# Process-Control Systems

# Process-Control Systems
## *Application / Design / Adjustment*

*Second Edition*

**F. G. SHINSKEY**
*Senior Systems Consultant, The Foxboro Company*

**McGRAW-HILL BOOK COMPANY**
New York   St. Louis   San Francisco   Auckland   Bogotá
Düsseldorf   Johannesburg   London   Madrid   Mexico
Montreal   New Delhi   Panama   Paris   São Paulo
Singapore   Sydney   Tokyo   Toronto

Library of Congress Cataloging in Publication Data

Shinskey, F Greg.
　Process-control systems.

　Includes bibliographical references and index.
　1.　Process control.　I.　Title.
TP155.75.S5 1979　　660.2'81　　78-12810
ISBN 0-07-0556891-X

1234567890　KPKP　7865432109

*The editors for this book were Jeremy Robinson and Patricia A. Allen,
the designer was Naomi Auerbach, and the production supervisor
was Tom Kowalczyk. It was set in VIP
by University Graphics, Inc.*

*Printed and bound by The Kingsport Press.*

*To all who have sought my help*

# Preface to the Second Edition

Since its publication in 1967, "Process-Control Systems" has been used extensively in industrial training courses at The Foxboro Company and in undergraduate and continuing education programs at many colleges and universities in the United States and Canada. While the time-domain approach to learning control-system behavior has been eminently successful, instructors have found it necessary to bring students very slowly through the first chapter. The concepts described therein are in some ways as unfamiliar to engineers trained in other methods as to students having no background in the subject at all. Gradually, we developed a curriculum in which topics of fundamental importance were expanded and elaborated upon while incidentals fell by the wayside. Definitions and examples were added where necessary, and illustrations modified to become more meaningful.

This second edition of the work has incorporated these improvements, to become a more effective tool both for instructional courses and for self-teaching. Chapter 1 has been almost completely rewritten. An example of the type of changes made is the factoring of steady-state and dynamic gains of both process and controller at the outset, rather than leaving the steady-state gains to be introduced later in the book. Similarly, distinction is now made between any period of oscillation and the *natural* period, which may differ, depending upon the phase shift introduced by the controller. Some of the nomenclature has also been changed in keeping with recently promulgated standards of the Instrument Society of America (ISA) and the Scientific Apparatus Manufacturers' Association (SAMA). For example, the term "reset" has been replaced by "integral", and "dead band" has been substituted where "hysteresis" formerly appeared.

The advent of highly accurate pocket calculators, capable of performing exponential and trigonometric functions, has also had an impact. Where some approximations were relied upon in the first edition, extensive and accurate calculations now appear, performed iteratively where necessary. This has also led to a reevaluation of the mode settings that provide optimum response and has given improved accuracy to many of the tables and curves. The first edition provided answers to the problems at the end of the chapters but not complete solutions. Because of many requests from instructors using the book, solutions are now included in the Appendix. In many cases, they illustrate problem-solving methods not specifically used in the text and therefore add another dimension to its instructional capability.

Experience gained in two important disciplines, pH control and distillation, has led to the development of new and more accurate process models. The mathematical representation of a mixed vessel given in the first edition has now been corrected and corroborated, permitting reliable predictions of performance under control. The relationship between energy inflow and separation in a distillation column has gone through four evolutionary stages, so that the version now presented is quite exact and broadly applicable.

Chapter 5, Nonlinear Control Elements, now includes accurate solutions for loops containing saturation, dead zone, dead band, velocity limiting, negative resistance, and two- and three-state controllers. Chapter 6 has two new subjects, multiple-output control systems, and selective systems that restructure themselves as conditions demand. Feedforward control has been moved from Chapter 8 to 7, as a natural extension of the cascade systems which precede it.

When the first edition was written, the concept of evaluating process interaction by relative gains was new. It was a subject that elicited great interest among both theoreticians and practitioners, evoking a vigorous exchange of ideas. Demonstrations were conducted using computer simulations, and decoupling experiments were tried, eventually on full-scale processes. A wealth of new information was generated, leading to a better understanding of the subject. Chapter 8, Interaction and Decoupling, has been completely rewritten to reflect this new level of comprehension and to report on the successes and risks of applying decoupling.

The last part of the book, concerned with applications, has been updated principally to include new experience gained since publication. Again, sections on pH control and distillation are the principal beneficiaries of this experience. The description of a new dynamic element, inverse response, is added under the heading of boiler drum-level control, where it is most frequently encountered. Finally, the section on control of dryers has been completely revised to incorporate mathematical models recently perfected and to describe systems which have been proved effective in controlling this very elusive process.

For their assistance in helping develop our very successful training program based on this book, let me express my appreciation to Stanley Burak and Lewis Gordon. My thanks also go to Marilyn Clark, who typed the manuscript with its many equations.

*F. G. Shinskey*

# Contents

*Preface*   *vii*

**Part One   Understanding Feedback Control**

**1. Dynamic Elements in the Control Loop   3**

*Negative Feedback   3*
*The Difficult Element—Dead Time   8*
*The Easy Element—Capacity   18*
*Combinations of Dead Time and Capacity   24*
*Summary   28*
*Problems   29*

**2. Characteristics of Real Processes   30**

*Multicapacity Processes   31*
*Steady-State Gain   42*
*Testing the Plant   52*
*References   56*
*Problems   56*

**3. Analysis of Some Common Loops   57**

*Flow Control   58*
*Pressure Regulation   63*

*Liquid Level and Hydraulic Resonance   66*
*Temperature Control   69*
*Control of Composition   75*
*Conclusions   80*
*References   80*
*Problems   80*


# Part Two    Selecting the Feedback Controller


## 4. Linear Controllers    85

*Performance Criteria   86*
*PI and PID Controllers   89*
*Complementary Feedback   100*
*Interrupting the Control Loop   103*
*Digital Control Systems   107*
*References   111*
*Problems   111*


## 5. Nonlinear Control Elements    112

*Nonlinear Elements in the Closed Loop   112*
*Nonlinear Phase-shifting Elements   116*
*Variations of the On-Off Controller   120*
*The Dual-Mode Concept   124*
*Nonlinear PID Controllers   132*
*References   136*
*Problems   136*


# Part Three    Multiple-Loop Systems


## 6. Improved Control through Multiple Loops    139

*Cascade Control   139*
*Multiple-Output Control Systems   147*
*Selective Control Loops   151*
*Adaptive Control Systems   156*
*Summary   164*
*References   164*
*Problems   164*


## 7. Feedforward Control    166

*The Control System as a Model of the Process   167*
*Ratio Control Systems   172*
*Applying Dynamic Compensation   178*
*Adding Feedback   186*
*Economic Considerations   190*
*Summary   194*
*References   194*
*Problems   194*

**8. Interaction and Decoupling    196**

*Relative Gain    197*
*Effects of Interaction    204*
*Decoupling    212*
*References    221*
*Problems    221*

# Part Four    Applications

**9. Energy Transfer and Conversion    225**

*Heat Transfer    225*
*Combustion Control    233*
*Steam-Plant Control Systems    235*
*Pumps and Compressors    239*
*References    245*
*Problems    245*

**10. Controlling Chemical Reactions    247**

*Principles Governing the Conduct of Reactions    248*
*Continuous Reactors    257*
*pH Control    263*
*Batch Reactors    272*
*References    275*
*Problems    276*

**11. Distillation    277**

*Factors Affecting Product Quality    277*
*Arranging the Control Loops    283*
*Composition Control    292*
*Batch Distillation    301*
*Summary    304*
*References    305*
*Problems    305*

**12. Other Mass-Transfer Operations    306**

*Absorption and Humidification    307*
*Evaporation and Crystallization    313*
*Extraction and Azeotropic Distillation    318*
*Drying Operations    322*
*References    325*
*Problems    325*

**Appendix A    Notation    327**

**Appendix B    Solutions to Problems    329**

**Index    343**

**Part One**

# Understanding Feedback Control

# Chapter One

# Dynamic Elements in the Control Loop

What makes control loops behave the way they do? Some are fast, some slow; some oscillate, others loll in stability. What determines how well a given variable can be controlled? How are the optimum controller settings related to the process? These questions must be answered before the reader can really comprehend the essence of the control problem. They will be answered in the pages that follow.

Negative feedback is the basic regulating mechanism of automatic systems, but it is not the only mechanism. Feedback has certain limitations which sometimes go unnoticed in the pursuit of better feedback controllers. Yet before progress can be made to more effective systems, the properties of simple feedback loops must be well defined.

Fortunately, a process need not be very complicated before the properties of the typical feedback loop make their appearance. A rapid introduction to loop behavior can be presented using the simplest dynamic element found in the process—dead time. This chapter is devoted exclusively to discussion of the control of simple dynamic elements which may never exist in the pure form. But these elements do exist in various proportions in every real process. Therefore a thorough familiarity with the parts is essential for estimating the behavior of the whole.

## NEGATIVE FEEDBACK

There are two kinds of feedback possible in a closed loop, positive and negative. Positive feedback is an operation which augments an imbalance, thereby precluding stability. If a temperature controller with positive feedback were used to heat a room, it would increase the heat when the temperature was above the set point and turn it off when it was below. Loops with positive feedback lock at one extreme or the other. Obviously this property is not conducive to regulation and therefore will be of no further concern at this time.

Negative feedback, on the other hand, works toward restoring balance. If the temperature is too high, the heat is reduced. The action taken—heating—is manipulated negatively, in effect, to the direction of the controlled variable—temperature. Figure 1.1 shows the flow of information in a feedback loop.

At this point in the development of the subject, consider a control loop to be divided into only two parts, the process and the controller. They are distinguished on the basis that the controller is adjustable whereas the process generally is not. In actual practice, the process comprises many elements: valve, piping, pump, vessel, measuring device, transmitter, etc. Yet for the purposes of this discussion all these are assumed to have fixed parameters. In later chapters, the characteristics of these elements are examined in detail; for the moment, they are simply lumped into a process whose characteristics are given.

The process gain is determined as the ratio of a change in its output $dc$ to the change in the input $dm$ which caused it

$$K_p\mathbf{g}_p = \frac{dc}{dm} \tag{1.1}$$

The gain is shown as consisting of two separate components. The steady-state component $K_p$ does not vary with the period of the exciting signal. The dynamic gain $\mathbf{g}_p$ does, however; it appears as a vector having a scalar component $G_p$ and a phase angle $\phi_p$.

The gain of the controller represents the change in its output $dm$ in response to a change in deviation $de$

$$K_c\mathbf{g}_c = \frac{dm}{de} \tag{1.2}$$

and is similarly divided into steady-state component $K_c$, which does not change with period, and a dynamic-gain vector $\mathbf{g}_c$, which does. The vector is ordinarily expressed in terms of its scalar gain $G_c$ and its phase angle $\phi_c$.

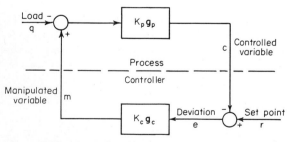

**FIG. 1.1** The controller manipulates $m$ to counteract the load $q$ and restore deviation $e$ to zero.

The negative sign applied to the controlled variable at the summing junction in Fig. 1.1 indicates that the controller has *increase-decrease action;* i.e., the output increases on a decrease in measured input. This action is necessary for negative feedback, with the process shown, because the sign applied to the manipulated variable entering the process is positive. Figure 1.1 might be representative of a heating system, where temperature $c$ increases on an increase in steam-valve position $m$ and decreases on an increase in heat loss $q$. For a cooling system, the controlled temperature $c$ would decrease on an increase in the position $m$ of the chilled-water valve and increase on an increase in heat gain $q$. When the signs of $m$ and $q$ are reversed, the signs applied to $c$ and $r$ also must be reversed; i.e., *increase-increase control action* is required. Most controllers operating in industry will have increase-decrease action.

The principal function of a controller is to provide regulation against changes in load. This is accomplished by making the gain $K_c \mathbf{g}_c$ of the controller as high as possible. As $K_c \mathbf{g}_c$ increases, the deviation $e$ required to drive $m$ to match a changing load becomes smaller. If $K_c \mathbf{g}_c$ could be set at a very high number, say 100 or more, very little deviation would appear on a change in load. Unfortunately, there is an upper limit which $K_c \mathbf{g}_c$ cannot exceed without leading to undamped oscillations. That limit of stability must be explored in order to determine how effective a feedback controller will be when it is applied to a given process.

## Oscillation in a Closed Loop

Oscillation in a closed loop is sustained much in the same way that a ball is bounced, by the periodic application of force at phase intervals of 360°. A ball struck repeatedly with the same force at its highest position will continue to cycle at a constant amplitude and period. But if the force is applied earlier in the cycle, i.e., less than 360° since the last impulse, the period will be shortened. Then a uniform oscillation requires a uniform force applied at exactly every 360° of phase.

Most controllers actually apply a force continuously rather than in pulses, with the magnitude of the force generally varying sinusoidally as the controlled variable changes sinusoidally. (The presence of nonlinear elements may alter the waveform as described in Chap. 5; for the moment, sinusoidal oscillation is assumed.)

Consider the controlled variable oscillating uniformly as depicted in Fig. 1.2. If the controller has increase-decrease action, as for a heating system, the deviation from the set point will cycle 180° out of phase with the controlled variable, as shown. This 180° shift in phase is the "negative" in negative feedback. If the control vector $\mathbf{g}_c$ has no phase shift, the manipulated variable will cycle in phase with the deviation; its different amplitude reveals the gain of the controller at the period of oscillation.

The controller has converted a rising controlled variable, e.g., temperature, into a falling manipulated variable, e.g., steam-valve position. Because there is assumed to be no phase shift in the controller (other than the 180° attributed to the negative sign), these events occur simultaneously. If there were also no phase shift in the process, a falling steam-valve position would cause temperature to fall *at the same time.* However, the figure shows temperature to be rising as the steam valve closes. Therefore the effect of the steam-valve closing does not appear as a falling temperature until the *next* half-cycle; in effect, the process has delayed the response of temperature to valve position by one-half cycle. This delay contributes the remaining 180° shift in phase necessary for oscillations to persist.

It is possible to introduce some phase shift into the controller by certain adjustments to be explained later in the chapter. In this case, the phase shift through the process will be altered. Yet whenever uniform oscillations are maintained, there must be a total of 180° of dynamic phase lag in the *loop*, i.e., combining that of the process and the controller (in addition to the 180° due to negative feedback)

$$\phi_p + \phi_c = -180°$$
(1.3)

The negative sign indicates a lag in phase.

### The Period of Oscillation

It has also been observed that the period of oscillation which a particular loop will exhibit is characteristic of that loop. The loop resonates at that period. Furthermore, any disturbance not periodic applied to the loop but containing components near the natural period will excite oscillations of the natural period. A pendulum is a good example of a feedback loop. The controlled variable is the angular position of the mass, and the set point is the vertical position. The mass of the pendulum, acted upon by gravity, is the manipulated variable, which tries to restore the angle to zero. Its natural period in seconds is

$$\tau_n = 2\pi \sqrt{\frac{L}{g}}$$

where $L$ is the length in feet and $g$ is the acceleration of gravity in feet per

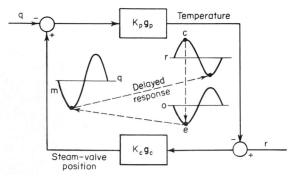

**FIG. 1.2**   The minimum position of the steam valve causes a minimum temperature one-half cycle later.

square second. A pendulum disturbed from rest by an impulse will proceed to oscillate at its own period. Impulse, step, and random disturbances contain a wide spectrum of periodic waves. The resonant system, however, responds principally to the component of its own natural period, rejecting the rest. For this reason, we are interested in the response of the loop to a wave of the natural period and at this point are unconcerned about the rest. The natural period of oscillation will be designated $\tau_n$ and will be recognized hereafter as a property peculiar to each control loop.

The period of any loop depends on the combination of all dynamic elements within it, including the controller. Since the amount of phase lag of most dynamic elements varies with the period of the wave passing through them, there is *one* particular period at which the total phase lag will equal 180°. This is the period at which the loop resonates. The period is a dependent variable. We can make use of its relation to the process dynamics in two ways: (1) if the characteristics of the elements in the process are known, the period under closed-loop control can be predicted; (2) if a process whose elements are largely unknown is under closed-loop control, the characteristics of these elements can be inferred by observing the period.

Because certain controllers can be adjusted to contribute varying degrees of phase shift to the loop, they can also change its period of oscillation. To avoid confusion, the term *natural period* and the symbol $\tau_n$ will refer to the period of oscillation observed in the absence of any phase shift introduced by the controller (other than the aforementioned always present 180°). When the controller contributes phase shift to the loop, the resulting period of oscillation $\tau_o$ will differ from the natural period $\tau_n$.

## Damping

During uniform oscillation, a signal passing through the loop will return to its starting point with exactly the same amplitude one complete cycle later. If it is attenuated more than it is amplified by the combination of elements through which it passes, the signal will gradually diminish in amplitude; the oscillation will be *damped*. For *undamped* oscillation to persist, the product of gains of all the elements in the loop must be unity

$$K_c G_c K_p G_p = 1.0 \tag{1.4}$$

*at the period of oscillation.* (This last consideration is important, because $G_c$ and $G_p$ usually vary with period.)

Should the gain product of the elements in the loop at the period of oscillation *exceed* unity, each successive cycle will exceed the last in amplitude until some natural limit is reached, possibly damaging equipment. Because of the inherent danger in an *expanding* cycle, gain products exceeding unity are scrupulously avoided.

When the gain product is less than unity, the oscillation will dampen and, in a linear system, eventually disappear. A gain product of unity is then the limit of stability. Yet to avoid exceeding unity, a value of 0.5 or less is usually desirable.

When the gain product is 0.5, each half-cycle is attenuated by one-half, with each full cycle being attenuated by one-quarter. This degree of damping, known as *quarter-amplitude damping,* has been found acceptable for most industrial control loops.

To summarize, a loop will oscillate uniformly (1) at a period at which the phase lags of all the elements in the loop total 180° and (2) when the gain product of all the elements at that period equals 1.0. The conditions for uniform oscillation will serve as a convenient reference on which to base rules for controller adjustment.

## THE DIFFICULT ELEMENT—DEAD TIME

### Identification

As the name implies, dead time is the property of a physical system by which the response to an applied force is delayed in its effect. It is the interval after the application of a force during which *no response* is observable. This characteristic does not depend on the nature of the applied force; it always appears the same. Its dimension is simply that of time.

Dead time occurs in the transportation of mass or energy along a particular path. The length of the path and the velocity of motion constitute the delay. Dead time is also called *pure delay, transport lag,* or *distance-velocity lag.* Like other fundamental elements, it rarely occurs alone in a real process, but there are few processes where it is not present in some form. For this reason, any useful technique of control system design must be capable of dealing with dead time.

An example of a process consisting of dead time alone is a weight-control system operating on a solids conveyor, as described in Fig. 1.3. The dead time between the action of the valve and the resulting change in weight is the distance between the valve and the cell (feet), divided by the velocity of the belt (feet per minute). Dead time is invariably a problem of transportation.

A feedback controller applies corrective action to the input of a process based on a present observation of its output. In this way the corrective action is moderated by its observable effect on the process. A process containing dead time produces no immediately observable effect; hence the control situation is

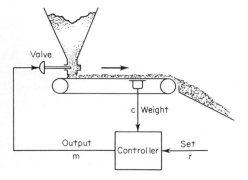

**FIG. 1.3** The response of the weigh cell to a change in solids flow is delayed by the travel of the belt.

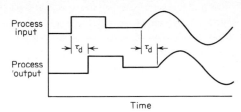

**FIG. 1.4**  Pure dead time transmits the input delayed by $\tau_d$.

complicated. For this reason, dead time is recognized as the most difficult dynamic element naturally occurring in physical systems. So that the reader may begin without illusions about the limitations of automatic controls in their influence over real processes, the difficult element of dead time is presented first.

The response of a dead-time element to any signal whatever will be the signal delayed by that amount of time. Dead time is measured as shown in Fig. 1.4.

Notice the response of the element to the sine wave in Fig. 1.4. The delay effectively produces a phase shift between input and output. Since one characteristic of feedback loops is the tendency toward oscillation, the property of phase shift becomes an essential consideration.

### The Phase Shift of Dead Time

We are primarily interested in phase characteristics of elements at the period of oscillation of the loop. Assume, to begin, that a closed loop containing dead time is already oscillating uniformly. The input to the process is the sine wave shown in Fig. 1.5

$$m = A \sin 2\pi \frac{t}{\tau_o} + m_0 \tag{1.5}$$

where  $m$ = manipulated variable whose average component is $m_0$
$\quad\quad\; A$ = amplitude
$\quad\quad\; t$ = time
$\quad\quad\; \tau_o$ = period

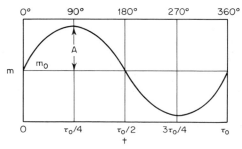

**FIG. 1.5**  The manipulated variable is cycling with an amplitude of $A$ at a period $\tau_o$.

Phase angles will be expressed both in degrees and in radians for reasons that will become clear later.

| | $2\pi t/\tau_o$ | | |
| --- | --- | --- | --- |
| $t/\tau_o$ | Degrees | Radians | $\sin(2\pi t/\tau_o)$ |
| 0 | 0 | 0 | 0 |
| $1/4$ | 90 | $\pi/2$ | $+1$ |
| $1/2$ | 180 | $\pi$ | 0 |
| $3/4$ | 270 | $3\pi/2$ | $-1$ |
| 1 | 360 | $2\pi$ | 0 |

This wave, passing through a dead time, will be delayed by an amount $\tau_d$ but will be undiminished, so that the output will be

$$c = K_p \left( A \sin 2\pi \frac{t - \tau_d}{\tau_o} + m_0 \right) \tag{1.6}$$

The gain of the process is the ratio of the output signal vector to the input vector at the period of oscillation

$$K_p \mathbf{g}_d = \frac{dc(\tau_o)}{dm(\tau_o)} = \frac{K_p A (\sin 2\pi(t - \tau_d)/\tau_o)}{A (\sin 2\pi t/\tau_o)} \tag{1.7}$$

The ratio of two vectors is a third vector whose magnitude is the ratio of the signal amplitudes and whose phase angle is the difference between their phase angles. Then $\mathbf{g}_d$ is calculated as a scalar gain

$$G_d = \frac{A}{A} = 1 \tag{1.8}$$

and a phase angle

$$\phi_d = 2\pi \frac{t - \tau_d}{\tau_o} - 2\pi \frac{t}{\tau_o} = -2\pi \frac{\tau_d}{\tau_o} = -360° \frac{\tau_d}{\tau_o} \tag{1.9}$$

**Proportional Control of Dead Time**

Having defined the process, our next step is to select a suitable controller. A proportional controller will be chosen first because of its simplicity. It contains no dynamic elements. Output and input are related by the expression

$$m = \frac{100}{P} e + b \tag{1.10}$$

where $P$ = proportional band, %
  $e$ = error or deviation of measurement from set point
  $b$ = output bias

The steady-state gain $K_c$ of the proportional controller is $100/P$; its dynamic gain $\mathbf{g}_c$ is 1.0, $\angle 0°$. The output of the controller equals the bias when there is no error.

Because there are no dynamic elements in the proportional controller, the entire 180° phase shift will take place in the dead-time element. This determines the natural period

$$\phi_d = -360° \frac{\tau_d}{\tau_n} = -180°$$

Solving for $\tau_n$ gives

$$\tau_n = 2\tau_d \tag{1.11}$$

The relationship is as plain as it appears. A 1-min dead-time process will cycle with a 2-min period under proportional control. This is not an approximation; it is exact.

Next it is important to estimate the proportional band necessary to sustain oscillation. Since dead time offers no gain contribution, if the loop-gain product is to be 1.0, the controller proportional band must be $100K_p$ percent. To dampen the oscillations, the band must be increased, thus attenuating the input cycle.

Figure 1.6 illustrates how a proportional band of $200K_p$ percent reduces the amplitude of each successive half-cycle by one-half, resulting in quarter-amplitude damping of each successive cycle.

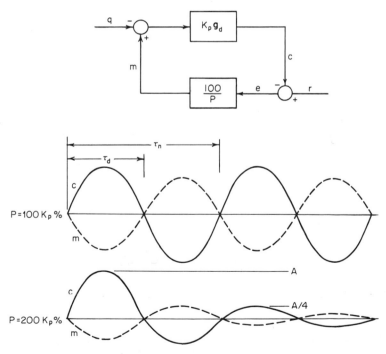

**FIG. 1.6**  A loop gain of 0.5 will provide quarter-amplitude damping; in this process, $K_p > 1.0$.

Notice that there is only one adjustment available, and it affects the damping. Given a process consisting of a 1-min dead time to be controlled by proportional action only, adjusted to quarter-amplitude damping, the natural period is fixed at 2 min, and the proportional band must be $200K_p$ percent. The nature of the process determines the results.

### Proportional Offset

The prime function of a controller is that of regulation. The controller is intended to change its output as often and as much as necessary to keep the controlled variable at the set point. Every process is subject to variations in load. In a well-regulated loop, the manipulated variable will be driven to balance the load. Consequently, the load is often measured in terms of the corresponding value of controller output.

In the equation describing the proportional controller, the bias $b$ equals the output when the error is zero. This bias may be fixed at the normal value of output, usually 50 percent, or it may be adjusted by hand to match the current load. This adjustment is called *manual reset*. But because of the proportional relationship between input and output, a change in output by any amount cannot be gained without a corresponding change in error. Should the output of the proportional controller have to change to meet a new load condition, a deviation will appear

$$e = \frac{P(m - b)}{100} \tag{1.12}$$

The deviation in this case is known as *offset*, and it increases with proportional band. With a proportional band of $200K_p$ percent, which was necessary for quarter-amplitude damping in the previous example, a 10 percent change in load would produce a $20K_p$ percent offset; depending on the value of $K_p$, this offset could be intolerable.

Consider a dead-time process having a $K_p$ of 1.0, at a steady state, with the controlled variable at the set point of 50 percent, as in the initial conditions of Fig. 1.7. A change in load will affect the controlled variable

$$dc = dm - dq \tag{1.13}$$

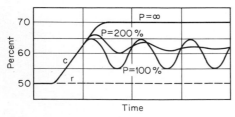

**FIG. 1.7** The response to a load change illustrates how the proportional band affects both damping and offset.

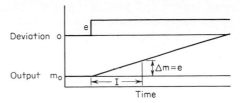

**FIG. 1.8** The output of an integrator will change by an amount equal to its deviation in time $I$.

In the absence of control action, $dm = 0$, and $dc = -dq$. The top curve in Fig. 1.7 describes the response of $c$ to a gradual load reduction of 20 percent in the absence of control ($P = \infty$).

The action of the controller changes $m$ with $c$

$$dm = \frac{100}{P} dc \qquad (1.14)$$

Then $dm$ from (1.14) can be substituted into (1.13) to evaluate the effect of a load change with control action

$$dc = \frac{-dq}{1 + 100/P} \qquad (1.15)$$

When $P$ is set at 200 percent, the deviation will be reduced from 20 to 13.3 percent, with quarter-amplitude damping, as shown in Fig. 1.7. Reducing $P$ further to 100 percent lowers the average deviation to 10 percent but sacrifices damping.

### Integral (Reset) Control of Dead Time

Proportional control is obviously rejected for most applications demanding a band wider than a few percent, and so another control mode is needed. An integral controller is a device whose output is the time integral of the deviation

$$m = \frac{1}{I} \int e \, dt \qquad (1.16)$$

where $I$ is the time constant of the controller, known as *integral* or *reset* time. As long as a deviation exists, this controller will change its output; hence it is capable of driving the deviation to zero. The rate of change of output is proportional to the deviation

$$\frac{dm}{dt} = \frac{e}{I} \qquad (1.17)$$

Response to a step input is shown in Fig. 1.8.

Before using an integral controller in a closed loop, its gain and phase characteristics must be defined. Again we are primarily interested in these properties at the period of oscillation of the loop, $\tau_o$. Introducing a sinusoidal input to the controller leads to

$$e = A \sin 2\pi \frac{t}{\tau_o} \tag{1.18}$$

and the controller output will be the time integral of the input

$$m = \frac{1}{I} \int e \, dt = \frac{1}{I} \int \left( A \sin 2\pi \frac{t}{\tau_o} \right) dt$$

Extraction of the appropriate item from a table of definite integrals enables us to solve the above equation:

$$m = \frac{A\tau_o}{2\pi I} \left( -\cos 2\pi \frac{t}{\tau_o} \right) + m_0$$

where $m_0$ is the output at time zero.

In order to evaluate phase and gain properties, the output must be reduced to the same form as the input. Using the trigonometric identity

$$-\cos \phi = \sin \left( -\frac{\pi}{2} + \phi \right)$$

we can convert $m$ into a sine function

$$m = \frac{A\tau_o}{2\pi I} \sin \left( -\frac{\pi}{2} + \frac{2\pi t}{\tau_o} \right) + m_0 \tag{1.19}$$

The phase shift of the integrator is the angle of the output minus the angle of the input:

$$\phi_I = \left( -\frac{\pi}{2} + \frac{2\pi t}{\tau_o} \right) - \frac{2\pi t}{\tau_o} = -\frac{\pi}{2} = -90° \tag{1.20}$$

An integrator exhibits a phase lag of 90° regardless of the period of the input.

The gain of an integrator is the amplitude of the output over the amplitude of the input:

$$G_I = \frac{A\tau_o/2\pi I}{A} = \frac{\tau_o}{2\pi I} \tag{1.21}$$

In closing the loop, the sum of the phase shifts of dead time and controller must equal $-180°$ at the period of oscillation

$$\phi_I + \phi_d = -180°$$

$$-90 + \left( -360 \frac{\tau_d}{\tau_o} \right) = -180°$$

Solving for $\tau_o$ gives

$$\tau_o = 4\tau_d \tag{1.22}$$

Notice that $\tau_o$ is twice the natural period $\tau_n$ because only 90° of phase shift is allowed to take place in the process.

To sustain oscillations, the loop gain must be 1.0. Because the process gain is $K_p$, the integrator gain for this condition must be $1/K_p$

$$G_I = \frac{\tau_o}{2\pi I} = \frac{1}{K_p}$$

Then the integral time for zero damping is

$$I = K_p \frac{\tau_o}{2\pi} = \frac{2}{\pi} K_p \tau_d \qquad (1.23)$$

A process having a dead time of 1 min would cycle with a period of 4 min under integral control, with oscillations sustained by an integral time constant of $2K_p/\pi$, or about $0.64K_p$ min. Quarter-amplitude damping will be achieved by reducing loop gain to 0.5, by doubling the integral time as shown in Fig. 1.9. The integral controller has but one adjustment, which affects only the damping. The period and the integral time required for damping are estab-

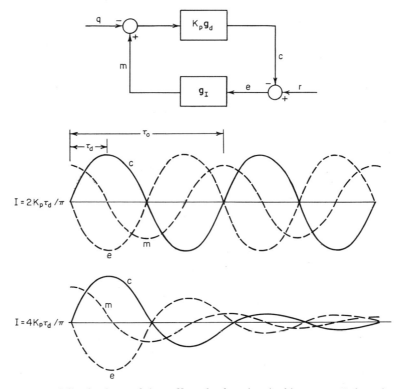

**FIG. 1.9**  Adjusting integral time affects the damping; in this process, $K_p$ is again greater than 1.

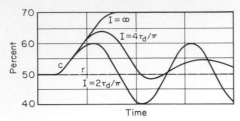

**FIG. 1.10**  Increasing integral time trades recovery for damping, although $\tau_o$ is unaffected.

lished by the characteristics of the process. While integrating action has succeeded in eliminating offset, the cost has been reduced speed of response.

Figure 1.10 shows the response of the loop to the same load change imposed on the process under proportional control. When integral time is too long, the rate of recovery is slow and the loop is overdamped.

### Proportional-plus-Integral Control

This controller combines the best features of the proportional and integral modes, in that proportional offset is eliminated with little loss of response speed. The controller is represented as

$$m = \frac{100}{P}\left(e + \frac{1}{I}\int e\,dt\right) \tag{1.24}$$

As with the proportional controller, its steady-state gain $K_c$ is $100/P$. Its dynamic-gain vector $\mathbf{g}_{PI}$ varies with period, between the limits established by the proportional and integral components. It is determined by vector summation as shown in Fig. 1.11

$$\mathbf{g}_{PI} = 1.0, \angle 0° + \frac{\tau_o}{2\pi I}, \angle -90° \tag{1.25}$$

$$G_{PI} = \sqrt{1 + \left(\frac{\tau_o}{2\pi I}\right)^2} \tag{1.26}$$

$$\phi_{PI} = -\tan^{-1}\frac{\tau_o}{2\pi I} \tag{1.27}$$

To achieve quarter-amplitude damping requires a loop gain of 0.5

$$K_p\frac{100}{P}\sqrt{1 + \left(\frac{\tau_o}{2\pi I}\right)^2} = 0.5 \tag{1.28}$$

Observe that the desired degree of damping can be attained with a variety of combinations of $P$ and $I$, each exhibiting a different phase angle and therefore developing a different period. For example, if the controller phase angle $\phi_{PI}$ is $-30°$, the phase shift remaining in the dead time is its difference from $-180°$

$$\phi_d = -180° - \phi_{PI} = -150°$$

**TABLE 1.1    Settings of Proportional and Integral Modes for a Dead-Time Process**

| $\phi_{PI}$, deg | $\phi_d$, deg | $\dfrac{\tau_o}{\tau_d}$ | $G_{PI}$ | $\dfrac{I}{\tau_d}$ | $\dfrac{P}{K_p}$, % |
|---|---|---|---|---|---|
| 0 | $-180$ | 2.00 | 1.000 | $\infty$ | 200 |
| $-15$ | $-165$ | 2.18 | 1.035 | 1.296 | 207 |
| $-30$ | $-150$ | 2.40 | 1.155 | 0.662 | 231 |
| $-45$ | $-135$ | 2.67 | 1.414 | 0.424 | 283 |
| $-60$ | $-120$ | 3.00 | 2.000 | 0.276 | 400 |
| $-75$ | $-105$ | 3.43 | 3.864 | 0.146 | 773 |
| $-90$ | $-90$ | 4.00 | 0.5† | 1.273 | † |

†The last row describes integral-only control.

Then the period of the loop is determined using Eq. (1.9)

$$\tau_o = -\frac{360}{\phi_d}\tau_d = 2.4\tau_d$$

The value of integral time required to produce that controller phase angle at the determined period can then be calculated from Eq. (1.27)

$$-\tan\,(-30°) = \frac{\tau_o}{2\pi I}$$

$$0.577 = \frac{2.4\tau_d}{2\pi I}$$

$$I = 0.662\tau_d$$

The proportional band required to produce a loop gain of 0.5 can then be estimated using Eq. (1.28)

$$P = 200\,K_p\,\sqrt{1 + (0.577)^2} = 231K_p$$

Table 1.1 compiles all the combinations of $P$ and $I$ which will produce quarter-amplitude damping at phase-angle increments of 15°, calculated in the above manner.

The selection of the optimum combination of $P$ and $I$ will be deferred until an analysis of controller performance is given in Chap. 4. At this point it is sufficient to note that extreme phase angles are to be avoided and that generally satisfactory performance will be found in the midrange, where the two modes

**FIG. 1.11**  The dynamic gain of the proportional-plus-integral controller is the vector sum of those of proportional and integral controllers.

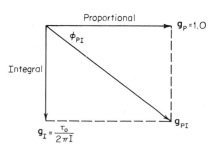

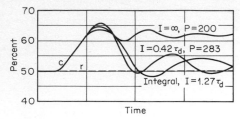

**FIG. 1.12** Various combinations of proportional and integral values can provide quarter-amplitude damping, but with different rates of recovery from a load change.

offer similar contributions to gain. Figure 1.12 shows the responses of a dead-time process having $K_p$ of 1 to a gradual load change, for proportional, integral, and a proportional-plus-integral controller having a phase angle of $-45°$.

## THE EASY ELEMENT—CAPACITY

### Identification

Capacity appears in many forms, but its properties are universal as far as automatic control is concerned. Capacity is a location where mass or energy can be stored. It acts as a buffer between inflowing and outflowing streams, determining how fast the level of mass or energy can change. In fluid systems, tanks have capacity to hold liquid or gas. In electrical systems, capacitors are used to store nominal amounts of charge. Heat capacity is a factor in thermal systems. And the mechanical measure of capacitance is inertia, which determines the amount of energy that can be stored in a stationary or a moving object.

Since our principal concern is with fluids, Fig. 1.13 is an appropriate introduction to capacity. In the system shown in the figure, the metering pump delivers a constant outflow, while inflow can be manipulated. The rate of change of tank contents equals the difference between inflow and outflow

$$\frac{dv}{dt} = F_i - F_o \tag{1.29}$$

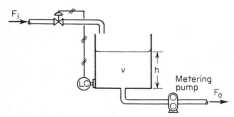

**FIG. 1.13** The rate of change of level is proportional to the difference between inflow and outflow.

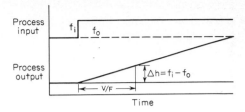

**FIG. 1.14** The percentage level change will equal the percentage flow change in time $V/F$.

Solving for $v$ gives

$$v = \int (F_i - F_o)\, dt \tag{1.30}$$

If the tank is vertical and of uniform inside area, its fractional liquid level $h$ will equal the fractional volume

$$h = \frac{v}{V}$$

where $V$ is the capacity of the tank. Since we are interested in tank level,

$$h = \frac{1}{V} \int (F_i - F_o)\, dt$$

In an effort to make the entire equation dimensionless, we can define $f_i$ and $f_o$ as fractions of the maximum flow $F$ which the valve can deliver. Then

$$F_i - F_o = F(f_i - f_o)$$

and

$$h = \frac{F}{V} \int (f_i - f_o)\, dt \tag{1.31}$$

This is called an *integrating process*. Notice its similarity to the integrating controller: $h$ is the output, $f_i - f_o$ is the input error, and $V/F$ is the time constant. The step response is given in Fig. 1.14.

The level in the tank could be controlled by manually adjusting the valve position, thereby setting inflow. But if inflow varied in the slightest from outflow, the tank would eventually flood or run dry. This characteristic is called *non-self-regulation*. It means that the integrating process cannot balance itself: it has no natural equilibrium or steady state. The non-self-regulating process cannot be left unattended for long periods of time without automatic control.

Most liquid-level processes are non-self-regulating; occasionally other processes will exhibit this characteristic. In general, it is not harmful as long as its peculiarities are taken into account. One of these peculiarities is its phase shift. Like the integrating controller, the non-self-regulating process exhibits a phase lag of 90° to any periodic wave. This has two consequences: (1) Under proportional control, the loop cannot oscillate because its phase lag never reaches 180°.

The proportional band therefore can be set to zero. (2) Under floating (integrating) control, the loop will *always* oscillate with uniform amplitude because the total phase shift of process and controller is 180° at all periods. The loop tends to oscillate at the period where the gain product is unity; the integral time then affects only the period and cannot change the damping.

The dynamic gain of an integrating process is like that of the integrating controller

$$\mathbf{g}_p = \frac{\tau_o}{2\pi\tau}, \ \angle -90° \tag{1.32}$$

where $\tau = V/F$. If an integrating controller is used to close the loop,

$$G_p G_I = \frac{\tau_o}{2\pi\tau}\frac{\tau_o}{2\pi I} = 1.0$$

Solving for $\tau_o$, we get

$$\tau_o = 2\pi\sqrt{I\tau} \tag{1.33}$$

### Self-regulation

Replace the metering pump in Fig. 1.13 with a valve. Then an increase in liquid level would inherently increase the outflow. This action works toward the restoration of equilibrium and is called *self-regulation*. It is as if a proportional controller were at work within the process. This is a natural form of negative feedback.

Although the relationship is in fact not linear, assume for the moment that flow out of the tank is proportional to the head of liquid above the valve

$$f_o = kh$$

The level will remain steady when $f_o = f_i$, which indicates that every condition of inflow will bring about a new steady-state level

$$h = \frac{f_i}{k}$$

In proceeding from one steady state to another, however, the level will vary with time. With a step increase in $f_i$, the level will start to change at the same rate as in the non-self-regulating case because outflow has not yet begun to increase. The rate of rise of level will then diminish with time as $f_o$ approaches $f_i$. As a result, the final level will only be reached in infinite time

$$\frac{dh}{dt} = \frac{F}{V}(f_i - f_o)$$

Substituting for $f_o$, we get

$$\frac{dh}{dt} = \frac{F}{V}(f_i - kh)$$

The next step is to solve for $h$, the controlled variable

$$h + \frac{V}{Fk}\frac{dh}{dt} = \frac{f_i}{k} \tag{1.34}$$

This is known as a *first-order differential equation*. The controlled variable $h$ is related to the manipulated variable $f_i$ both in the steady state and with respect to time.

The general form of the first-order differential equation is

$$c + \tau_1\frac{dc}{dt} = K_p m \tag{1.35}$$

where $\tau_1$ is the time constant and $K_p$ the steady-state gain. By equating $c$ to $h$ and $m$ to $f_i$ in Eq. (1.34) it can be shown that

$$\tau_1 = \frac{V}{Fk} \qquad \text{and} \qquad K_p = \frac{1}{k}$$

While $V$ and $F$ are constants, $k$ will vary with the opening of the valve(s) draining the tank and also with level. Therefore *neither* the time constant nor the steady-state gain of this process is constant. Indeed, this characteristic is typical of many if not most self-regulating processes. While it may be the source of some confusion, in most cases the simultaneous variation of both parameters is compensatory, leaving a constant process gain, as will be demonstrated.

The solution of Eq. (1.35) for a step change $\Delta m$ is exponential with time

$$\Delta c = K_p\,\Delta m(1 - e^{-t/\tau_1}) \tag{1.36}$$

The step response of the self-regulating level process is shown in Fig. 1.15. The rate of rise of $c$ varies with its position, as can be seen by rearranging Eq. (1.35)

$$\frac{dc}{dt} = \frac{K_p m - c}{\tau} \tag{1.37}$$

The instant after introducing the input change $\Delta m$, the numerator in (1.37) is $K_p\,\Delta m$. Therefore the initial rate of rise of $c$ is $K_p\,\Delta m/\tau$. The initial rate of rise of the level process is

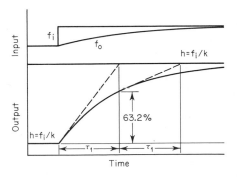

**FIG. 1.15** The slope of the response curve equals the departure from steady state divided by $\tau_1$.

$$\left(\frac{dh}{dt}\right)_{t=0} = \frac{\Delta f_i}{V/F} \tag{1.38}$$

Note the absence of $k$.

As the level begins to respond, so does the outflow, reducing the rate of change of level from this initial maximum value. After an elapsed time equal to the time constant, $\Delta c/(K_p \Delta m)$ equals $1 - e^{-1}$ or 0.632; that is, the controlled variable has traversed 63.2 percent of the distance to the next steady state. During the next time interval $\tau_1$, 63.2 percent of the remainder will be traversed, etc., as shown in Fig. 1.15.

The dynamic gain of the liquid-level process can be calculated by putting Eq. (1.35) in terms of $m$ alone

$$m = \frac{1}{K_p}\left(c + \tau_1 \frac{dc}{dt}\right) \tag{1.39}$$

Because this expression has the process input in terms of two components of the output, it is necessary first to calculate the *inverse* of the dynamic gain $\mathbf{g}_1$ of the first-order lag

$$\frac{1}{\mathbf{g}_1} = K_p \frac{dm(\tau_o)}{dc(\tau_o)}$$

and later invert it to obtain $\mathbf{g}_1$. The value of the *steady-state* component of $1/\mathbf{g}_1$ is simply 1.0 at a phase angle of zero; the *derivative* component produces a vector whose gain and phase are the inverse of the integral gain described in Eq. (1.32)

$$\frac{1}{\mathbf{g}_1} = 1, \angle 0° + \frac{2\pi\tau_1}{\tau_o}, \angle +90° \tag{1.40}$$

The vector summation is shown in Fig. 1.16 and below:

$$\frac{1}{\mathbf{g}_1} = \sqrt{1 + \left(\frac{2\pi\tau_1}{\tau_o}\right)^2}, \angle \tan^{-1} \frac{2\pi\tau_1}{\tau_o} \tag{1.41}$$

The next step is the inversion of (1.41) to produce the true dynamic gain of the first-order lag

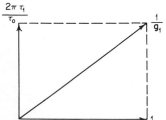

$$\mathbf{g}_1 = \frac{1}{\sqrt{1 + (2\pi\tau_1/\tau_o)^2}}, \angle -\tan^{-1}\frac{2\pi\tau_1}{\tau_o} \tag{1.42}$$

Since the phase lag of the self-regulating process cannot exceed 90°, it cannot oscillate under proportional control. This was also true of the non-self-regulating process. Then we can make a general statement that single-capacity processes can be controlled without oscillation at zero proportional band. This means that the valve can be driven full range in response to an infinitesimal

**FIG. 1.16** The vector sum is the gain of $m$ with respect to $c$, that is the inverse of the process dynamic gain.

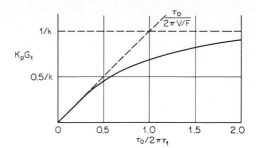

**FIG. 1.17** In most instances, the period of oscillation is less than the time constant, such that the product of the steady-state and dynamic gains is the same as that of a non-self-regulating process.

error, allowing no offset to develop. Single-capacity processes are therefore the easiest to control.

Substitution for the parameters of the liquid-level processes into the scalar component of Eq. (1.42) and inclusion of the steady-state gain gives

$$K_p G_1 = \frac{1}{\sqrt{k^2 + (2\pi V/F\tau_o)^2}} \approx \frac{\tau_o}{2\pi V/F} \qquad (1.43)$$

A plot of $K_p G_1$ vs. the ratio of $\tau_o$ to $\tau_1$ appears in Fig. 1.17. Observe that at periods shorter than about $3\tau_1$, that is, $\tau_o/2\pi\tau_1 < 0.5$, $K_p G_1$ follows the asymptote $\tau_o/(2\pi V/F)$. This is nearly always true when $\tau_1$ is the largest or dominant time constant in the loop. Under these conditions, $k$ disappears from consideration altogether.

Parameter $k$ may be looked upon as an index of self-regulation. If $k = 1.0$, the process is completely self-regulating in that liquid level would remain within the range of the measuring device for all rates of flow, without control. Very few liquid-level processes approach this degree of self-regulation. In most, the total head across the drain valve(s) will be much higher than the level measuring range. Furthermore, flow is proportional to the square root of head in the usual turbulent regime, further complicating the issue. Let $k$ be redefined over a small operating range

$$k = \frac{df_o}{dh} \qquad (1.44)$$

The actual outflow depends on the flow coefficient of the drain valve(s) $C_v$, and the head $h_v$ across them

$$f_o = C_v\sqrt{h_v} \qquad (1.45)$$

Differentiating (1.45) will yield $k$

$$k = \frac{df_o}{dh} = \frac{C_v}{2\sqrt{h_v}} = \frac{f_o}{2h_v} \qquad (1.46)$$

Since $h$ is expressed as a fraction of the liquid-level measurement range, $h_v$ must also be expressed in those terms, although in most cases it will exceed that range.

Consider the case where $f_o$ and $h$ are both 0.5, and the drain valve(s) are located at the bottom of the range; $h_v$ will then equal $h$ at 0.5, and from Eq. (1.46) $k$ will be 0.5. Should $h_v$ be ten times the liquid-level range, $k$ will be 0.05. In summary, $k$ will nearly always be much less than unity, further decreasing its contribution to process gain as described in Eq. (1.43). This factor, combined with the comparatively large value of the other term under the radical in (1.43), allows $k$ to disappear from consideration in almost every process of this sort. This is a fortunate circumstance, in that Eq. (1.46) shows $k$ to vary with both the liquid level in the vessel and the opening $C_v$ of the drain valve(s). Although both the time constant and steady-state gain of the process are variable, the gain product is not.

## COMBINATIONS OF DEAD TIME AND CAPACITY

Occurrences of either pure dead-time or ideal single-capacity processes are rare. Virtually all physical processes have some capacity to store mass and energy and involve some transportation between input and output. Between the most and the least difficult processes lies a broad spectrum of moderately difficult processes. Although most are dynamically complex, their behavior can be modeled with reasonable faithfulness by a combination of dead time and a single capacity. Consequently the next topic for consideration is the control of a process comprising these two now-familiar elements.

### Proportional Control

The dynamic characteristics of proportional control of a process composed of dead time and an integrating capacity have already been developed. Figure 1.18 contains the same phase-shifting elements as Fig. 1.9, although the integrator has been moved from the controller to the process, where its time constant is no longer adjustable. Because of the phase lag of the integrator, only 90° remains for the dead-time element. As a consequence, the natural period of this process, i.e., under proportional control, is

$$\tau_n = 4\tau_d \tag{1.47}$$

For quarter-amplitude damping, the product of all gains in the loop must be 0.5

$$K_c \mathbf{g}_c \mathbf{g}_I K_p \mathbf{g}_d = 0.5$$

The controller gain is $100/P$, and the integrator gain is given in (1.32)

$$\frac{100}{P} \frac{\tau_o}{2\pi\tau} K_p = 0.5$$

Substituting $4\tau_d$ for $\tau_o$ and solving for $P$ gives

$$P = \frac{400}{\pi} K_p \frac{\tau_d}{\tau} = 127 K_p \frac{\tau_d}{\tau} \tag{1.48}$$

The salient feature of this relationship is that the allowable proportional band, which is an index of the difficulty of control, varies directly with the ratio of dead time to time constant. As the dead time existing in the process approaches zero, controllability improves; therefore processes should be designed in such a way as to minimize dead time insofar as practicable.

Observe also that as the time constant approaches zero, the proportional band required for damping approaches infinity. This is due to the dynamic gain of the integrating capacity increasing with the ratio of period to time constant. A *self-regulating* capacity has an upper limit of unity for its dynamic gain; therefore as its time constant approaches zero, the process behaves as if it had only dead time.

Because the phase shift of a self-regulating capacity varies with period, the period of oscillation under proportional control must be determined by trial and error. As an example, let the ratio of dead time to time constant for a self-regulating capacity be

$$\frac{\tau_d}{\tau_1} = 0.1$$

The phase shift in the capacity was given in (1.42). Although $\tau_o$ is unknown, the approximate phase angle of the first-order lag can be calculated assuming $\tau_o = 4\tau_d$

$$\phi_1 = -\tan^{-1}\frac{2\pi}{4(0.1)} = -86.4°$$

Subtracting this phase angle from $-180°$ leaves $\phi_d$

$$\phi_d = -180 -(-86.4) = -93.6°$$

Then the second approximation of the period is

$$\tau_o = \frac{-360}{-93.6}\tau_d = 3.84\tau_d$$

Using this value for $\tau_o$, a new $\phi_1$ is calculated

$$\phi_1 = -\tan^{-1}\frac{2\pi}{3.84(0.1)} = -86.5°$$

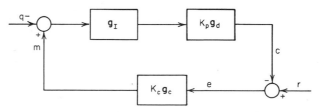

**FIG. 1.18** The combination of a single capacity with dead time is representative of the dynamics of most industrial processes.

The second approximation of $\phi_1$ is so close to the first that no further trials are needed.

Next, the dynamic gain of the first-order lag can be calculated from Eq. (1.42)

$$G_1 = \frac{1}{\sqrt{1 + (\tan \phi_1)^2}} = 0.0611$$

Because the phase angle is so close to $-90°$, the dynamic gain approximates that of an integrator

$$G_1 \approx \frac{\tau_o}{2\pi\tau_1}$$

To demonstrate,

$$G_1 \approx \frac{3.84\tau_d}{2\pi\tau_1} = \frac{3.84}{2\pi}(0.1) = 0.0611$$

In most cases, the approximation is sufficient.

The range of ratios of $\tau_d/\tau_1$ over which the approximation of a non-self-regulating process is valid can be perceived from Fig. 1.19. The dynamic gain of a self-regulating capacity is plotted against $\tau_d/\tau_1$, along with the asymptote representing the non-self-regulating capacity. For $\tau_d/\tau_1$ ratios of 0.5 and below, the approximation is usually adequate, and this includes most processes encountered in industry. On the right-hand scale appears the proportional band required for quarter-amplitude damping.

### Proportional-plus-Integral Control

For the case where the process capacity is non-self-regulating, the settings of proportional and integral modes necessary for quarter-amplitude damping may readily be determined. The procedure also applies to any self-regulating capacity whose time constant is greater than $\tau_d/\pi$, in which case its dynamic gain can be approximated as that of an integrator, as described in connection with Fig. 1.17. As was done for the dead-time process in Table 1.1, Table 1.2 has been compiled starting with a selected controller phase angle, then determining the period from the phase lag remaining in the dead-time element. The gain of the process capacity increases with period, as does the dynamic gain of the

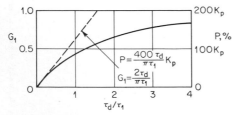

**FIG. 1.19**  The proportional band required for quarter-amplitude damping for any combination of dead time and capacity can be selected from this chart.

**TABLE 1.2    Settings of Proportional and Integral Modes for a Process with Dead Time and Capacity**

| $\phi_{Pb}$ deg | $\phi_d$, deg | $\dfrac{\tau_o}{\tau_d}$ | $G_1$ | $G_{PI}$ | $\dfrac{I}{\tau_d}$ | $P, \%$ |
|---|---|---|---|---|---|---|
| 0 | −90 | 4.00 | $0.637\tau_d/\tau_1$ | 1.000 | ∞ | $127K_p\tau_d/\tau_1$ |
| −15 | −75 | 4.80 | 0.764 | 1.035 | 2.85 | 158 |
| −30 | −60 | 6.00 | 0.955 | 1.155 | 1.65 | 220 |
| −45 | −45 | 8.00 | 1.273 | 1.414 | 1.27 | 360 |
| −60 | −30 | 12.00 | 1.910 | 2.000 | 1.10 | 764 |
| −75 | −15 | 24.00 | 3.820 | 3.864 | 1.02 | 2951 |

controller. As a result, the proportional band must be expanded greatly as more phase lag is introduced into the controller by integral action.

Observe that the integral time seems to approach the process dead time in the limit. This would seem reasonable, in that corrective action cannot be applied at a rate faster than the process can respond. As discussed in connection with Table 1.1, there is an optimum combination of proportional and integral settings, which will be determined after examining performance criteria in Chap. 4.

At this point, it is sufficient to note that the integral time required to produce a particular phase shift in the controller varies directly with dead time and that the proportional band necessary for damping varies with the ratio of dead time to process time constant. In essence, the characteristics of the process determine how well it can be controlled.

### Proportional-plus-Derivative Control

Integral action was able to eliminate offset but at the expense of increasing the period of the loop with its phase lag. Derivative action is the opposite of integration, contributing a phase lead. A proportional-plus-derivative controller can be described mathematically as

$$m = \frac{100}{P}\left(e + D\,\frac{de}{dt}\right) \tag{1.49}$$

where $D$ is the derivative time constant.

Its dynamic gain is calculated as the vector sum of the proportional and derivative components at the period of oscillation

$$\mathbf{g}_{PD} = 1,\angle\, 0° + \frac{2\pi D}{\tau_o},\angle\, +90°$$

$$\mathbf{g}_{PD} = \sqrt{1 + \left(\frac{2\pi D}{\tau_o}\right)^2},\angle\, \tan^{-1}\frac{2\pi D}{\tau_o} \tag{1.50}$$

The gain of derivative action approaches infinity as $\tau_o$ approaches zero, making the controller hypersensitive to noise and other high-frequency disturbances. In actual practice, $\mathbf{g}_{PD}$ is limited to a value between 10 and 20 to stabilize the

**TABLE 1.3   Settings of Proportional and Derivative Modes for a Process with Dead Time and Capacity**

| $\phi_{PD}$, deg | $\phi_d$, deg | $\dfrac{\tau_o}{\tau_d}$ | $G_1$ | $G_{PD}$ | $D/\tau_d$ | $P$, % |
|---|---|---|---|---|---|---|
| 0 | −90 | 4.00 | $0.637\tau_d/\tau_1$ | 1.000 | 0.00 | $127K_p\tau_d/\tau_1$ |
| +15 | −105 | 3.43 | 0.546 | 1.035 | 0.146 | 113 |
| +30 | −120 | 3.00 | 0.477 | 1.155 | 0.276 | 110 |
| +45 | −135 | 2.67 | 0.424 | 1.414 | 0.424 | 120 |
| +60 | −150 | 2.40 | 0.382 | 2.000 | 0.662 | 153 |
| +75 | −165 | 2.18 | 0.347 | 3.864 | 1.296 | 268 |

controller itself and reduce noise sensitivity. This limit is discussed in detail in Chap. 4. At this point, the effectiveness of the ideal derivative as described by Eq. (1.50) is to be evaluated for the process with dead time and capacity.

Table 1.3 lists the proportional and derivative settings for quarter-amplitude damping at selected values of controller phase lead. The procedure followed to generate the table is the same as for Table 1.2, although the results are markedly different. Increasing controller phase angle reduces the period and therefore reduces the process dynamic gain. At the same time, however, the controller gain is increasing. The result of this combination of events is that the proportional band required for quarter-amplitude damping passes through a minimum. If this is accepted as a measure of control effectiveness, 30° appears to be the optimum phase angle for the controller.

Derivative action does not seem to bring about a substantial improvement in control over this process. This factor is borne out by actual tests. In practice, derivative will be found more effective on multicapacity processes, described next, and in canceling the phase lag of integral action in three-mode controllers. This last point is explored in Chap. 4.

## SUMMARY

Control over a given process will be most effective when the gain of the controller is as high as possible. A point is soon reached, however, beyond which controller gain cannot be increased without developing expanding oscillations. This maximum allowable controller gain is determined by the nature of the process, i.e., its steady-state and dynamic response characteristics.

Dead time is the dynamic element principally responsible for limiting controllability. The allowable mode settings and speed of response of the loop are directly related to the value of the dead time therein. A process should be designed for a minimum dead time, insofar as practicable, to maximize controllability. Given a process dead time, there is relatively little a control engineer can do to improve feedback loop response; integral control action can eliminate steady-state offset but sacrifices response speed; derivative action can improve speed somewhat but not greatly. In effect, how well a process can be controlled depends more on the process than on the controller.

## PROBLEMS

**1.1**   The belt speed of the process described in Fig. 1.3 is 12 ft/min, and the weigh cell is located 4 ft from the valve. Estimate the natural period under integral control and the integral time required for quarter-amplitude damping. Is this setting likely to be conservative? Why?

**1.2**   The same process is to be controlled with a proportional-plus-integral controller, adjusted for a phase lag of 60°. Calculate the settings required for quarter-amplitude damping, and check your answer against Table 1.1.

**1.3**   Figure 1.16 is an inverse vector diagram of a first-order lag. Construct a true vector diagram, indicating the magnitude and phase angle of each vector.

**1.4**   Construct a vector diagram for the proportional-plus-derivative controller described by Eq. (1.50). Indicate the magnitude and phase angle of each vector.

**1.5**   Calculate the gain of a dead-time-plus-single-capacity process whose natural period under proportional control is $3.0\tau_d$. What is the ratio of $\tau_d/\tau_1$? Does this point fall on the curve of Fig. 1.19?

**1.6**   A certain process consists of a 1-min dead time and a 30-min lag. Estimate the period and settings for quarter-amplitude damping under proportional-plus-derivative control. Repeat for a proportional-plus-integral controller. Assume 45° phase angle in the controllers.

# Chapter Two

# Characteristics of
# Real Processes

As pointed out in Chap. 1, it is doubtful whether any real process consists exclusively of dead time or single capacity or even a combination of the two. But having become familiar with the properties of these elements, we now can proceed to identify their contributions to complex processes. Some processes are difficult to control, particularly where dead time is dominant. But many processes are poorly controlled because their needs are not understood and therefore not satisfied.

Real processes consist of a combination of dynamic elements and steady-state elements. When many dynamic elements are present, their combined effect is hard to visualize. Even worse, one or more of these elements may be variable. The same is true for steady-state elements. In fact, one could venture to say that many engineers have less comprehension of the steady-state relationships in a complex process than of the dynamic properties. This chapter is devoted to identifying these characteristics for the general case and to putting them into a form in which they can be readily recognized and handled.

Nonlinearities naturally occurring in the process are cause for grave concern. Most processes are nonlinear in some respect. Identification of the source and nature of a nonlinearity is of the utmost significance. Whether it is severe enough to be troublesome and how the trouble can be corrected are important questions which will be answered in the pages that follow. General rules and methods will be stipulated, with a concrete example to illustrate each point. Many more cases will be cited in later chapters as part of specific applications.

It is especially important to keep in mind the prominence of nonlinear characteristics when studying an unfamiliar process; the engineer must know what to look for and what to expect. Tests improperly conducted can give results that are meaningless, confusing, or altogether misleading. The full significance of the "characteristics of real processes" must be appreciated before an intelligent program of testing and evaluation can be undertaken.

## MULTICAPACITY PROCESSES

Many industrial processes consist of two or more capacities through which material or energy must flow on the way from the manipulated to the controlled variable. A distillation column is a familiar example; liquid cascades across a series of trays on its way from the top of the column to the base. Each tray has a capacity to hold liquid, and its content changes with flow rate. In order to evaluate the response of this and similar processes, it is necessary first to examine the effect of a second capacity added to the first and then how connected capacities interact.

### Two-Capacity Processes

Consider the already familiar non-self-regulating process of Chap. 1 having its liquid level measured in a side chamber connected as shown in Fig. 2.1. In the steady state, the level in the chamber will be the same as that in the tank, and so the steady-state gain of the chamber is unity. However, a changing level will require flow to pass between tank and chamber, forcing their levels to differ.

The rate of change of volume $v_2$ in the chamber is proportional to the flow $F_2$ into it

$$\frac{dv_2}{dt} = F_2 \qquad (2.1)$$

Let the controlled variable $c$ be the ratio of contained volume $v_2$ to chamber capacity $V_2$. Then

$$\frac{dc}{dt} = \frac{F_2}{V_2} \qquad (2.2)$$

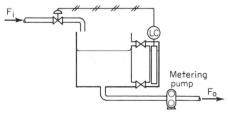

**FIG. 2.1** Because the displacement chamber cannot fill instantaneously, it introduces a second capacity.

For small changes in level, the flow into the chamber is proportional to the difference between tank level $h$ and chamber level $c$

$$F_2 = k_2(h - c) \tag{2.3}$$

Note that $F_2$ will be negative when $h < c$. Combining (2.2) and (2.3) gives a first-order differential equation with a steady-state gain of 1

$$c + \tau_2 \frac{dc}{dt} = h \tag{2.4}$$

where $\tau_2 = k_2/V_2$, a function of chamber volume and resistance to flow.

The response of this second capacity to a step input is of no consequence, since $h$ cannot change stepwise. However, $h$ can change in a ramp, resulting from a step in inflow to the vessel. The difference between measured and actual level is proportional to the rate of change of measured level

$$h - c = \tau_2 \frac{dc}{dt} \tag{2.5}$$

which can be related to the rate of change of actual level and the time $t$ since the start of the ramp

$$h - c = \frac{dh}{dt} \tau_2 (1 - e^{-t/\tau_2}) \tag{2.6}$$

The difference $h - c$ caused by a step change in inflow $f_i$ is found by substituting the derivative of Eq. (1.31) for $dh/dt$

$$h - c = (f_i - f_o) \frac{\tau_2}{\tau_1} (1 - e^{-t/\tau_2}) \tag{2.7}$$

The dynamic gain of the second capacity is simply that of the first-order lag, as described in Eq. (1.42). Since its phase lag only approaches 90°, when it is operating in conjunction with an integrating capacity and proportional controller, the loop cannot oscillate uniformly. It is theoretically possible to control this process with a zero proportional band and obtain a damped response. The response of such a loop to a step in set point when outflow is fixed at 50 percent is shown in Fig. 2.2.

Because the proportional band is zero, inflow is at 100 percent when $c < r$ and zero when $c > r$. Since ramp rates are thereby fixed, the period of oscillation must decrease as amplitude decreases, as Fig. 2.2 shows. This is a characteristic of two-capacity processes.

In actual practice, there will always be other dynamic elements scattered through the loop. In Fig. 2.1, level is measured by a displacer which acts as a pendulum having its own natural period. The liquid-level loop cannot cycle faster than the natural period of the pendulum. This then fixes the lower limit of the dynamic gains of the two capacities and hence the proportional band needed for damped response. The period of the level loop will actually be

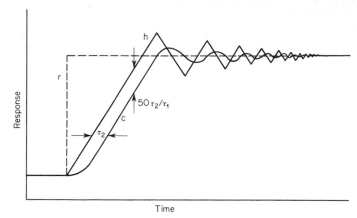

**FIG. 2.2** Zero proportional band will cause a two-capacity process to overrun the set point.

longer than that of the pendulum because the phase lags in the three elements are cumulative. The phase contributions of resonant elements like a pendulum are more complex than a simple lag and are considered in detail in the next chapter.

The control valve also has a limited dynamic response, which may have to be considered. Adding its lag to that of the other two capacities will permit the loop phase to shift beyond 180° even without the pendulum. This combination of dynamic elements is so common that virtually every loop contains multiple capacities. However the individual capacities do not necessarily respond independently. Consequently, the study of multicapacity processes must include an examination of interaction.

### Interacting Lags

The principal distinction to be made in multicapacity processes is in how the capacities are joined. If they are said to be isolated or noninteracting, the capacities behave exactly as they would alone. But if coupled, they interact with one another, in which case the contribution of each is altered by the interaction. Figure 2.3 compares the two forms.

In the upper left, the two tank levels do not interact because the flow from the first to the second is independent of the level in the second. The lower picture, however, illustrates the case where *both* inflow and outflow are a function of tank level. The levels interact because any change in the downstream level will affect the upstream level. An electrical analog of each process appears on the right. The amplifier in the upper figure isolates the two lags by preventing the voltage on the second from affecting that on the first. The lower right figure, without the amplifier, is a two-stage ladder network. A multistage ladder is often used to simulate a transmission line.

The significance of interaction is that it changes the effective time constants of the individual capacities. The magnitude of the change is striking. The solution of the equation for determining the effective time constants is irrational [1], but for the special case where two equal capacities with equal time constants $\tau$ interact, their combined response is that of two noninteracting lags with the values

$$\tau_{1,2} = \frac{3 \pm \sqrt{5}}{2} \tau \qquad \tau_1 = 2.618\tau \quad \text{and} \quad \tau_2 = 0.382\tau$$

The following general rules apply to the principle of interaction:

1. The degree of interaction is proportional to the ratio of the smaller to the larger capacity (not time constant). Where this ratio is low ($<0.1$), the capacities may be assumed not to interact.

2. Interaction always works toward increasing the larger time constant and decreasing the smaller one.

3. Specifically with regard to the behavior of systems with equal time constants $\tau$, of equal capacity, the effect is a combination of one large and the rest small time constants whose normalized sum is

$$\sum_{i=1}^{i=n} \frac{\tau_i}{\tau} = \sum_{i=1}^{i=n} i = \frac{n^2 + n}{2} \tag{2.8}$$

where $i$ is each time constant and $n$ is the number of capacities, and whose normalized product is

$$\prod_{i=1}^{i=n} \frac{\tau_i}{\tau} = 1.0 \tag{2.9}$$

A case in point is the two-capacity process cited above

$$\frac{\tau_1 + \tau_2}{\tau} = 2.618 + 0.382 = 3$$

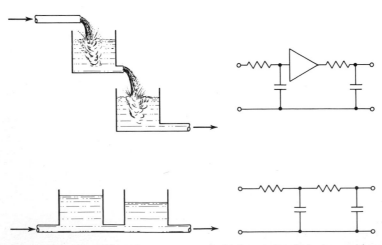

**FIG. 2.3**   Noninteracting (above) compared with interacting (below) capacities.

Since $n = 2$,

$$\frac{n^2 + n}{2} = \frac{4 + 2}{2} = 3 \quad \text{and} \quad \frac{\tau_1 \tau_2}{\tau \ \tau} = (2.618)(0.382) = 1.00$$

With three capacities of time constant $\tau$,

$$\frac{\tau_1}{\tau} = 5.0505$$

$$\frac{\tau_2}{\tau} = 0.6405$$

$$\frac{\tau_3}{\tau} = \underline{0.3090}$$

$$6.0000$$

and $(5.0505)(0.6405)(0.3090) = 1.00$.

The reasons for interaction can be visualized to some extent. For example, in the interacting tanks of Fig. 2.3, the flow entering the first tank must ultimately fill both tanks, whereas that entering the second fills the second. The sum of the time constants then becomes 3.

In the absence of interaction, each capacity in a multicapacity system develops its own phase lag at the period of oscillation. For the case of $n$ equal time constants in a proportional control loop, each contributes a phase lag of $180\%/n$

$$\frac{180}{n} = \tan^{-1} \frac{2\pi\tau}{\tau_n}$$

Solving for $\tau_n$ gives

$$\tau_n = \frac{2\pi\tau}{\tan (180/n)} \tag{2.10}$$

The dynamic gain of the combination is that of the individual time constant raised to the $n$th power:

$$\prod_{i=1}^{i=n} G_i = \left\{ \frac{1}{\sqrt{1 + [\tan(180/n)]^2}} \right\}^n = \left[ 1 + \left( \tan \frac{180}{n} \right)^2 \right]^{-n/2} \tag{2.11}$$

The natural period $\tau_n$ and the dynamic-gain product calculated in this manner are listed for selected numbers of equal noninteracting lags in Table 2.1.

Observe that as $n$ increases, the natural period and dynamic gain approach those of dead time. Step responses bear out the resemblance. The step response for a series of $n$ equal noninteracting lags is

$$\Delta c = K_p \, \Delta m \left\{ 1 - \left[ 1 + \frac{t}{\tau} + \frac{(t/\tau)^2}{2!} + \cdots + \frac{(t/\tau)^{n-1}}{(n-1)!} \right] e^{-t/\tau} \right\} \tag{2.12}$$

Responses for $n = 2$, 10, and $\infty$ are shown in the upper half of Fig. 2.4.

The response characteristics of equal interacting lags are considerably different. Step responses for two and ten equal interacting lags are shown in the

**TABLE 2.1    Natural Period and Gain Product of a Series of Equal, Noninteracting Lags**

| $n$ | $\dfrac{\tau_n}{n\tau}$ | $\Pi G_i$ |
|-----|------|-------|
| 3   | 1.21 | 0.125 |
| 5   | 1.73 | 0.347 |
| 10  | 1.93 | 0.605 |
| 20  | 1.98 | 0.781 |
| 50  | 2.00 | 0.906 |
| 100 | 2.00 | 0.952 |

lower half of Fig. 2.4, normalized to the sum of all the effective lags. Characteristically, a number of equal interacting lags resolves into noninteracting lags, one large and the rest small. The large lag becomes the dominant time constant, while the remaining small lags combine to form what is effectively dead time.

As a general rule, multicapacity processes contain a natural interaction, responding like the lower set of curves in Fig. 2.4. This form of response is evident both in processes consisting of a large number of discrete stages and in those embodying a continuum of distributed particles. Examples of multistage processes are plate columns for distillation, extraction, and absorption. Counterflow of the two phases produces the interaction. Packed columns, on the other hand, are *distributed* systems, which behave similarly. Diffusive processes such as heat transfer by conduction, mixing in pipes and vessels, and flow through porous media react in much the same manner. More attention will be devoted to these operations when specific applications are investigated.

From Fig. 2.4 it can be seen that the interacting multicapacity process differs from the dead-time-plus-single-capacity process in the smooth upturn at the beginning of the step response. This curvature indicates that the dead time is

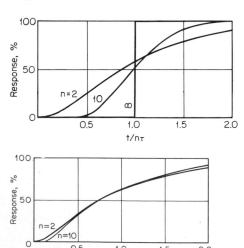

**FIG. 2.4** The difference in step response between isolated (above) and interacting (below) lags becomes more pronounced as $n$ increases.

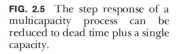

**FIG. 2.5** The step response of a multicapacity process can be reduced to dead time plus a single capacity.

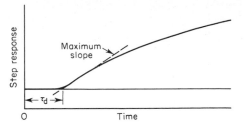

not pure but is the result of many small lags and therefore the process will be somewhat easier to control. By the same token, derivative action will be of more value than it was in the case of dead time and a single capacity. Nonetheless, if we choose to estimate the necessary controller settings on the basis of a single-capacity-plus-dead-time representation, we will err on the safe side.

The natural period of the loop can be predicted with surprising reliability by noting where the maximum slope of the step-response curve intersects the time base. This intersection, marked in Fig. 2.5, identifies the effective dead time of the process. The effective dead time plus the effective time constant equals the total lag in the process

$$\tau_d + \tau_1 = \tau \frac{n^2 + n}{2} \tag{2.13}$$

Equation (2.13) requires that the step response of any number of equal interacting lags reach 63.2 percent at time $\tau(n^2 + n)/2$, which is corroborated by Fig. 2.4.

Ziegler and Nichols [2] noted that the natural period of oscillation will be 4 times the effective dead time, whether the process is interacting or not. So the technique of dealing with single capacity plus dead time takes on added value in being applicable to these examples of complex dynamics. This is an important insight; without it, numerical methods would have to be rejected for use on any process containing more than two dynamic elements. And when on-the-spot analysis must be made, the shortcut numerical method is invaluable. Fortunately, a single-capacity-plus-dead-time process can be made to represent any degree of difficulty from one extreme to the other, simply by varying the ratio $\tau_d/\tau_1$. Thus its application is universal, if approximate.

As an example, the 10-capacity interacting process of Fig. 2.4 has an effective dead time of about 0.15 of the total lag. Since the balance is the dominant time constant, the ratio $\tau_d/\tau_1 = 0.15/0.85 = 0.18$. The proportional band needed for quarter-amplitude damping of this process can be found by referring to Fig. 1.19.

Figure 2.6 is a correlation of the ratio of effective dead time to effective lag against the number of interacting stages. Data from tests on systems of 2 to 10 capacities fall in a straight line on semilogarithmic coordinates. This relationship is extremely useful in predicting the dynamic behavior of any process with a discrete number of interacting stages.

The dynamic gain of three equal interacting lags at the natural period of the loop can readily be calculated, but beyond that it becomes increasingly difficult. A period must first be assumed and the phase angles of the lags calculated and summed; if they do not total $-180°$, another period must be used. Table 2.2 lists the phase angles and gains for three equal interacting lags at their natural period.

Observe that the natural period is about the same as for three equal noninteracting lags, $2.57\tau$ vs. $2.42\tau$. However, the dynamic gain is lower by the factor of 0.034/0.125, indicating how much more easily interacting lags can be controlled. When $n$ is increased to 10, the effective dead time of the noninteracting process increases to about equal the effective time constant, while that for the interacting process is only about 18 percent of the effective time constant.

At this point it is worthwhile to review from the aspect of interaction the examples of two-capacity processes that have already been presented. The capacities appearing in Fig. 2.1 are definitely interacting because changes in chamber level can cause changes in tank level. But the capacity of the chamber is so much smaller than that of the tank that the response is effectively noninteracting. This is generally true for lags in measuring elements; a temperature bulb draws heat from the fluid surrounding it, but its heat capacity is so much smaller that noninteraction can be assumed. However, when the process stream flows *through* connected lags of similar time constant, their capacities also tend to be similar in that they carry the same flow. Then interaction is probable unless the capacities are decoupled, as in Fig. 2.3.

### Pneumatic Transmission Lines

Transmission lines are true distributed lags; i.e., resistance and capacity are distributed uniformly throughout their entire length. Their step response, as indicated by data in Ref. 3, can be represented by a dead time and single capacity whose ratio is approximately 0.24. This could be modeled by a network of about 20 equal interacting time constants, according to Fig. 2.6.

Because pneumatic tubing is, in effect, a series of interacting lags, doubling its length triples its total lag. Conversely, half the length produces one-third the lag. Table 2.3 shows this relationship to hold between 200 and 2000 ft, while below 200 ft, the relationship between lag and length is more nearly linear.

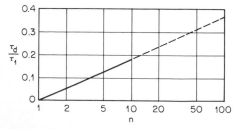

**FIG. 2.6** The ratio of effective dead time to effective lag of $n$ equal interacting capacities varies with the logarithm of $n$.

**TABLE 2.2  Phase Angles and Loop Gain for Three Equal Interacting Lags at $\tau_n = 2.57\tau$**

| Effective lag | $\phi_i$, deg | $G_i$ |
|---|---|---|
| $\tau_1 = 5.0505\tau$ | 85.4 | 0.081 |
| $\tau_2 = 0.6405\tau$ | 57.4 | 0.538 |
| $\tau_3 = 0.3090\tau$ | 37.1 | 0.798 |
| $\Sigma\tau_i = 6\tau$ | $\Sigma\phi_i = 179.9$ | $\Pi G_i = 0.034$ |

Tubing of 0.305-in-ID (⅜-in-OD) offers nearly a 50 percent reduction in both dead time and lag over 0.188-in-ID (¼-in-OD) tubing. Being more costly, however, it is generally reserved for shorter runs terminating in a large-volume device such as a valve motor.

The terminating volume can substantially increase both the dead time and time constant. Where speed of response is important, an isolating amplifier should be inserted between the tubing and a valve operator. The amplifier may be a simple repeating device or a valve positioner which closes a feedback loop around the valve. The choice depends on other considerations, discussed under valve positioners in Chap. 6.

### Predicting the Behavior of a Process

The appearance of a piece of processing equipment often reveals the nature of its dynamic characteristics. If all the dimensions are similar, as in a cylindrical tank where the height is of the same magnitude as the diameter, capacity will predominate, dead time, if any, being short. But if the vessel has one dimension much greater than the others, dead time may be dominant, though not without some capacity. Thus a shell-and-tube heat exchanger will exhibit considerable dead time, compared with a heated tank, whose principal elements would be lags. Just the appearance of a tower—distillation, absorption, or whatever—indicates the presence of dead time.

One could almost generalize to the extent of relating controllability to dimension:

$$\frac{\tau_d}{\tau_1} = f\left(\frac{\text{length}}{\text{diameter}}\right)$$

**TABLE 2.3  Time Lags for Pneumatic Tubing**
(0.188-in (4.8 mm) ID, bellows termination)

| Length | | | | |
|---|---|---|---|---|
| ft | m | $\Sigma\tau$, s | $\tau_d$, s | $\tau_1$, s |
| 50 | 15 | 0.20 | 0.04 | 0.16 |
| 100 | 30 | 0.40 | 0.08 | 0.32 |
| 200 | 61 | 0.90 | 0.17 | 0.73 |
| 500 | 152 | 4.0 | 0.77 | 3.2 |
| 1000 | 305 | 12 | 2.3 | 9.7 |
| 2000 | 610 | 38 | 7.4 | 31 |

Of course such an expression could only be written to apply within a specific system because many more factors are involved. Nonetheless, if dead time is related to length, the natural period is similarly related to length, as with a simple pendulum.

**Variable Time Constants**

In Chap. 1 a self-regulating process was described whose principal time constant varied with valve opening. The reader should regard the paradox of a *variable* time *constant* more as the rule than the exception. Fortunately, that process had a constant gain product because its time constant and steady-state gain varied together. This will not always be the case.

A very common process in industry is the tubular heat exchanger, shown in Fig. 2.7, which responds as a distributed lag. Step-response tests indicate that the point where tube outlet temperature reaches 63.2 percent of its final value occurs at a time roughly equivalent to the residence time of the fluid in the tubes. However, the residence time, i.e., tube volume divided by flow rate, varies with flow rate. As a consequence, both dead time and time constant vary inversely with flow.

In this process, the steady-state gain, i.e., the final change in outlet temperature for a given change in heat input, also varies inversely with flow. This is reasonable; the same increment of heat applied to half the flow will produce twice the temperature rise. Response curves for outlet temperature resulting from a step in heat input at two different flow rates appear in Fig. 2.8. The variation of steady-state gain and time constant with flow mirrors the behavior of the self-regulating liquid-level process exactly and is characteristic of self-regulating processes in general. Composition processes also illustrate this behavior. However, the variability of dead time causes the period (and hence the gain product of the process) to vary inversely with flow. A 50 percent reduction in flow will double the period and loop gain. If no compensation is applied for these parameter changes, the temperature control loop will tend to be overdamped at high flow rates and unstable at low rates. If instability cannot

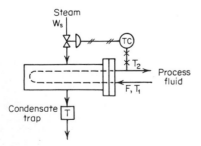

**FIG. 2.7** The tubular heat exchanger responds as a distributed lag with variable parameters.

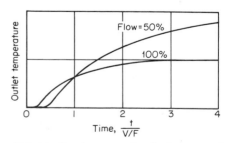

**FIG. 2.8** Steady-state gain, dead time, and time constant all vary with flow through the heat exchanger.

be tolerated, the control modes must be adjusted to favor the worst-case conditions, i.e., the lowest anticipated flow.

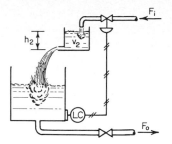

Fortunately, a number of techniques are available for compensating this variability. It can be achieved to some degree by valve selection, by adapting the controller settings, and by applying feedforward control, as presented at appropriate points later in the book.

**FIG. 2.9** Because the volume in the head tank varies with flow squared, its time constant varies directly with flow.

Another element with variable time constant is the head tank operating in turbulent flow. The process is described in Fig. 2.9. The time constant in the head tank is essentially the change in volume caused by a small change in flow

$$\tau_2 = \frac{dv_2}{dF_i} \qquad (2.14)$$

However, flow and volume are related nonlinearly in the steady state

$$F_i = k_2\sqrt{h_2} = k_2 \sqrt{\frac{v_2}{A_2}} \qquad (2.15)$$

where $h_2$ is the head of liquid above the nozzle and $A_2$ is the inside area of the tank in the horizontal plane. Differentiating (2.15) gives

$$\frac{dv_2}{dF_i} = \frac{2A_2}{k_2^2} F_i$$

Therefore

$$\tau_2 = \frac{2A_2}{k_2^2} F_i \qquad (2.16)$$

If $A_2$ and $k_2$ are constant, $\tau_2$ varies directly with $F_i$. Because the secondary time constant varies directly with flow, the period of the level loop and hence the dynamic gain of the principal capacity tend also to vary directly with flow. The compensation required for this process is the opposite of what was needed for the heat exchanger. It is achievable with proper valve selection, as described later in this chapter, and also through the use of a cascade loop, as described in Chap. 6.

Another not uncommon characteristic is a period of oscillation that increases with time. This can be caused by fouling of a surface with a growing thickness of crust. It can develop in heat-transfer processes, particularly polymerization reactors, and in pH loops, where solids accumulate on the surface of the measuring electrode. After cleaning, normal response is usually achieved.

Continuous cleaning, like that provided with ultrasonic devices, may be required to eliminate this variability.

## STEADY-STATE GAIN

Process steady-state gain $K_p$ has been shown in the various loops examined but without much discussion. At this point, it is to be flanked by two other steady-state gains, that of the valve $K_v$ and that of the transmitter $K_t$. The product of the three terms must be dimensionless, in that it relates controller input to its output. The input to the valve is flow. The process converts flow into the units of the controlled variable, which the transmitter converts into a signal again. For a loop where temperature is controlled by manipulating a steam valve

$$K_v K_p K_t = \frac{\text{lb/h}}{\%} \frac{°\text{F}}{\text{lb/h}} \frac{\%}{°\text{F}}$$

leaving a dimensionless product.

### Transmitter Gain

In the liquid-level process presented in Chap. 1, the measurement $h$ was defined as representing the fractional contents of the tank. This trick enabled us to find the time constant of the vessel in terms of its capacity $V$ and its nominal throughput $F$. When instrumenting a plant, however, it is not necessary that every liquid-level transmitter be scaled to measure the entire volume of the vessel. If, instead, the transmitter span represents only a small percentage of the vessel volume, the vessel will have effectively shrunk to the span of the transmitter. To state it another way: for control purposes, those parts of the vessel beyond the range of the transmitter do not exist.

Reducing the span of a transmitter is equivalent to reducing the proportional band of the controller. If a particular damping, hence a particular loop gain, is to be achieved, the proportional-band estimate must take into account the span of the transmitter.

In order to facilitate the evaluation of systems more complex than the liquid-level process, the transmitter gain will be explicitly defined

$$K_t = \frac{100\%}{\text{span}} \tag{2.17}$$

The numerator in Eq. (2.17) is the output that will be produced for a full-span change in input. $K_t$ is not a pure number; it has the dimensions of the measurement. Suppose a level transmitter were calibrated to a range of 20 to 100 in of water. Its gain would then be 100%/80 in, or 1.25%/in.

It is entirely possible that $K_t$ is not constant. This would be the case if the transmitter were nonlinear. Few transmitters are sufficiently nonlinear to show any marked effect on control-loop stability. A change in gain of at least 1.5/1 would be necessary to cause difficulty. Some temperature measurements are nonlinear but seldom to this extent. The most notable case of a nonlinear transmitter is the differential flowmeter, whose output varies with the square of flow through the primary element.

Each flow transmitter has its own particular span. But in addition, the differential flow transmitter has the nonlinear relationship. $K_t$ can be determined on the basis of transmitter span, with the nonlinearity applied as a coefficient. Let $h$ = fractional differential pressure and $f$ = fractional flow. Then

$$h = f^2 \tag{2.18}$$

and

$$\frac{dh}{df} = 2f \tag{2.19}$$

The derivative $dh/df$ is the dimensionless gain of the transmitter. Its dimensional gain is then

$$K_t = 2f\frac{100\%}{\text{span}} \tag{2.20}$$

As an example, look at a differential flowmeter whose scale is 0 to 500 gal/min. $K_t = 2f[0.2\%/(\text{gal/min})]$. At full-scale flow, $K_t = 0.4\%/(\text{gal/min})$; at 50 percent flow, $K_t = 0.2\%/(\text{gal/min})$; at zero flow, $K_t = 0$.

The result of the nonlinearity is that the control loop will not perform consistently at different rates of flow. If the proportional band is adjusted for acceptable damping at 50 percent flow, the loop will be undamped at 100 percent flow and sluggish near zero flow. The problem can be readily resolved, however, by inserting a square-root extractor, whose output would be linear with flow.

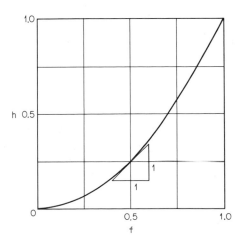

**FIG. 2.10**  The gain of a differential flow transmitter is directly proportional to flow.

## Valve Gain

Referring again to the liquid-level process of Chap. 1, the time constant of the vessel was based upon the rated flow $F$ which the control valve was capable of delivering. The time constant thus depends on valve size; consequently, the proportional band is a function of valve size. Looking at it another way, an oversize valve would be operated only over part of its travel; the span of stem travel would be less than 100 percent. Therefore the proportional band must be wider to compensate.

The gain of a valve can be defined as the change in delivered flow vs. percent change in stem position. The gain of a linear valve is simply the rated flow under nominal process conditions at full stroke

$$K_v = \frac{F_{max}}{100\%} \tag{2.21}$$

If a linear valve were able to deliver 500 gal/min fully open at stipulated process conditions, $K_v$ would be 5 (gal/min)/%. Notice that valve gain has dimension, as did transmitter gain, but now the percent sign is in the denominator. The valve is at the output of the controller, whereas the transmitter is at the input. Controller gain is therefore in terms of %/%, hence dimensionless.

Since valves cannot be manufactured to the same tolerance as transmitters, there is no such thing as a truly linear valve. But perfect linearity is not essential because a control loop does not demand it. Some valves are deliberately characterized to particular nonlinear functions, in order to carry out certain specific duties better. The most commonly used characterized valve is the equal-percentage type.

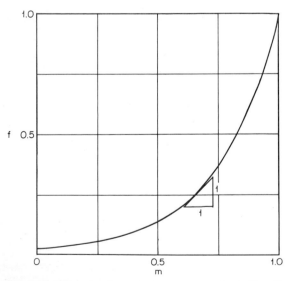

**FIG. 2.11**  The gain of an equal-percentage valve is directly proportional to flow.

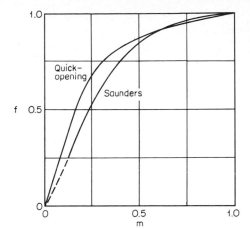

**FIG. 2.12**  Characteristics of quick-opening and Saunders valves.

The equal-percentage valve has an exponential characteristic

$$f = R^{m-1} \qquad (2.22)$$

where $R$ is the rangeability of the valve and $m$ is its fractional stem position. Rangeability is defined as the ratio of maximum to minimum controllable flow. Figure 2.11 gives the characteristic curve for an equal-percentage valve with 50:1 rangeability, under constant pressure drop. Notice that $f = 0.02$ at $m = 0$; this is the minimum controllable flow; below this point, the valve tends to close completely.

The slope of the characteristic curve is the derivative of (2.22)

$$\frac{df}{dm} = f \ln R \qquad (2.23)$$

This is the source of the valve's name: the flow varies an equal percentage of its value, that is, $df/f$, for each increment in stem position $dm$.

Most common equal-percentage valves have a rangeability of 50, exhibiting a slope at full flow of 3.9, which is ln 50. Combining the slope with the dimensional component of gain gives

$$K_v = \ln R \, f \frac{F_{max}}{100} \qquad (2.24)$$

Observe however that the product $fF_{max}$ is the actual flow being delivered. The gain of an equal-percentage valve is then not a function of valve size as long as operation is confined to the range where the characteristic remains undistorted.

Ball valves and butterfly valves also tend to exhibit equal-percentage characteristics because of the way port opening varies with stem rotation. Other common valves include the quick-opening and the Saunders-style diaphragm valve. Their characteristics have the opposite curvature to the equal-percentage type, as shown in Fig. 2.12. The lower 10 percent of the Saunders characteristic is shown as a broken line because results are generally not reproducible in that region.

The equal-percentage valve is used to compensate for gain variations elsewhere in the loop. One such instance is the inverse variation of process dynamic gain with flow through the tubular heat exchanger, described in Figs. 2.7 and 2.8. If an equal-percentage valve were used to deliver steam to the heat exchanger, its direct-gain variation with steam flow would cancel the inverse-gain variation of the process. Although it would have no influence over the period of oscillation, the loop gain at least would be compensated.

### Linearizing an Equal-Percentage Valve

There are instances when a linear characteristic is desirable but unavailable. This is often the case where butterfly or ball valves must be used because of mechanical considerations but linear performance is needed. One method of moderating the equal-percentage characteristic is through the use of a divider, as shown in Fig. 2.13. Most dividers have adjustable scaling factors, providing calculations of a form

$$X = \frac{yY}{z + (1 - z)Z} \tag{2.25}$$

where $Y, Z$ = input signals
$y$ = constant multiplying factor
$z$ = adjustable zero elevation
When used as shown in Fig. 2.13, with $y$ fixed at 1.0, the formula becomes

$$f(m) = \frac{m}{z + (1 - z)m} \tag{2.26}$$

This function passes through points $(0, 0)$ and $(1, 1)$ for all values of $z$, with the

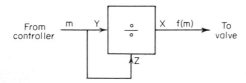

**FIG. 2.13** A divider can generate a nonlinear function suitable for compensating an equal-percentage valve characteristic.

severity of the curvature determined by $z$. Several such curves are shown in Fig. 2.14. The slope of the function is the derivative of (2.26)

$$\frac{df(m)}{dm} = \frac{z}{[z + (1 - z)m]^2} \tag{2.27}$$

When evaluated at $m = 0$, the slope is $1/z$; evaluated at $m = 1$, the slope is $z$.

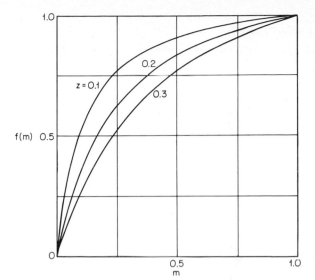

**FIG. 2.14** The divider can be used as a function generator to compensate the gain variation of an equal-percentage valve.

Therefore the gain variation for the function over its entire range is $1/z^2$. When $z$ is 0.10, for example, the slope of the curve varies from 10 to 0.1, changing by a factor of 100.

The gain of an equal-percentage valve varies directly with flow. Therefore its gain variation is also its rangeability. A divider used to compensate for the curve of an equal-percentage valve should have its gain changed by the same amount. This provides a quick estimate of the value of $z$ needed to match the valve

$$z = \sqrt{\frac{1}{R}} \tag{2.28}$$

For the valve with a rangeability of 50, $z$ should be set at 0.141.

**Variable Pressure Drop**

The inherent characteristics of the valves described are subject to distortion due to variations in pressure drop with flow. The actual flow $F$ of liquid passing through a valve is related to pressure drop $\Delta p$ as

$$F = aC_v \sqrt{\frac{\Delta p}{\rho}} \tag{2.29}$$

where $a$ = fractional opening of valve
$C_v$ = flow coefficient of valve
$\rho$ = density of liquid
(Standard values of $C_v$ are used with $\Delta p$ expressed in pounds per square inch, $\rho$ in specific-gravity units, and flow in gallons per minute.) In a linear valve,

fractional opening $a$ equals fractional stem position $m$, while in an equal-percentage valve $a = R^{m-1}$.

In a flowing system, pressure drop across the control valve usually varies with flow due to losses elsewhere in the piping, pump, and vessels. When the valve is shut, these losses are zero and therefore the pressure drop across the valve is at its maximum $\Delta p_{max}$. As flow increases, pressure drop is absorbed elsewhere, reducing that across the valve

$$\Delta p = \Delta p_{max} - kF^2 \tag{2.30}$$

where $k$ is a factor representing fixed resistances in the system. At maximum flow, $\Delta p$ is minimum

$$\Delta p_{min} = \Delta p_{max} - kF^2_{max} \tag{2.31}$$

Rearranging (2.30) and (2.31) and dividing gives

$$f^2 = \frac{F^2}{F^2_{max}} = \frac{\Delta p_{max} - \Delta p}{\Delta p_{max} - \Delta p_{min}} \tag{2.32}$$

Then if (2.29) is solved first for $F$ and then for $F_{max}$ and their quotient is obtained, we have

$$f^2 = \frac{a^2 \, \Delta p}{\Delta p_{min}} \tag{2.33}$$

When (2.32) and (2.33) are combined by eliminating $\Delta p$, we get

$$f = \frac{1}{\sqrt{1 + (1/a^2 - 1)\Delta p_{min}/\Delta p_{max}}} \tag{2.34}$$

Figure 2.15 is a plot of fractional flow vs. fractional valve opening as found by Eq. (2.34) for three $\Delta p$ ratios. These are known as the *installed* characteristics of a linear valve.

The derivative of (2.34) is the slope of the installed characteristic curve

$$\frac{df}{da} = \frac{\Delta p_{min}}{\Delta p_{max}} \left[ a^2 + (1 - a^2) \frac{\Delta p_{min}}{\Delta p_{max}} \right]^{-3/2} \tag{2.35}$$

The maximum slope occurs at $a = 0$

$$\left( \frac{df}{da} \right)_0 = \frac{1}{\sqrt{\Delta p_{min}/\Delta p_{max}}} \tag{2.36}$$

and the minimum slope lies at $a = 1$

$$\left( \frac{df}{da} \right)_1 = \frac{\Delta p_{min}}{\Delta p_{max}} \tag{2.37}$$

The change in gain experienced across the entire travel of the valve opening is the ratio of (2.36) to (2.37)

$$\frac{(df/da)_0}{(df/da)_1} = \left(\frac{\Delta p_{\min}}{\Delta p_{\max}}\right)^{-3/2} \tag{2.38}$$

The slopes of the curves shown in Fig. 2.15 change by factors of 31.6, 11.2, and 6.1, for pressure-drop ratios of 0.1, 0.2, and 0.3, respectively.

If the remaining elements in the control loop all have constant gains or gains varying in the same direction, stability will vary substantially with flow. However, the type of curve shown in Fig. 2.15 can be useful in at least two applications. It can complement the curve of the flowmeter shown in Fig. 2.10 quite well. The flowmeter curve has a gain of 2 at full scale; an ideal match at full flow would be a valve characteristic with a slope of ½ at full flow, achieved when $\Delta p_{\min}/\Delta p_{\max} = 0.5$. If a match is desired at a lower flow rate, a lower pressure-drop ratio would be preferred.

Another process whose gain increases with flow was the two-capacity process shown in Fig. 2.9. Like the differential flowmeter, this liquid-level process exhibited a gain increasing directly with flow and can therefore be compensated in the same way.

In many instances, a valve gain varying inversely with flow is undesirable and must be compensated. The equal-percentage characteristic is best able to offset the effect of variable pressure drop on valve gain. Figure 2.16 compares the installed characteristic of an equal-percentage valve for three pressure-drop ratios with its inherent characteristic. Observe that extreme pressure-drop ratios cause the installed characteristic to become more nearly linear. Consequently, when these conditions are encountered, an equal-percentage valve is usually chosen, perhaps explaining its predominant use over linear valves.

Variable pressure drop reduces valve rangeability, as shown in the curves of Fig. 2.16. Rangeability is the ratio of maximum to minimum controllable flow.

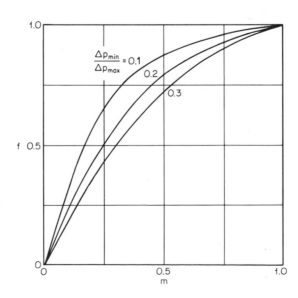

**FIG. 2.15** Line resistance distorts a linear characteristic toward that of a quick-opening valve.

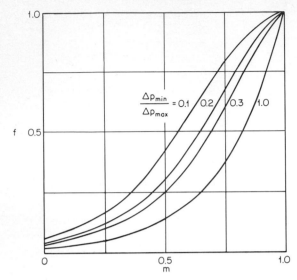

**FIG. 2.16** An equal-percentage valve is able to remove most of the effect of line drop.

Since maximum flow occurs at minimum pressure drop and vice versa, the pressure-drop ratio determines the effective rangeability of a valve

$$R_{eff} = R \sqrt{\frac{\Delta p_{min}}{\Delta p_{max}}} \qquad (2.39)$$

The rangeability of the valve in Fig. 2.16 is reduced from 50 to 15.8 when the pressure drop changes by 10:1 with flow.

### Process Gain

The output of a valve is flow; the process accepts this flow and converts it into the controlled variable. If the controlled variable is also *flow,* as in a flow-control loop, the process gain is *unity.* But for any other controlled variable, the process gain has a dimensioned value.

If the controlled variable is an integral of flow, as liquid level or pressure, process gain may be taken into account in its integrating time constant $V/F$. If the process is self-regulating, there will be a finite steady-state gain, although it may also appear in the time constant as well. These situations have already been described as being approximated by an integrating element having a constant $V/F$.

Processes producing a controlled variable that is a property of a flowing stream, such as temperature or composition, have a readily calculable (although not necessarily constant) steady-state gain. A case in point is the heat exchanger described in Fig. 2.7. A heat balance across the exchanger gives

$$Q = W_s \, \Delta H_s = FC(T_2 - T_1) \qquad (2.40)$$

where $Q$ = heat-transfer rate
$\quad W_s$ = steam mass flow
$\quad \Delta H_s$ = latent heat of vaporization
$\quad F$ = feed rate
$\quad C$ = heat capacity of feed
$\quad T_1$ = inlet temperature
$\quad T_2$ = outlet temperature

The steady-state gain is the derivative of outlet temperature with respect to steam flow

$$K_p = \frac{dT_2}{dW_s} = \frac{\Delta H_s}{FC} \tag{2.41}$$

having the dimensions of degrees temperature per unit mass flow. Observe that the steady-state gain for this process varies inversely with flow; this point was made in discussing the step response of the heat exchanger in Fig. 2.8. It is a common property in temperature and composition-control loops.

A composition-control loop noted for its variable gain is pH. The nature and characteristics of pH curves are developed in detail in Chap. 10. At this point we need examine only how their gain is determined. Figure 2.17 describes a typical curve where measured pH is related to the ratio of acid reagent added to the flow of an influent stream. Observe first the existence of this ratio; it indicates that process gain, i.e., the change in pH per unit acid flow, changes with the flow of influent to be treated.

The controller should respond to a given deviation $e$ with a change in output $\Delta m$ that will reduce $e$ to zero. The controller gain required to accomplish this is

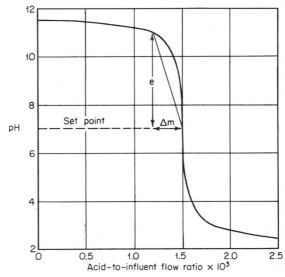

**FIG. 2.17** The gain of the pH curve is the pH deviation caused by a change in reagent flow.

inversely related to the process gain, which can be defined as the ratio of the deviation to the change in $m$ which will eliminate it

$$K_p \equiv \frac{e}{\Delta m} \tag{2.42}$$

It is represented by the slope of a straight line connecting a value of deviation to zero, as shown in Fig. 2.17. The slope of the pH curve is equal to its gain only at the control point. The slope at pH 11, for example, is unrelated to the control point and is much less than $K_p$.

The controller must be adjusted to provide damping for small values of deviation if steady-state stability is to be realized. Therefore the value of $K_p$ at the set point is used to determine loop gain. The fact that $K_p$ decreases on an increasing deviation reduces loop gain to larger upsets, retarding recovery. Special nonlinear components are presented in Chap. 5 to cope with this situation.

## TESTING THE PLANT

In the late 1950s there was much talk of extensive tests on processes using frequency-response analysis. In fact some tests were conducted on reactors, heat exchangers, and distillation columns. Although a certain amount of information was obtained using this method, two major objections stand out: (1) the tests are unbelievably time-consuming, and (2) they assume that the process is linear and invariant.

The first objection rules out testing in most plants because of the unwillingness of operating personnel to tolerate upsets for long intervals and because of the expense of manpower and equipment. The second objection indicates that the results of tests on a process with nonlinear elements may be not only invalid but also misleading. Frequency response is suitable only for fast, linear devices like instruments, controllers, amplifiers, etc.

The author has been called upon many times to investigate a process which was in trouble. In these instances it was impossible to bring an extensive array of test equipment or to spend days gathering information. In most cases the process was nonlinear in some respect and not well understood by the operating people; otherwise it would not have been in trouble. A simple test procedure was decided on, independent of linearity, from which the dominant properties of the system could be determined. A properly conducted test should pinpoint problem areas with a minimal upset to the process.

To keep testing to a minimum, all available knowledge of the process must be employed. The volume of vessels and flow rates is always available, from which time constants can be calculated. The length and diameter of piping runs can serve to locate dead-time elements. By identifying all the known or knowable elements in this way, any tests will be of more value in defining the unknown elements which make up the balance of the loop.

The author has always reacted strongly to any test procedure that is based upon knowing nothing about a process. Many things about an unfamiliar process can be learned by observing the vessels and piping, examining the chemistry and physics involved, and talking to the operators. Preliminary information like this is of inestimable help in indicating what to look for and where to find it. It is surprising how much can often be learned about a particular process without even making a test. Occasionally the tests will not substantiate the expectations, which provides a challenging opportunity to learn.

### A Simplified Test Procedure

Before describing how to conduct a test, it is important to point out how not to conduct a test, in order to avoid some serious pitfalls.

1. Do *not* test for steady-state gain. In Chap. 1 it was pointed out that the steady-state gain of a single-capacity liquid-level process is not constant. It varies with both flow and level. Yet the dynamic gain changes too. Because the process is in a control loop, their product, the loop gain at $\tau_o$, is of real consequence.

2. Do *not* test for time constants. There are several methods available for finding the time constants in a linear system. But, as in the single-capacity level process, the time constant may vary with flow without affecting loop gain. The likelihood of a nonlinear element in a troublesome process is extremely high, rendering these tests meaningless. The tests also require the process to come to rest after a disturbance. A non-self-regulating process will not come to rest and therefore cannot be treated in this way. Furthermore these tests require the control loop to be open until a new steady state has been reached, which could be a long time.

Fortunately there is a quick and easy method for obtaining enough information to suggest corrective measures in most instances. The method consists of one open-loop and one closed-loop test. In the latter case, the proportional mode of a controller serves as the test instrument. The procedure is as follows:

1. With the controller in manual, step or pulse the control valve sufficiently to produce an observable effect. Measure the time elapsed between the disturbance and the first indication of a response. This is the dead time $\tau_d$.

2. Transfer control to automatic, with minimum derivative and maximum integral time. Adjust the proportional band to develop nearly undamped oscillation. Note the natural period of oscillation $\tau_n$ and the proportional band setting. (A disturbing pulse may be necessary to start the cycle.)

In this test, it is only necessary to leave the loop open (manual control) long enough to measure the dead time. Any other type of open-loop test would consume more time. The closed-loop test places the process under those conditions which are of greatest significance, i.e., at the natural period. Two complete cycles are enough to measure $\tau_n$. If it is not practical to induce uniform oscillations, damped oscillations will suffice, although the proportional band reading should be corrected for the damping.

From the data obtained, a representation of the dynamic elements in the process can be constructed:

If $\tau_n/\tau_d = 2$, the process is pure dead time.
If $2 < \tau_n/\tau_d < 4$, dead time is dominant.
If $\tau_n/\tau_d = 4$, there is a single dominant capacity.
If $\tau_n/\tau_d > 4$, more than one capacity is present.

Furthermore, the setting of proportional band responsible for uniform oscillation equals the gain product of the other elements in the loop at $\tau_n$.

When these bits of information are combined with the characteristics of the known elements, a remarkably accurate picture of the process can be assembled. For example, if the process is known to contain one principal capacity and $\tau_n/\tau_d = 4$, no other time constants need be sought. If the time constant of this capacity is known, its dynamic gain $G_1$ at $\tau_n$ can be calculated. Combining this with known values of transmitter and valve gain, together with the controller proportional band, yields the process gain

$$K_p = \frac{P}{100 G_1 K_t K_v} \tag{2.43}$$

Since these tests are made only at one operating point, they will not disclose any nonlinear properties. Closed-loop response should be observed at other flow conditions to detect any change in damping. If the period changes with flow, a variable dynamic element is present. An extremely nonlinear measurement, such as pH, can be identified by the distorted waveform it produces. A less severe nonlinear measurement may not be detected without changing the set point. In short, if a thorough analysis is to be made, the closed-loop test should be repeated at other values of flow and set point.

There is an oft-quoted axiom to the effect that "if you can't control a process manually, you can't control it automatically." There are exceptions to this rule, as in very fast processes, but it is largely true. When difficulties are encountered in attempting to control a process automatically, try to control it manually. If the process contains an inherently unstable element such as a negative resistance or backlash, you may not be successful. But by approaching the control point first from one direction and then the other, the region of instability may be outlined, to help locate the problem.

In neutralization processes it is wise to withdraw a sample of the influent from time to time and run a laboratory titration. The general shape of the pH curve and its variability can then be determined. If the process responds faster than influent composition is likely to change, a pH curve may be generated online by moving the controller output incrementally and allowing a steady pH reading to be obtained between changes. This test should also be conducted in both directions, in case an irreversible reaction occurs, which could create an unstable loop.

## Testing a Neutralization Process

This is an actual case history of the process upon which the above test procedure was first tried. It was a neutralization process in which a reagent was being added to bring the effluent leaving a reactor to pH 7. The pH controller was in manual, simply because automatic control was unsatisfactory.

The open-loop test gave a dead time of 40 s. The volume of the sample piping divided by the sample flow was 15 s. The remainder was due to incomplete mixing in the reactor.

With a proportional band of 150 percent, the loop sustained uniform oscillation with a period of 2.8 min. The ratio $\tau_n/\tau_d$ = 2.8/0.67, or 4.2, indicated essentially a single capacity along with the measured dead time.

The reaction vessel contained 200 gal of material, flowing at 2.5 gal/min. Therefore $V/F$ = 200/2.5, or 80 min. The dynamic gain of an 80-min capacity at a 2.8-min period is

$$G_1 = \frac{2.8}{2\pi 80} = 0.004$$

Yet the proportional band for zero damping was 150 percent. This can mean only one thing, extremely high process gain. Dividing $G_1$ into $P/100$ yields the gain product of valve, process, and transmitter

$$K_p K_v K_t = \frac{150}{100(0.004)} = 375$$

The pH was measured over a range of 2 to 12, giving a span of 10 units. Then

$$K_t = \frac{100\%}{10 \text{ pH}} = 10\%/\text{pH}$$

The product of process and valve gains is

$$K_p K_v = \frac{375}{10\%/\text{pH}} = 37.5 \text{ pH}/\%$$

This high sensitivity is not unusual in neutralization processes. Repiping the sample line reduced its dead time to 5 s, bringing the total dead time to 30 s. This reduced the period to 2.1 min and the proportional band by the same factor of ¾. So the controller was adjusted for damping at the new conditions. (A procedure for adjusting three-mode controllers is described in Chap. 4.)

A later observation revealed that the loop had become more heavily damped. The only noticeable change since the controller was adjusted was a lower value of output. The loop gain apparently had decreased with load. An inquiry about the valve characteristic produced the answer: reagent was being delivered through an equal-percentage valve under constant pressure drop. The loop gain therefore varied directly with flow, as did the valve gain.

Although a pH process is nonlinear, its characteristic curve cannot be corrected with an equal-percentage valve because the valve acts on the output of the controller, not on the input [4]. The valve characteristic, in fact, made matters worse. Not only did the loop gain become variable, but it was higher than it would have been with an equivalent linear valve. The gain of an equal-percentage valve is 4 times the fractional flow; fractional flow in excess of 0.25 will cause the gain to exceed unity. If the normal flow is 50 percent of the valve's capacity, the equal-percentage characteristic will contribute twice the gain of a linear valve. This necessitates a proportional band twice as wide.

The time required to test this process at one operating point was only a few minutes. Yet together with known facts about the plant and one subsequent observation, the process was thoroughly defined and two recommendations made to improve control. Any other test procedure would have taken longer and might not have achieved comparable results.

## REFERENCES

1. Caldwell, W. I., G. A. Coon, and L. M. Zoss: "Frequency Response for Process Control," pp. 76–80, McGraw-Hill, New York, 1959.
2. Ziegler, J. G., and N. B. Nichols: Optimum Settings for Automatic Controllers, *Trans. ASME,* December 1941.
3. Bradner, M.: Pneumatic Transmission Lag, *ISA Pap.* 48-4-2.
4. Shinskey, F. G.: Controls for Nonlinear Processes, *Chem. Eng.,* Mar. 19, 1962.

## PROBLEMS

**2.1**  Given two processes consisting of 15 identical time constants; in one process they interact, but in the other they do not. Estimate the dynamic gain and natural period of each in terms of the time constant $\tau$ of the individual capacities.

**2.2**  Control of the liquid level in the base of a 50-tray distillation column by manipulating the flow of reflux into the top is being considered. Maximum reflux flow is 100 gal/min, and the volume of the base is 200 gal. There is a pressure drop of 100 lb/in² across the drain valve. Each tray has a time constant of about 6 s; the flow cascades from one tray to the next through a downcomer. Estimate the proportional band required for quarter-amplitude damping of the level loop. Comment on the effectiveness of this loop.

**2.3**  Flow measured by an orifice meter is controlled using a linear valve at constant pressure drop. If the controller is adjusted for quarter-amplitude damping at 60 percent flow, at what flow will the loop be undamped? If an equal-percentage valve is used, at what flow will the loop be undamped?

**2.4**  It is desired to linearize an equal-percentage valve with a rangeability of 50 by throttling a manual valve in series with it. How is the manual valve to be adjusted in relation to the pressure drop across the control valve? How much must the control valve be oversized? What is its installed rangeability?

**2.5**  It is desired to linearize an equal-percentage valve with a rangeability of 50 by using a divider at the controller output. Estimate the optimum value of $z$ for the divider and the maximum departure from linearity for the system.

**2.6**  A process exhibits a dead time of 23 s at 50 percent flow; proportional control with a 40 percent band causes damped oscillations of 1.5 min period. At 25 percent flow, the loop is undamped, and its natural period increases to 2.9 min. Draw some conclusions about the process and make recommendations about improving control.

# Chapter Three

# Analysis of
# Some Common Loops

Classification of processes into broad areas with certain common characteristics is both desirable and informative. We know, for example, that a temperature-control loop behaves very differently from a level-control loop. Why it does so is the essence of the classification.

The first control loop to be considered is flow. It has the distinction that the manipulated variable and the controlled variable are the same. They may not have the same range or the same linearity, but nevertheless they are the same variable. For this reason the flow loop is the easiest to understand, as far as steady-state characteristics are concerned.

We will next analyze the control of variables that are the integral of flow. Liquid level is the integral of liquid flow, whereas the integral of gas flow in a constant-volume system is pressure. These loops have certain features not common to other classifications. For example, they can be non-self-regulating. This is never true of flow and rarely true of other variables. Second, the rate of change of measurement is a function of the *difference* between inflow and outflow; either inflow or outflow will be load-dependent, while the other is manipulated. Furthermore these processes are dominated by capacity; dead time will rarely be found because pressure waves travel through the process at the velocity of sound.

The third group includes energy- and mass-transfer processes, where control is exercised primarily over temperature and composition. The controlled variable here is always a property of the flowing stream, as opposed to being the flowing stream or its integral. These processes ordinarily have a steady state in which the controlled variable is a function of the *ratio* of the manipulated flow to the load. (Note the abscissa of Fig. 2.17, expressed in terms of this ratio.) Because the controlled property travels with the fluid, it must be transported to the measuring element. Transportation involves dead time. Hence loops in this category are usually dominated by dead time, which makes control difficult and response slow.

In this chapter, five typical control loops will be analyzed: flow, level, pressure, temperature, and composition. The principal dynamic elements of each process will be derived and will be related to the closed-loop response. Constraints and nonlinearities will be included, as well as means for coping with them. A few additional comments will serve to distinguish control problems which are not typical or which appear to cross into other areas.

## FLOW CONTROL

Since flow is the manipulated variable as well as the controlled variable, it seems as though the process were unity. But this is not the case. Opening a valve does admit flow, but the response is not quite instantaneous. If the fluid is gaseous, it is subject to expansion upon a change in pressure; therefore the contents of a pipe vary somewhat with pressure drop, hence with flow. In a liquid stream, inertia is significant; flow cannot be started or stopped without accelerating or decelerating. To demonstrate the dynamic character of inertia, the time constant of a column of liquid in a pipe will be derived.

### Inertial Lag of a Flowing Liquid

In the steady state, the velocity of flow in a pipe varies with pressure drop

$$u^2 = 2gC^2 \frac{\Delta p}{\rho}$$

where $u$ = velocity, ft/s
$\quad C$ = flow coefficient
$\quad g$ = gravity, ft/s$^2$
$\quad \Delta p$ = pressure drop, lb/ft$^2$
$\quad \rho$ = density, lb/ft$^3$

But velocity is proportional to flow

$$u = \frac{F}{A}$$

where $F$ is the flow in cubic feet per second and $A$ is the inside area in square feet. Therefore the pressure drop due to flow in the steady state is

$$\Delta p = \frac{u^2 \rho}{2gC^2} = \frac{F^2 \rho}{2gA^2C^2}$$

If the applied force $A\,\Delta p$ exceeds resistance to flow, acceleration takes place.

An equation can be written for the unsteady state; net force equals mass times acceleration:

$$A \, \Delta p - \frac{AF^2\rho}{2gA^2C^2} = M \frac{du}{dt} = \frac{M}{A} \frac{dF}{dt}$$

where $M$ is the mass in pounds and $t$ is the time in seconds. The mass of fluid in the pipe is

$$M = \frac{LA\rho}{g}$$

where $L$ is the length in feet. Rearranging, we get

$$\frac{F^2\rho}{2gAC^2} + \frac{L\rho}{g} \frac{dF}{dt} = A \, \Delta p \qquad (3.1)$$

Equation (3.1) is a nonlinear differential equation. To extract the gain and time constant from it, both sides must be differentiated to give a linear equation

$$\frac{F\rho}{gAC^2} dF + \frac{L\rho}{g} \frac{d(dF)}{dt} = A \, d(\Delta p)$$

Isolating $dF$ puts the equation in its familiar form

$$dF + \frac{LAC^2}{F} \frac{d(dF)}{dt} = \frac{gA^2C^2}{\rho F} d(\Delta p)$$

The time constant is then

$$\tau = \frac{LAC^2}{F} \qquad (3.2)$$

Observe that the time constant varies inversely with flow when $\Delta p$ is the manipulated variable. This case would be represented by the manipulation of speed of the prime mover with constant line resistance. The more common case would be the manipulation of a valve opening $C$ with constant $\Delta p$. The time constant then varies directly with flow. To demonstrate, flow coefficient $C^2$ can be replaced by its steady-state equivalent

$$C^2 = \frac{F^2\rho}{2gA^2 \, \Delta p}$$

leaving

$$\tau = \frac{LF\rho}{2gA \, \Delta p} \qquad (3.3)$$

**example 3.1**    To test the significance of the last expression, a numerical example is presented. Consider a 200-ft length of 1-in Schedule 40 pipe, containing water flowing at 10 gal/min with a drop of 20 lb/in²;

$$L = 200 \text{ ft}$$
$$F = 10 \text{ gal/min} = 0.0223 \text{ ft}^3/\text{s}$$
$$\rho = 62.4 \text{ lb/ft}^3$$
$$g = 32.2 \text{ ft/s}^2$$
$$A = 0.006 \text{ ft}^2$$
$$\Delta p = 20 \text{ lb/in}^2 = 2880 \text{ lb/ft}^2$$
$$\tau = \frac{(200)(0.0223)(62.4)}{2\,(32.2)(0.006)(2880)} = 0.25 \text{ s}$$

Notice that the time constant varies with both flow and pressure drop, because of the square relation between the two. Nevertheless, the derivation permits evaluation of the dynamic response at a nominal flow and at least a qualitative indication of the response elsewhere. As may have been anticipated, the time constant is small but not zero, except at zero flow.

### Dynamic Elements Elsewhere in the Loop

This time constant is fundamentally the only dynamic element in the process. But its response is of the same order of magnitude as the instruments in the control loop, and therefore the entire loop must be analyzed.

Figure 3.1 describes a pneumatic flow-control loop consisting of transmitter (2), controller (4), valve (6), and two transmission lines (3, 5). The flow transmitter contains an amplifier with certain dynamic properties. Because of the amplifier, the lag of the transmission line is isolated from that of the flowing fluid. The transmission line is terminated by a controller, isolating it from the

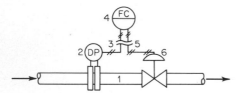

**FIG. 3.1**   At least six elements contribute to the dynamic response of the flow-control loop.

second transmission line. The figure shows no isolating amplifier between the output line and the valve, however, allowing interaction there.

> **example 3.2**   Analyze the response of the flow loop whose inertial lag was estimated in Example 3.1. The flowmeter has a range of 0 to 15 gal/min; both transmission lines are 100 ft of ¼-in tubing; the valve is a 1-in size with a linear characteristic. The combination of transmission line and valve motor produces a time constant of about 3 s, according to Ref. 1. The dead time in this line is not to be neglected.
>
> A closed loop of this description will be found to oscillate with a natural period of about 3 s. Each of the dynamic elements contributes some amount of phase shift at that period, as Table 3.1 indicates.
>
> To determine the proportional band necessary for damping, the steady-state

**TABLE 3.1    Dynamic Elements in the Flow Loop**

|  | $\tau$, s | $-\phi$, deg | $G$ |
|---|---|---|---|
| Process | 0.25 | 27.6 | 0.886 |
| Transmitter | 0.16 | 18.5 | 0.948 |
| Transmission line: | | | |
| $\tau$ | 0.32 | 33.8 | 0.831 |
| $\tau_d$ | 0.08 | 9.1 | 1.0 |
| Transmission line $\tau_d$ | 0.08 | 9.1 | 1.0 |
| Valve $\tau$ | 3.0 | 80.9 | 0.157 |
| | | $\Sigma = 180.1$ | $\Pi = 0.110$ |

gains of valve and transmitter need to be multiplied by the dynamic-gain product of 0.110 appearing in Table 3.1. Flow through the linear valve is

$$F = mC_v\sqrt{\Delta p}$$

Valve gain is then the derivative

$$K_v = \frac{dF}{dm} = C_v\sqrt{\Delta p}$$

In this example, 200 ft of pipe and fittings produce a drop of about 8 lb/in² at 10 gal/min, leaving 12 lb/in² across the valve. A 1-in valve has a $C_v$ of about 10. The valve gain is then

$$K_v = 10\sqrt{12} = \frac{34.6 \text{ gal/min}}{100\%} = 0.346 \text{ (gal/min)/\%}$$

At 10 gal/min, fractional flow is 10/15 or 0.667. The gain of the orifice meter is then

$$K_t = 2(0.667)\frac{100\%}{15 \text{ gal/min}} = 8.89\%/\text{(gal/min)}$$

Loop gain is then

$$G_pK_vK_t = 0.110(0.346)(8.89) = 0.338$$

Quarter-amplitude damping will require a proportional band of 200(0.338), or about 68 percent.

With a wide proportional band, integral action is always necessary to eliminate offset. This will increase the period somewhat above $\tau_n$ and require a still wider proportional band for stability, perhaps about 100 percent. This is quite typical for a flow controller. Using Table 1.2 as a guide to determining settings for the proportional-plus-integral controller, we see that a 30° controller phase lag ought to increase the period of the flow loop by about 50 percent to 4.5 s. The integral time required to do this would be 1.2 s.

The series of similar noninteracting lags and the dead time in transmission can all be treated as dead time, to obtain a quick estimate of natural period and dynamic gain. The valve plus its transmission line would remain as the dominant time constant.

**example 3.3**  Estimate the period and dynamic gain of the flow loop in Example 3.2 using the above procedure.

$$\Sigma\tau \text{ (less valve)} = 0.89 \text{ s}$$
$$\tau_n \text{ (estimated)} = \dfrac{\times 4}{3.56 \text{ s}}$$
$$G_v \text{ (approximate)} = \dfrac{\tau_o}{2\pi\tau_v} = \dfrac{3.56}{2\pi 3} = 0.189$$

Observe that the simplification afforded by the above procedure is justified. The estimates of period and gain are reasonable and err on the safe side; the process is somewhat easier to control than the approximation indicates.

Some improvements can be made in the response of the flow loop. For example, the transmission line could be decoupled from the control-valve motor with an isolating amplifier. This would separate the single 3-s time constant into two smaller lags, 0.32 s for the transmission line and about 1.0 s for the valve. While the period of the loop would thus be reduced somewhat, the dynamic gain of the valve will increase, requiring a wider proportional band for stability. However, if the transmission lines are eliminated by locating the controller at the valve, the period and the valve lag would *both* be reduced to about 1.0 s, so that response speed would be improved *without* a sacrifice in proportional band.

## Flow Noise

In an equivalent electronic flow loop, absence of the transmission lines reduces the natural period to the vicinity of 1 s. Noise, however, becomes more prominent. *Noise* means disturbances, either periodic or random, occurring at frequencies too high for control action. Figure 3.2 is a record of noise in an electronic flow loop. Turbulence in the stream and vibration from pumps are the chief sources of this noise. Even in pneumatic systems, flow noise is invariably present in sufficient magnitude to prevent the use of derivative. Phase lead is useful, but unfortunately the increase in high-frequency gain which accompanies it actually explodes the loop into instability.

## Summary

The purpose of the analysis is not to show how an analysis should be made but to explain why a flow loop behaves the way it does. Because many dynamic

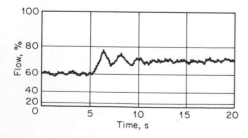

**FIG. 3.2**  Flow noise precludes the use of derivative.

**TABLE 3.2 The Significance of Pressure as a Measurement of Specific Volume and Enthalpy of Steam and Water at 100 lb/in² Absolute (6.8 atm)**

| System | Spec. vol. change, %<br>Pressure change, % | Enthalpy change, %<br>Pressure change, % |
|---|---|---|
| Superheated vapor at 1000°F (538°C) | −1.006 | −0.00163 |
| Saturated vapor | −0.945 | +0.0158 |
| Compressed liquid at 100°F (38°C) | −0.0003 | +0.00395 |

elements are present, all of the same order of magnitude, dynamic gain is high. The proportional band of a flow controller is rarely less than 100 percent, making integral action mandatory. Where the valve and transmitter are in the same line, the period of oscillation will invariably fall within 1 to 10 s. The presence of noise precludes the use of derivative. As long as these factors are appreciated, there is little reason to spend time analyzing flow loops.

## PRESSURE REGULATION

The thermodynamic state of a system can be defined from its pressure, enthalpy, and volume. If a gas phase alone is present, pressure and volume are inversely proportional, with enthalpy playing a relatively minor role. When a vapor is in equilibrium with its liquid, however, a change in enthalpy of the system will produce a pronounced pressure change, while volume variations will have less effect. Liquids, moreover, are virtually incompressible, with the result that neither pressure nor enthalpy has much influence over system volume.

The thermodynamic properties of gas, vapor, and liquid systems have been brought out expressly to establish that the properties of system pressure are decidedly a function of state. It is extremely important to attach the correct significance to the pressure measurement if acceptable performance of a control loop is to be gained. Table 3.2 gives an example of each of the three states listed above, where water is the substance under pressure. It indicates the conditions under which pressure is a suitable measurement of the material content (specific volume) and energy content (enthalpy) of the system.

The table points out that pressure is an adequate measurement of the material content of a system which contains only gas. Enthalpy of a gas, on the other hand, is more a function of temperature than of pressure. Consequently gas pressure should be controlled by manipulating the material content of the system, i.e., inflow or outflow. But in a system where vapor and liquid are in equilibrium, pressure can be controlled by adjusting the flow of either material or heat. Finally, pressure is a poor measure of either heat or mass content of a liquid, so that another approach must be taken in stipulating its control.

### Gas Pressure

The perfect-gas law states that

$$pV = MRT$$

where $p$ = system pressure
   $V$ = volume
   $M$ = mole content
   $R$ = gas constant
   $T$ = absolute temperature

The rate of change of pressure in a constant-volume system is related to the change in material content of the system

$$\frac{dp}{dt} = \frac{dM}{dt}\frac{RT}{V}$$

If $R$ and $T$ are both constant, the rate of change of mole content of the system is the difference between mole inflow and outflow

$$\frac{dp}{dt} = \frac{F}{V}(f_i - f_o)$$

where $F$ = nominal mole flow
   $f_i$ = fractional inflow
   $f_o$ = fractional outflow

Integration of the last equation places pressure in terms of flow

$$p = \frac{1}{V/F}\int (f_i - f_o)\ dt \tag{3.4}$$

For dimensional conformity, $p$ would be in units of atmospheres, $V$ in cubic feet, and $F$ in standard cubic feet per minute, i.e., at 1.0 atm. Thus the time constant $V/F$ is expressed in minutes.

Just as level control was used to close a liquid material balance around a tank, pressure control is used to close a gas material balance. The gas-pressure process is ordinarily self-regulating, except at zero flow, because pressure always influences inflow and outflow. The process is fundamentally single-capacity, although the pressure transmitter and valve can add very small secondary lags. If there is no transmitter, as with a self-contained regulator, one secondary lag is eliminated.

Pressure of a gas is easy to control, even when the volume of the system is small, e.g., only piping. In fact, the narrow-band proportional action of self-contained regulators is sufficient for most applications. They are, for the most part, as sensitive as their simple construction will allow, indicating that loop gain is not a problem. Pressure acting on the diaphragm compresses the spring, moving the plug within the valve. Each position of the seat corresponds to a given pressure on the diaphragm. Initial compression of the spring sets the pressure at which the valve begins to open.

Because pressure will vary with flow, as in Fig. 3.3, a regulator is said to exhibit *droop*. Regulators differ, but a typical proportional band would be 5 percent. Near zero flow, extra pressure is needed for shutoff; at the other extreme, the valve is wide open and acts as a fixed resistance.

### Vapor Pressure

In a system containing liquid and vapor in equilibrium, the difference between inflow and outflow of vapor changes the pressure, from a material-balance standpoint,

$$F_i - F_o = V \frac{dp}{dt}$$

But if the enthalpy of inflow and outflow differ, flow of material between the vapor and liquid phases will also affect system pressure. An energy balance shows the relationship

$$F_i H_i - F_o H_o + Q_i - Q_o = V H_v \frac{dp}{dt} \tag{3.5}$$

where $H_i$, $H_o$ = enthalpy of inflow and outflow, respectively

$\qquad Q_i$, $Q_o$ = transfer of heat in and out

$\qquad H_v$ = heat of vaporization

Both mass flow and heat flow affect pressure. But where the net change of enthalpy across a process is zero, mass flow alone is sufficient for control. An example of this situation is pressure reduction of saturated or wet steam; there is no change in enthalpy across the reducing valve.

In a boiler, or distillation column, or evaporator, transfer of heat is an integral part of the operation, and system pressure can be used to close the heat balance. In this role, the pressure controller has much the same type of dynamic and steady-state relationships as a temperature controller normally does. Therefore the properties of this sort of pressure-control loop will be covered for the most part under considerations of temperature control.

### Liquid Pressure

Pressure control of a liquid stream is exactly like flow control. The pressure at the origin of a pipeline, for example, is directly related to flow in the line. The process's only dynamic contribution is that of inertia of the flowing fluid.

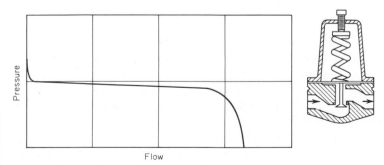

FIG. 3.3   The characteristic curve for a pressure regulator indicates proportional action.

The process gain $K_p$ in a flow loop is, by definition, 1.0. But in a pressure loop there must be a conversion from flow into units of pressure. Liquid pressure upstream of a resistance $C_R$, as in Fig. 3.4, varies with flow squared

$$p = p_0 + \frac{F^2}{C_R^2} \tag{3.6}$$

The intercept $p_0$ is the static pressure at no flow. Differentiating, we obtain the process gain

$$K_p = \frac{dp}{dF} = \frac{2F}{C_R^2} \tag{3.7}$$

Ordinarily pressure moves less than full scale for full-scale change in valve position, resulting in a lower proportional band than for a flow loop. Other characteristics, including noise, are similar.

Self-contained regulators are sometimes used for liquid pressure and perform moderately well on quiet streams. When we recall that the dynamic elements which caused most of the problems in the flow loop were instruments and transmission lines, the application makes good sense. But where accurate regulation and tight shutoff are important, these simple devices are insufficient.

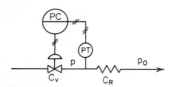

**FIG. 3.4** Pressure drop across a fixed resistance varies with the square of flow.

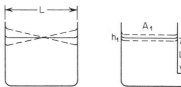

**FIG. 3.5** The period of hydraulic resonance varies with the distance between the bounded surfaces.

## LIQUID LEVEL AND HYDRAULIC RESONANCE

Control of liquid level is not as easy as the examples given in Chap. 1 indicate. The descriptions of Figs. 1.13 and 2.1 were intentionally oversimplified to aid understanding of single- and two-capacity processes. But the existence of waves in any body of water as large as a bay or as small as a cup gives rise to the speculation that any liquid with an open surface is capable of sustaining oscillation. While average level responds to flow as an integrator, level responds to level in a resonant manner. Consequently the liquid-level process is not single-capacity, even with a directly connected measuring element.

### The Period of Hydraulic Resonance

To analyze this resonance, let us take the case of the vessel with a measuring chamber shown in Fig. 3.5, neglecting resistance to flow. If the level in the measuring chamber momentarily exceeds that in the tank, the differential force developed causes a downward acceleration in that leg

$$\rho h_2 A_2 - \rho h_1 A_1 = -M_2 \frac{du_2}{dt} - M_1 \frac{du_1}{dt} \tag{3.8}$$

where $h_2$ = head of fluid in measuring chamber

       $A_2$ = area of fluid in measuring chamber

       $M_2$ = mass of fluid in measuring chamber

       $u_2$ = velocity of fluid in measuring chamber

As before, $\rho$ is the fluid density. Furthermore

$$u_1 = \frac{A_2}{A_1} u_2 \quad \text{and} \quad M_1 = \rho \frac{L_1 A_1}{g} \quad M_2 = \rho \frac{L_2 A_2}{g}$$

Substituting for $u_1$, $M_1$, and $M_2$ in Eq. (3.8) yields

$$h_2 A_2 - h_1 A_1 = -\frac{L_2 A_2}{g} \frac{du_2}{dt} - \frac{L_1 A_2}{g} \frac{du_2}{dt} \tag{3.9}$$

Level in the measuring chamber $h_2$ is related to average level $h$ by

$$h - h_1 = (h_2 - h) \frac{A_2}{A_1}$$

Including this in Eq. (3.9) yields the response of measured level $h_2$ to average level $h$

$$2h_2 = h \left( 1 + \frac{A_1}{A_2} \right) - \frac{L_1 + L_2}{g} \frac{du_2}{dt}$$

But velocity $u_2$ is the rate of change of level, $dh_2/dt$. Therefore a differential equation can be written eliminating $u_2$

$$h_2 + \frac{L_1 + L_2}{2g} \frac{d^2 h_2}{dt^2} = \frac{h}{2} \left( 1 + \frac{A_1}{A_2} \right) \tag{3.10}$$

     This differential equation is descriptive of a second-order undamped system. The U-tube resonates at a natural period established by the square root of the coefficient of the differential

$$\tau_0 = 2\pi \left( \frac{L_1 + L_2}{2g} \right)^{1/2} = 0.78\sqrt{L_1 + L_2} \tag{3.11}$$

where the period is in seconds and length is in feet.

Notice that the period is unaffected by density, area, or any property other than the distance $L_1 + L_2$ between the bounded surfaces. Compare it with that of a pendulum, also a function of length and gravity only.

     Liquid in a vessel may also oscillate without the benefit of a U-tube. The period of oscillation of the surface of diameter $L$ is

$$\tau_0 = 2\pi \left( \frac{L}{2g} \right)^{1/2} = 0.78\sqrt{L} \tag{3.12}$$

Rectangular vessels can oscillate at two different periods. Vessels with an attached measuring chamber can oscillate with at least two different periods.

The natural period of any control loop containing a resonant element cannot exceed that of the resonant element. The phase shift of a resonant element is exactly $-90°$ at its natural period, no matter how heavily damped it may be. Since the integration of flow into average level represents an inherent phase shift of $-90°$, the process, from flow to measured level, will exhibit $-180°$ at the natural period of the vessel. To damp the measuring chamber by throttling its connecting valves will not change this period but will only reduce the gain of the resonance.

**example 3.4**   As an example of a liquid-level control problem, consider a vessel with a measuring chamber of the following description:

Volume $V$ = 100 gal     Maximum flow $F$ = 50 gal/min
Diameter $L$ = 2.0 ft        Normal level $L_1$ = 3.6 ft     Chamber $L_2$ = 4.4 ft

The liquid can oscillate on the surface and in the U-tube, but since the largest resonant period is always the limiting one, only the period of the U-tube is important.

$$\tau_o = 2\pi \left[ \frac{3.6 \text{ ft} + 4.4 \text{ ft}}{2(32.2 \text{ ft/s}^2)} \right]^{1/2} = 2.2 \text{ s}$$

The dynamic gain of the integrator is

$$G = \frac{\tau_o}{2\pi V/F} = \frac{2.2/60}{(6.28)(100/50)} = 0.003$$

This control problem can be accommodated with a proportional band of $200G$ = 0.6 percent.

Since dimensions of process vessels generally fall between 2 and 200 ft, liquid resonance lies principally in the region from a 1- to 10-s period. Hence it is of serious consequence, from the standpoint of control-loop stability, only in vessels with time constants of less than 1 min.

## Liquid-Level Noise

Measurement of liquid level is usually noisy, because of splashing and turbulence of fluids entering the vessel. As we have seen, loops that resonate respond

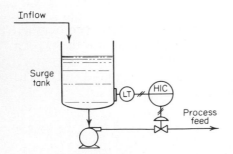

FIG. 3.6  A transmitter will give 100 percent proportional control action, which utilizes the entire surge volume to dampen fluctuations in inflow.

to random disturbances by oscillating at their natural period. As a result, level measurements are rarely quiet, often fluctuating 20 or 30 percent of scale. This is particularly true in vessels containing boiling liquids, where turbulence is high.

Although a narrow proportional band, like the one determined in the example, may be sufficient for control-loop stability, random fluctuations of only a few percent will drive the control valve to its limits. This may be unobjectionable in some cases but too severe in others. Often the liquid level in a tank is used to control flow into another part of the process. It is certain that wide fluctuations in feed rate are not tolerated in most operations. To provide steady flow in these instances, the proportional band is widened substantially.

In many applications, exact regulation of liquid level is not important. In fact, a surge tank does not fulfill its purpose if tight control is imposed on it. As a result, control adjustments are often relaxed, and the process is sometimes left to be operated manually if its time constant is long enough. A preferred practice is to connect the level transmitter to the valve through an automanual station, as shown in Fig. 3.6. Tank level will then swing full scale with changes in inflow, maximizing its capacity to absorb disturbances.

In some applications, a special controller whose proportional band changes with deviation is warranted. This type of controller is devised to deliver smooth flow while level is normal but to change flow radically if high or low limits are approached. Chapter 5 discusses more details of this function.

### Boiling Liquids and Condensing Vapors

Whenever level control is to be effected on a boiling liquid or condensing vapor, properties more typical of thermal processes appear. Transfer of both heat and mass is involved, which, combined with the integration of flow into level, renders control surprisingly difficult. Level control in boilers and distillation columns is sufficiently problematic to warrant special consideration, which is given in Chaps. 7, 9, and 11.

## TEMPERATURE CONTROL

Temperature-control problems are really heat-transfer problems, whether the mechanism is radiation, conduction, or convection. Although an entire chapter is devoted exclusively to energy control, it is important at this time to assay the general features of the temperature loop in order to establish its place in the classification that has been made.

### Example of a Constant-Parameter System

Because most heat-transfer processes have variable parameters (heat-transfer coefficient, dead time, etc.) which vary with flow, care has been taken to choose an example free of these complications, to better introduce the subject. The example chosen is that of a stirred-tank reactor cooled by a constant flow of liquid circulating through its jacket.

The temperature controller, as shown in Fig. 3.7, adds cold water to the circulating coolant, in order to remove the heat of reaction. There are five important dynamic elements in the process:

1. Heat capacity of the contents of the reactor
2. Heat capacity of the wall
3. Heat capacity of the contents of the jacket
4. Lag in the temperature bulb
5. Dead time of circulation

Because all the heat leaving the reactor flows through the walls and into the coolant, the capacities of reactants, walls, and coolant interact. But in view of the slight heat capacity of the bulb, its time constant does not significantly interact with the others. Basically the process is four-capacity plus dead time.

### Finding the Time Constants

To determine the values of the time constants, an unsteady-state heat balance must be written across each heat-transfer surface. The equation takes the following form: Heat in = heat out, + heat capacity × rate of temperature rise. If we assume a constant rate of heat evolution (the case of a variable rate will be taken up later), the heat balance at the surface of the reactor wall is

$$Q = k_1 A(T - T_1) + W_1 C_1 \frac{dT}{dt} \tag{3.13}$$

where $Q$ = rate of heat evolution, Btu/h
$k_1$ = heat-transfer coefficient, Btu/(h)(ft$^2$)(°F)
$A$ = heat-transfer area, ft$^2$
$T$ = reactor temperature, °F
$T_1$ = wall temperature, °F
$W_1$ = weight of reactants, lb
$C_1$ = specific heat of reactants, Btu/(lb)(°F)

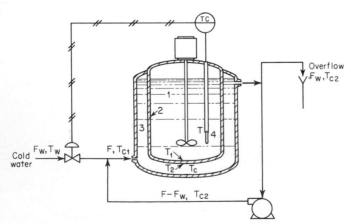

**FIG. 3.7** The thermal process contains four interacting lags.

Rearranging in the standard form, we get

$$T + \frac{W_1 C_1}{k_1 A} \frac{dT}{dt} = T_1 + \frac{Q}{k_1 A} \tag{3.14}$$

The thermal time constant is

$$\tau_1 = \frac{W_1 C_1}{k_1 A} \tag{3.15}$$

Reactor temperature responds to wall temperature with a time constant of $\tau_1$ and a steady-state gain of 1. If $k_1 A$ is not directly known, $Q/(T - T_1)$ may be substituted

$$\tau_1 = \frac{W_1 C_1}{Q}(T - T_1) \tag{3.16}$$

By the same token, the temperature of the outside wall of the reactor responds to that of the inside wall with a time constant of

$$\tau_2 = \frac{W_2 C_2 l}{k_2 A} = \frac{W_2 C_2}{Q}(T_1 - T_2) \tag{3.17}$$

where $W_2$ = weight of wall, lb
$\quad C_2$ = specific heat of wall, Btu/(lb)(°F)
$\quad k_2$ = thermal conductivity, Btu/(h)(ft²)(°F/in)
$\quad l$ = wall thickness, in
$\quad T_2$ = outside wall temperature

Next, outside wall temperature responds to coolant temperature with a time constant of

$$\tau_3 = \frac{W_3 C}{k_3 A} = \frac{W_3 C}{Q}(T_2 - T_c) \tag{3.18}$$

where $W_3$ = weight of jacket contents
$\quad C$ = specific heat of coolant
$\quad k_3$ = heat-transfer coefficient
$\quad T_c$ = average coolant temperature

The lag of the temperature bulb can be calculated in the same way as that of the other time constants

$$\tau_4 = \frac{W_4 C_4}{k_1 A_4} \tag{3.19}$$

where $W_4$ = weight of bulb
$\quad C_4$ = specific heat of bulb
$\quad A_4$ = surface area of bulb

For most types of thermal systems and heat-transfer conditions, data on bulb response are already available [2].

## Process Gain

The steady-state response of reactor temperature to cooling-water flow can be broken into two steps for ease of calculation. First, the effect on coolant outlet temperature $T_{c2}$ is considered, from a heat balance. The net flow of heat from the vessel is essentially the flow of cooling water multiplied by the difference in temperature between supply and exit

$$Q = F_W C(T_{c2} - T_W) \tag{3.20}$$

Solving for $T_{c2}$ gives

$$T_{c2} = T_W + \frac{Q}{F_W C} \tag{3.21}$$

The steady-state gain of $T_{c2}$ with respect to coolant flow is the derivative of (3.21)

$$\frac{dT_{c2}}{dF_W} = -\frac{Q}{CF_W^2} \tag{3.22}$$

Next the effect of $T_{c2}$ on reactor temperature $T$ is to be evaluated as a function of overall heat-transfer coefficient $U$ (which includes $k_1$, $k_2$, $l$, and $k_3$) and the logarithmic mean temperature difference

$$Q = \frac{UA\,[(T - T_{c1}) - (T - T_{c2})]}{\ln\,[(T - T_{c1})/(T - T_{c2})]} \tag{3.23}$$

Equation (3.23) can be equated to a heat balance

$$Q = FC(T_{c2} - T_{c1}) = \frac{UA(T_{c2} - T_{c1})}{\ln\,[(T - T_{c1})/(T - T_{c2})]}$$

Observe that the coolant temperature differences cancel, leaving

$$\frac{T - T_{c1}}{T - T_{c2}} = e^{UA/FC} \tag{3.24}$$

Coolant inlet temperature can be eliminated by substituting the following expression obtained from the heat balance:

$$T_{c1} = T_{c2} - \frac{Q}{FC}$$

Substituting into (3.24) gives

$$T - T_{c2} + \frac{Q}{FC} = (T - T_{c2})e^{UA/FC}$$

Solving for $T$, we have

$$T = T_{c2} + \frac{Q}{FC(e^{UA/FC} - 1)} \tag{3.25}$$

The derivative of $T$ with respect to $T_{c2}$ is seen to be unity; therefore process gain is

$$K_p = \frac{dT}{dF_W} = -\frac{Q}{CF_W^2} \tag{3.26}$$

By combining this expression with (3.20) we see that process gain varies directly with $T_{c2}$ and inversely with $F_W$

$$K_p = -\frac{T_{c2} - T_W}{F_W} \tag{3.27}$$

The variation with $F_W$ can be compensated exactly using an equal-percentage valve under conditions of constant pressure drop. (Pressure drop is likely to be nearly constant in this case, in that water is drawn from a supply header and overflows to an open drain.) The direct variation with $T_{c2}$ will require that the temperature controller be adjusted for stability when $T_{c2}$ is maximum, which would occur at maximum reactor temperature and minimum heat-transfer rate.

**example 3.5**   If a reactor contains 40,000 lb of material of specific heat of 0.8 Btu/(lb)(°F), evolving 20,000 Btu/min at 200°F with a wall temperature of 170°F,

$$\tau_1 = \frac{(40,000)(0.8)(200 - 170)}{20,000} = 48 \text{ min}$$

$\tau_2$ can be estimated from the weight of the reactor wall, 8000 lb, of specific heat 0.15 and a temperature gradient of 10°F

$$\tau_2 = \frac{(8000)(0.15)(10)}{20,000} = 0.6 \text{ min}$$

Jacket contents of 500 gal (4160 lb) of water at an average temperature of 140°F exhibit a time constant of

$$\tau_3 = \frac{(4160)(1.0)(160 - 140)}{20,000} = 4.2 \text{ min}$$

A typical value for lag in a temperature well is $\tau_4 = 0.5$ min. Finally, circulation through the jacket at a rate of 250 gal/min yields a dead time

$$\tau_d = \frac{500}{250} = 2 \text{ min}$$

It happens that a reactor of this description will oscillate at a period of about 35 min in a closed loop. Even if all the secondary elements consisted of pure dead time, they could not cause the period to exceed 29 min. Therefore, some secondary element remains hidden, and the only place it could hide is in the reaction mass. The assumption has been made, in calculating its time constant, that the reaction mass was perfectly mixed, that it was all at the same temperature. This, of course, is a false premise, because it is impossible to transport fluid, hence heat, from the wall of the vessel to the temperature bulb in zero time. Heat is transferred both by convection and by conduction; conduction would be the mechanism if the fluid were motionless. It has been pointed out that heat transfer by conduction is a distributed process, involving some effective dead time. So it does not seem unreasonable that a small percentage of the 48-min primary time constant is dead time due to imperfect mixing. An examination of the mechanism of mixing will be taken up under composition control.

**example 3.6**    The dynamic gain of the process is principally that of the primary time constant

$$G_1 = \frac{\tau_o}{2\pi\tau_1} = \frac{35}{6.28(48)} = 0.116$$

As pointed out in Chap. 2, the gain of an equal-percentage valve is its actual flow multiplied by the natural logarithm of its rangeability. For a 50:1 valve, $\ln R$ is 3.91. Coolant valves are always reverse-acting, having a negative gain

$$K_v = -3.91\frac{F_W}{100\%}$$

When multiplied by process gain from Eq. (3.27), this becomes

$$K_p K_v = \frac{3.91(T_{c2} - T_W)}{100\%}$$

Given that $T_{c2}$ is 145°F and $T_W = 80°F$, we have

$$K_p K_v = \frac{254°F}{100\%}$$

Let the temperature range be 100 to 300°F. Then $K_t = 100\%/200°F$, and

$$K_p K_v K_t = \frac{254°F}{200°F} = 1.27$$

The proportional band required for quarter-amplitude damping is 200 times the gain product

$$P = 200(0.116)(1.27) = 29.5\%$$

The period of this loop and the proportional band needed for damping can both be cut in half through cascade control of coolant outlet temperature; the reactor temperature controller would set $T_{c2}$, whose controller would in turn manipulate $F_W$. Cascade control is described in detail in Chap. 6.

## Summary

The most important points to be grasped from this analysis of a simple heat-transfer process are as follows:

  1. Time constants in a temperature-control loop are not easy to identify, and they interact.

  2. The presence of distributed lags makes the exact performance of the loop difficult to predict.

  3. Processes involving heat transfer are always nonlinear in at least one

respect. Each process ought to be evaluated on its own merits to be sure correct compensation is applied.

If the rate of heat evolution in the example had been made a function of temperature, as it is in a real reactor, a second nonlinearity would have made its appearance. Obviously much further consideration must be given to each individual heat-transfer application as it is encountered. Although certain characteristics are common, many others are not. In short, there is no such thing as a "typical" temperature-control loop.

## CONTROL OF COMPOSITION

By far the greatest single contributor to the problems of a control engineer is the composition loop. Since composition is a property of a flowing stream, it travels with the stream. This means that dead time is always in the loop. Further, sampling difficulties, incomplete mixing, and intermittent analyses lend the measurement a certain amount of random character, often making tight controller adjustment inadvisable. Most significant of all, the composition of a stream is a function of the performance of the processing equipment producing it, which many control engineers do not fully understand.

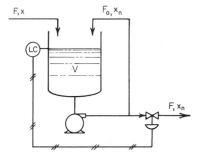

**FIG. 3.8** The dynamic characteristics of incomplete mixing can be evaluated using a dead-time model with recirculation.

Like temperature control, the process within a composition loop may be extremely complex. In fact, most mass-transfer operations require multiple control loops to cope with the number of variables which affect product quality. But for the moment it is important to examine the properties of a composition loop apart from the intrigues of mass transfer. Therefore a simple blending system will be analyzed.

### The Problem of Mixing

Although most agitators are mounted internally, the dynamic properties of mixing are most easily demonstrated using the plug-flow model with external recirculation shown in Fig. 3.8. Assume that no mixing whatever takes place within the vessel itself. Consider the system initially in a steady state with the concentration of component X in the feed and in the discharge at $x_0$. Then let the feed concentration be changed to $x_f$ stepwise.

The composition of the product will remain at $x_0$ until the dead time within the vessel elapses. That dead time is the volume $V$ divided by the internal flow rate, $F + F_a$. During this time, the feed is being diluted with recycled product having initial composition $x_0$. The blend of the two has a composition $x_1$

$$x_1 = \frac{Fx_f + F_a x_0}{F + F_a} \tag{3.28}$$

which, when factored, gives

$$x_1 = x_f \frac{F}{F + F_a} + x_0 \frac{F_a}{F + F_a} \tag{3.29}$$

To simplify this statement, let $\alpha$ represent the completeness of mixing

$$\alpha = \frac{F_a}{F + F_a} \tag{3.30}$$

Without mixing, $\alpha = 0$; with complete mixing, $\alpha = 1.0$. Combining the last two expressions yields

$$x_1 = (1 - \alpha)x_f + \alpha x_0 \tag{3.31}$$

When time $t = V/(F + F_a)$, product composition will step from $x_0$ to $x_1$. At this point, feed at $x_f$ will start being mixed with product at $x_1$ to produce a blend $x_2$

$$x_2 = (1 - \alpha)x_f + \alpha x_1 \tag{3.32}$$

Substitution for $x_1$ from (3.31) yields

$$x_2 = (1 - \alpha^2)x_f + \alpha^2 x_0 \tag{3.33}$$

The operation can be repeated $n$ times

$$x_n = (1 - \alpha^n)x_f + \alpha^n x_0 \tag{3.34}$$

where $n$ is the number of dead times having elapsed since initiation of the step

$$n = \frac{t}{V/(F + F_a)} = \frac{t}{(1 - \alpha)V/F} \tag{3.35}$$

Equation (3.35) can be more conveniently expressed as

$$\frac{x_n - x_0}{x_f - x_0} = 1 - \alpha^n = 1 - \alpha^{t/[(1 - \alpha)V/F]} \tag{3.36}$$

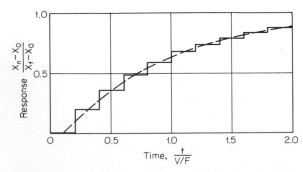

**FIG. 3.9**  Connecting the midpoints of all the steps produces a dead time followed by a first-order lag *(From F. G. Shinskey, "pH and pIon Control in Process and Waste Streams" Fig. 7.2, Wiley-Interscience, New York, 1973, with permission.)*

**TABLE 3.3   Step Response of a Mixed Vessel**

| $\dfrac{t}{V/F}$ | $\dfrac{x_t - x_0}{x_f - x_0}$ | $\dfrac{x_n - x_0}{x_f - x_0}$ | | | |
|---|---|---|---|---|---|
| | | $\alpha = 0.95$ | 0.9 | 0.8 | 0.5 |
| 0.05 | 0.049 | 0.050 | 0     | 0     | 0     |
| 0.1  | 0.095 | 0.097 | 0.100 | 0     | 0     |
| 0.2  | 0.181 | 0.185 | 0.190 | 0.200 | 0     |
| 0.4  | 0.330 | 0.337 | 0.344 | 0.360 | 0     |
| 0.5  | 0.393 | 0.401 | 0.410 | —     | 0.500 |
| 1.0  | 0.632 | 0.641 | 0.651 | 0.672 | 0.750 |
| 2.0  | 0.864 | 0.871 | 0.878 | 0.892 | 0.938 |
| 4.0  | 0.982 | 0.983 | 0.985 | 0.988 | 0.996 |

In this way, response can be compared with that of a simple first-order lag

$$\frac{x_t - x_0}{x_f - x_0} = 1 - e^{-t/(V/F)} \tag{3.37}$$

Table 3.3 lists the step response for various values of $\alpha$. Equation (3.36) can be solved only at integers of $n$, which explains the zeros and the blank space in the table. Observe that as $\alpha$ approaches 1.0, Eq. (3.36) approaches (3.37).

The step response for the process with $\alpha = 0.8$ is plotted in Fig. 3.9. It is an exponential series of steps spaced at intervals of time $V/(F + F_a)$. In a real vessel, this staircase response will not be seen because of internal diffusion. Some particles will be retained in the vessel longer than the average, while some will escape earlier. As a result, the steps will be rounded into a smooth curve running through their midpoints, as shown. This smooth curve is exactly represented by a first-order lag with dead time, where

$$\tau_d = \frac{V}{2(F + F_a)} = \frac{V}{F}\frac{1 - \alpha}{2} \tag{3.38}$$

and

$$\tau_1 = \frac{V}{F} - \tau_d = \frac{V}{F}\frac{1 + \alpha}{2} \tag{3.39}$$

As in the case of interacting lags described in Chap. 2, all the rounded step-response curves will pass through the point 0.632, that is, $1 - e^{-1}$, at $t = V/F$, regardless of the value of $\alpha$. As an example, the midpoint of the step at $t = V/F$ in Fig. 3.9 is 0.6314. From Table 3.3, the midpoint of the step at $t = V/F$ for $\alpha = 0.5$ is 0.625.

This theory is seen to fit the distributed-lag concept particularly well. Furthermore, its faithfulness has been proved in actual plant tests. Whether the mixing takes place through external circulation, as shown in Fig. 3.8, or by internal agitation, makes no difference. Although it is difficult to measure the true pumping rate of an agitator, these devices are often specified by that parameter.

The index of control difficulty $\tau_d/\tau_1$ varies only with $F$ and $F_a$:

$$\frac{\tau_d}{\tau_1} = \frac{1}{1 + 2F_a/F} = \frac{1 - \alpha}{1 + \alpha} \tag{3.40}$$

Notice that this ratio is *not* a function of volume; a large vessel may be controlled as easily as a small one, given equivalent agitation. Horsepower requirements as a function of vessel volume for constant agitation are given in Ref. 3.

## The Analyzer

Dynamics associated with the analysis play an important role in the performance of the loop. The foremost limitation in the speed of analysis is generally that of transporting the sample to the detector. Fortunately some composition measurements can be made without withdrawing a sample: electrolytic conductivity, density, and pH are notable examples. But any analysis requiring the withdrawal of a sample, particularly if that sample must undergo a certain amount of preparation, results in a significant accumulation of dead time (see the example cited at the close of Chap. 2). Naturally any effort spent in minimizing the sampling time will be rewarded both by tighter control and by faster response.

Some analyzers are discontinuous. They produce only one analysis in a given time interval. This characteristic is worthy of much more attention, because it periodically interrupts the control loop. Process chromatographs are the principal, but not sole, constituents of this group. The response of this kind of control loop will be given extensive coverage in Chap. 4, and methods for coping with it will be presented.

A few analyzers exhibit a time lag in addition to the dead time associated with sample transport. Normally this property is of little consequence, except when the process itself consists of nothing but the volume of a pipeline, whose time constant may be less than that of the analyzer. Measurements which are fast are by the same token subject to noise. Conductivity and pH are usually in this category, because they are fast enough to react to an incompletely mixed solution or particles of an immiscible phase.

Dead time in sample lines is understandably constant. Dead time in a pipe carrying the main stream varies with flow. Dead time within a stirred tank is slightly affected by flow, to the extent of $F/F_a$; in most systems this variation would not be significant. The natural period of the composition loop would therefore be virtually constant, producing constant dynamic gain, except for a process whose dominant element is a pipeline.

Most analyzers are not so far from being linear that they materially affect the gain of the control loop. But analyzers are generally given a high order of sensitivity because of the importance placed on quality control. As a result, the gain of a composition-control loop is invariably high. Objectively, composition is not as difficult to control as flow, for example, but the specifications placed on product quality are so stringent that ordinary performance is seldom acceptable. The impurity of a product stream leaving a fractionator, for example, may be specified at $1.0 \pm 0.2$ percent. It is virtually impossible to regulate flow within $\pm 1$ percent in the unsteady state, yet the composition controller is asked to perform 5 times as well. This is perhaps the greatest single reason why composition control has the distinction of being a problem area. Because quality can be measured to 0.1 percent is apparently reason enough to expect it to be controlled to the same tolerance.

**Process Gain**

Consider a blending system where concentrate X is added to diluent F to produce a mixture having composition $x$. A material balance gives

$$x = \frac{X}{F + X} \tag{3.41}$$

The steady-state gain of this process is

$$K_p = \frac{dx}{dX} = \frac{F}{(F + X)^2} = \frac{(1 - x)^2}{F} \tag{3.42}$$

If $X \ll F$, which is often the case, process gain reduces to

$$K_p \approx \frac{1}{F} \tag{3.43}$$

If $x$ is closer to 1 than to 0, $K_p$ will vary significantly with it.

The time constant of the mixed vessel has been identified in Eq. (3.40). The product of steady-state and dynamic gains is then

$$K_p G_1 = \frac{1}{F} \frac{\tau_o}{2\pi(V/F)(1 + \alpha)/2} = \frac{\tau_o}{\pi V(1 + \alpha)} \tag{3.44}$$

Flow then disappears from consideration. Period $\tau_o$ is proportional to dead time, which varies slightly with F according to (3.38) but not significantly where agitation is provided. Therefore this process tends to exhibit a constant loop gain.

Should flow F be manipulated for control, steady-state gain would be

$$K_p = \frac{dx}{dF} = -\frac{x(1 - x)}{F} \tag{3.45}$$

Variation with F would be canceled by the dynamic gain as before, but now $K_p$ varies with $x$, a problem which may require special nonlinear compensation, as described in connection with distillation in Chap. 11.

In the above derivations, the controlled variable was given as a fractional concentration. If its actual measurement is a *function* of concentration, such as pH or density, the gain of that function must also be included. The pH curve shown in Fig. 2.17 represents that type of function.

**TABLE 3.4    Properties of Common Loops**

| Property | Flow and liquid pressure | Gas pressure | Liquid level | Temperature and vapor pressure | Composition |
|---|---|---|---|---|---|
| Dead time | No | No | No | Variable | Constant |
| Capacity | Multiple | Single | Single | 3–6 | 1–100 |
| Period | 1–10 s | 0–2 min | 1–10 s | Minutes to hours | Minutes to hours |
| Linearity | Square | Linear | Linear | Nonlinear | Either |
| $K_pK_t$ | 2–5, 0.5–1† | Integrating | Integrating | 1–2 | 10–1,000 |
| Noise | Always | None | Always | None | Often |
| Proportional | 100–500% 50–200%† | 0–5% | 5–50% | 10–100% | 100–1,000% |
| Integral | Essential | Unnecessary | Seldom | Yes | Essential |
| Derivative | No | Unnecessary | No | Essential | If possible |
| Valve | Linear | Linear | Linear | Equal-percentage | Linear |

†Applies to liquid pressure.

## CONCLUSIONS

The purpose of this chapter has been to acquaint the reader with the properties of common process control loops and the reasons for these properties. Analysis served as a useful tool to present the case while at the same time demonstrating how to identify the significant elements in a loop. Rarely will a flow or level loop need analysis, but when composition-control problems arise, this procedure can be of inestimable value.

Much of what has been derived and weighed and discussed in the foregoing pages is summarized in Table 3.4.

Nothing that has not been already covered is presented in the table, yet gathering all this information together discloses some interesting features. Notice, for example, the similarity between level and flow loops, with respect to both natural period and the presence of noise. Without any doubt, however, each of the five groups above is separate and distinct from the rest.

## REFERENCES

1. Catheron, A. R.: Factors in Precise Control of Liquid Flow, *ISA Pap.* 50-8-2.
2. Considine, D. M.: "Process Instruments and Controls Handbook," Chap. 2, McGraw-Hill, New York, 1957.
3. Shinskey, F. G.: "pH and pIon Control in Process and Waste Streams," pp. 156–162, Wiley-Interscience, New York, 1973.

## PROBLEMS

**3.1**  Estimate the proportional band, integral time, and period for controlling the flow process of Example 3.2 with a controller having a phase lag of 30°.

**3.2**  A volume booster (isolating amplifier) is installed at the inlet to the valve motor of the flow process in Example 3.2, reducing its time constant to 1.0 s. Estimate the resulting natural period and dynamic gain of the process.

**3.3** Let pressure downstream of the valve in Example 3.2 be controlled instead of flow. At no flow, there is a static head of 5 lb/in² gage, while 10 gal/min will raise the pressure to 13 lb/in² gage; the range of the pressure transmitter is 0 to 25 lb/in² gage. Estimate what the proportional band of the controller will be for quarter-amplitude damping with the period used in the example.

**3.4** A mercury manometer capable of reading 15-in differential pressure is used to indicate the flow in a gas stream. What is its natural period? How would it affect the control of flow?

**3.5** Two fluids are blended in a pipeline 20 ft upstream of where the mixture is sampled. The pipe contains 0.4 gal per foot of length, and the flow rate of the blend varies from 10 to 80 gal/min. Dead time in the sample line to the analyzer is 15 s. A circulating pump is installed to maintain 100 gal/min flow through that 20-ft section of pipe without affecting the throughput. Compare the natural period with and without the pump in operation. What else does the pump provide?

**3.6** In the same process, the flow of additive is manipulated through a linear valve whose maximum flow is 1.2 gal/min. The range of the analyzer is 0 to 1 percent additive concentration. Estimate the proportional band required for at least quarter-amplitude damping if the integral time is set for 60° phase lag with the pump operating.

# Part Two

# Selecting the Feedback Controller

# Chapter Four

# Linear Controllers

Now that the characteristics of typical processes have been presented, it is possible to look more closely into various means for controlling them. The range of process difficulty has been seen to vary from zero to several hundred, as measured by the proportional band needed for damping. The very existence of such a range of control problems suggests the possibility of a variety of means for their control. The first distinction to be made is between linear and nonlinear control methods.

A linear device is one whose output is directly proportional to its input(s) and any dynamic function thereof. This definition includes not only proportional controllers, but those with integral action, derivative, lag, dead time—in short, any time function of a linear variable. To be sure, a device is linear only over a specified range. A pneumatic controller, for example, ceases to operate linearly when its output falls to zero or reaches full supply pressure. All linear devices are similarly limited, and their proper use demands an appreciation of these limitations.

In the earlier chapters certain nonlinear charactcristics were dealt with, both in processes and in the measuring devices and valves. An attempt was made in every case to compensate for process nonlinearities so as to obtain constant loop gain. This assures uniformity of performance under all conditions of operation. In general, compensation is effected outside the controller, leaving the controller as a linear device.

But within the domain of linear controllers, a variety of dynamic elements exists. Each dynamic element, such as integral or derivative, has certain undesirable properties along with those which are beneficial. A thorough understanding of the assets and liabilities of each control mode is prerequisite to their intelligent selection.

## PERFORMANCE CRITERIA

If selection between various control configurations is to be made, some basis must be established for their comparison. For example, a given process may be controlled in a number of ways. One way will be better than the others from the standpoint of performance, i.e., how it responds to a set-point or load change. The three load-response curves in Fig. 1.12 show the performance of three different controllers on the same process.

The shape of the load-response curve depends to a considerable degree on the type of control action used and the settings of the parameters involved. Furthermore, the penalty ascribed to a typical response curve is determined by the specifications of the process. Several means of weighing the response curve suggest themselves:

1. *Integrated error:* since the error $r - c$ can be either positive or negative, an integrated error of zero could be obtained in a continuously oscillating loop. Integrated error is therefore not, of itself, a measure of stability.

2. *Error magnitude:* this criterion allows the possibility of offset (a small permanent error), which is generally undesirable in any loop.

3. *Integrated absolute error* (IAE): this is a measure of the total area under the response curve on both sides of zero error. It is one of the generally accepted performance criteria. Since the error following a load change eventually disappears, the IAE approaches a finite value for any stable loop.

4. *Integrated square error* (ISE): the instantaneous error is first squared and then summed (integrated). Squaring prevents a negative error from canceling a positive one (as does absolute value) and also weighs large errors more heavily than small ones.

5. *Root-mean-square (rms) error:* this index is the standard deviation of the error. If the error reduces to zero with time, so does the rms error; this criterion is therefore applicable only to systems without a steady state.

Technically, the IAE and the ISE are the only all-encompassing indexes of performance. The principal distinction between them is the weight placed on large errors. Two response curves with the same IAE would have different values of ISE if there were a difference in error magnitude. For this reason, the ISE criterion is seen to be a combination of error magnitude and IAE.

For the case where the response curve lies wholly on one side of zero error, the integrated error equals the IAE. But this is not the limit of usefulness of the integrated error, for it represents the average error that has existed over a particular time span. The average error or integrated error is a valid basis for comparing response curves with equal damping, like the comparisons shown in Fig. 1.12. By specifying the damping, the objection raised in item 1 above is overruled. Integrated error will therefore be used as a performance index throughout the balance of the book, and in every case quarter-amplitude damping will be meant unless otherwise indicated.

### Integrated Error

The choice of integrated error as a performance index has a very practical aspect, in that it can be readily calculated from controller settings. In a proportional-plus-integral controller

$$m = \frac{100}{P}\left(e + \frac{1}{I}\int e\, dt\right)$$

Before a load change at time $t_1$ the output will be stationary at a level $m_1$, and the error will be zero. After the transient from a load change has subsided, i.e., at time $t_2$, the output will come to rest at a new level $m_2$, at which the error will again be zero. Then, subtracting the two outputs gives

$$m_2 - m_1 = \frac{100}{P}\left(\frac{1}{I}\int_0^{t_2} e\, dt - \frac{1}{I}\int_0^{t_1} e\, dt\right)$$

Reducing the last expression yields

$$\Delta m = \frac{100}{P}\left(\frac{1}{I}\int_{t_1}^{t_2} e\, dt\right) \tag{4.1}$$

Let the integrated error resulting from the load change $\Delta m$ be designated $E$

$$E = \int_{t_1}^{t_2} e\, dt$$

Then the load response of a given control loop can be assessed on the basis of integrated error per unit load change

$$\frac{E}{\Delta m} = \frac{PI}{100} \tag{4.2}$$

Again this is integrated error and not IAE; the damping of the loop must be assured before this index can be used.

The load-response criterion $E/\Delta m$ depends on the proportional band and integral time, which, in turn, depend on the characteristics of the plant. This is another way of illustrating the difficulty of control which was described in Chap. 1. If the proportional band can be made to approach zero because of the ease with which the process can be controlled, or the integral time because of its speed of response, $E/\Delta m$ will approach zero. The integrated error will be found useful in evaluating not only the difficulty of a process but also the effectiveness of the means used in its control.

### Error Magnitude

Figure 1.1 showed that changes in the controlled variable $dc$ are induced by changes in load $dq$ and controller output $dm$

$$dc = K_p \mathbf{g}_p (dm - dq) \tag{4.3}$$

In turn, the controller output changes by an amount $dm$ as a function of $dc$

$$dm = -K_c \mathbf{g}_c \, dc \tag{4.4}$$

Load response of the closed loop can be found by eliminating $dm$

$$\frac{dc}{dq} = -\frac{K_p \mathbf{g}_p}{1 + K_p \mathbf{g}_p K_c \mathbf{g}_c} \tag{4.5}$$

Dynamic components $\mathbf{g}_p$ and $\mathbf{g}_c$ are vectors whose phase and gain vary with period. Therefore Eq. (4.5) has a different solution at each period $\tau_q$ of the disturbing input $dq$

$$\frac{dc(\tau_q)}{dq(\tau_q)} = -\frac{K_p G_p}{1, \angle\, 0 + K_p G_p K_c G_c, \angle\, (\phi_p + \phi_c)} \tag{4.6}$$

Consider a process consisting of dead time and an integrating capacity such that at $\tau_q$

$$G_p = \frac{\tau_q}{2\pi\tau_1} \qquad \phi_p = -90 - 360\frac{\tau_d}{\tau_q} \tag{4.7}$$

Let the controller have proportional-plus-integral action, adjusted for quarter-amplitude damping and $30°$ phase shift at loop period $\tau_o$, according to Table 1.2. Then $I = 1.65\tau_d$,

$$G_c = \sqrt{1 + \left(\frac{\tau_q}{2\pi 1.65\tau_d}\right)^2} \qquad \phi_c = -\tan^{-1}\frac{\tau_q}{2\pi 1.65\tau_d} \tag{4.8}$$

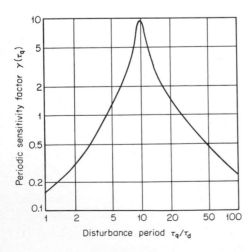

**FIG. 4.1** The sensitivity of a closed loop to disturbances for dead-time plus capacity process under PI control.

**TABLE 4.1  Sensitivity of the Controlled Variable to Sinusoidal Load Changes under PI Control**

| $\dfrac{\tau_q}{\tau_d}$ | $C_p$ | $G_c$ | $-\dfrac{dc(\tau_q)}{dq(\tau_q)}$ |
|---|---|---|---|
| 1 | $0.159\tau_d/\tau_1$ | 1.000 | $0.159K_p\tau_d/\tau_1$ |
| 2 | 0.318 | 1.018 | 0.306 |
| 4 | 0.637 | 1.072 | 0.885 |
| 6 | 0.954 | 1.156 | 1.908 |
| 10 | 1.592 | 1.390 | 9.107 |
| 20 | 3.183 | 2.174 | 1.470 |
| 40 | 6.366 | 3.988 | 0.604 |
| 100 | 15.92 | 9.704 | 0.230 |

and

$$K_c = \frac{1}{2.2K_p\tau_d/\tau_1} \tag{4.9}$$

Process gain can be placed in terms of the ratio $\tau_q/\tau_d$

$$K_pG_p = \frac{K_p}{2\pi}\frac{\tau_q}{\tau_d} \tag{4.10}$$

as can loop gain

$$K_pG_pK_cG_c = \frac{1}{2\pi 2.2}\frac{\tau_q}{\tau_d}\sqrt{1 + (\tan \phi_c)^2} \tag{4.11}$$

Table 4.1 lists the solution of Eq. (4.6) for load changes occurring at several periods expressed as a ratio to process dead time. At short periods attenuation is provided by the capacity of the process and proportional control. At long periods, attenuation is provided by integral control action, as indicated by increasing $G_c$. In the range of periods around 10 dead times, the sensitivity of the loop to load changes is maximum, owing to the oscillatory nature of the loop itself. From Table 1.2 the period of the loop is seen to be $6\tau_d$.

The numerical coefficient appearing in the last column of Table 4.1, designated *periodic sensitivity factor* $\gamma(\tau_q)$, is plotted against the ratio $\tau_q/\tau_d$ in Fig. 4.1, exposing the region of maximum sensitivity. Figure 4.1 indicates the type of problem that may be encountered when a number of identical processes under closed-loop control are connected in series. An oscillation developed in the first can readily propagate through the entire series.

## PI AND PID CONTROLLERS

In Chap. 1 the characteristics of proportional-plus-integral (PI) and proportional-plus-derivative (PD) controllers were presented as ideal devices. Now that their principal functions are understood, it is appropriate to investigate their limitations and nonidealities as they affect both normal and abnormal operation. Furthermore, the combination of the three modes into a PID controller introduces the prospect of mode interaction. While common to most controllers, it is not widely appreciated. However, effects of mode interaction must be evaluated before a generally applicable tuning procedure can be developed.

### Limitations of Derivative Action

As was pointed out earlier, derivative action must be limited in gain to avoid controller instability and noise propagation. Gain limits in the range of 10 to 20 are typical, applied either by a limited-gain amplifier or by adding a first-order lag. In either case, the effect is the same, producing derivative action with a lag having a time constant 10 to 20 times smaller than the derivative setting. The contribution of the lag is easily included in the loop. Table 4.2 shows the effect of a lag of $D/10$ on the dynamic-gain vector of the PD controller.

As the table indicates, the gain limit is noticeable only at values of $\tau_o/2\pi D$ of 0.2 and less, which are typically outside the control region. At a value of 1.0, which is the region of normal operation for a PID controller, the gain is not significantly altered, and the phase angle is reduced less than 6°. Consequently, the gain limitation has little effect at $\tau_n$ and even less when the phase lead is adjusted to about 30°, as found optimum in Table 1.3. It is doubtful whether an increase in gain limit to 20 would be discernible, although a reduction to 5 definitely reduces control effectiveness.

Older controllers, and certain field-mounted pneumatic controllers, have but a single amplifier. Derivative action is accomplished by imposing a lag in the feedback loop around the amplifier. These styles have two undesirable characteristics: (1) there is no derivative action when the controller output is saturated, and (2) derivative action is applied to the set point as well as to the controlled variable. The first characteristic can cause overshoot upon startup or recovery from a condition which caused saturation because the derivative function is disabled. The second relates to the different requirements of the two inputs to the controller.

Applied to the controlled variable, derivative is in the loop and therefore contributes to stability and responsiveness. Applied to the set point, derivative is outside the loop and can cause unnecessary disturbances. Since set points are

**TABLE 4.2   Dynamic Gain of a PD Controller with a Gain Limit of 10**

| $\dfrac{\tau_o}{2\pi D}$ | Ideal | | Limited | |
|---|---|---|---|---|
| | $G_{PD}$ | $\phi_{PD}$, deg | $G_{PD}$ | $\phi_{PD}$, deg |
| 0.02 | 50 | 88.9 | 9.81 | 10.2 |
| 0.05 | 20 | 87.1 | 8.96 | 23.7 |
| 0.10 | 10 | 84.3 | 7.11 | 39.3 |
| 0.20 | 5.1 | 78.7 | 4.55 | 52.1 |
| 0.50 | 2.23 | 63.4 | 2.19 | 52.1 |
| 1.00 | 1.41 | 45.0 | 1.41 | 39.3 |
| 2.00 | 1.12 | 26.6 | 1.12 | 23.7 |

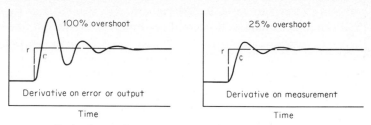

FIG. 4.2 Derivative makes some controllers hypersensitive to set-point changes.

usually introduced stepwise by an operator, derivative action will be applied to its gain limit, overdriving the final actuator and causing a severe overshoot. Most modern controllers apply derivative action to the controlled variable alone, avoiding this hypersensitivity to set-point changes. Figure 4.2 compares the response to a step change in set point with and without derivative action on the set point. Set-point changes can be introduced by another controller in a cascade configuration, as described in Chap. 6. The controller being adjusted cannot tolerate derivative action on its set point, particularly if the adjusting controller also has derivative.

## Integral Windup

The integral mode of a controller causes its output to continue changing as long as there is a deviation from set point. Often that deviation cannot be eliminated when the load exceeds the range of the manipulated variable because manipulation is obstructed by a closed valve, failed pump, etc. Nonetheless, the controller will attempt to correct for the deviation and, given enough time, will saturate its integral mode; this condition is called *windup*.

The effect of windup can be visualized by examining a common configuration of PI controller described in Fig. 4.3. Integration is accomplished by positive feedback of the output of the controller through a first-order lag having a time constant $I$. The feedback connection is shown as a broken line to indicate the desirability of breaking that loop to avoid windup. However, if that loop is closed and $e \neq 0$, $m$ will be driven continuously by positive feedback to the saturation limit. Windup is not complete, however, till the output of lag $I$, here designated $b$, is also at the saturation limit. Then, even if $e$ should

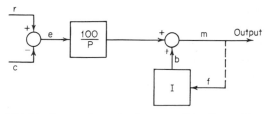

FIG. 4.3 Integrating action is accomplished by positive feedback of the controller output through lag $I$.

suddenly return to zero, $m$ will remain at $b$, which is at the saturation limit. (Saturation is usually well beyond the ends of the output scale; in a pneumatic controller with an air supply of 20 lb/in² gage, maximum pressure is about 42 percent above the 3 to 15 lb/in² gage scale, and minimum pressure is 25 percent below it.) Control action cannot begin until the controller output returns on scale, which requires a reversal of the error signal. Therefore windup *always* results in overshoot before control is restored.

The effect of windup can be moderated by placing limits on feedback signal $f$ entering the integral time constant. However, they cannot be set *within* the 0 to 100 percent control range without producing offset during normal operation. A preferred arrangement is to add a controller to the feedback circuit, as shown in Fig. 4.4. This device has been called a *batch unit,* in that it is normally required for controllers used to start up batch processes automatically.

The batch unit shown in Fig. 4.4 is designed for high limiting. When controller output $m$ reaches set high limit $m_h$, the high-gain amplifier drives feedback signal $f$ downward; $m$ is then prevented from exceeding $m_h$. During normal operation, $m$ is below $m_h$, in which case the amplifier drives upward; however the low selector ($<$) preferentially accepts the lower $m$ as feedback, producing normal integrating action.

Upon a large deviation, the batch amplifier may drive $f$ to the low limit of saturation and still be unable to keep $m$ at $m_h$. Then, when $e$ decreases, $m$ may remain below zero for an extended period of time. This has the opposite effect of windup, resulting in an excessively slow approach to the set point. This sluggishness can be avoided by limiting the output of the batch amplifier so that it cannot fall below "preload" setting $m_q$. Should $e$ return to zero in this mode of operation, $m$ will equal $m_q$; this preload is set roughly equal to the anticipated process load when control is to be restored. It allows the operator to adjust the amount of overshoot or undershoot obtainable upon startup. Figure 4.5 compares control of temperature of a batch reactor without batch control, with that achieved with batch control but without preload, and with both.

Some processes will require a batch unit acting on a low limit. It is configured in the same manner as in Fig. 4.4, although the actions of the two selectors are

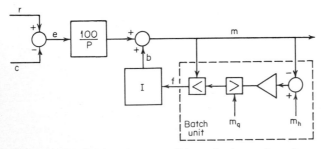

**FIG. 4.4** The batch unit preconditions the integral mode to resume control when the deviation returns toward zero.

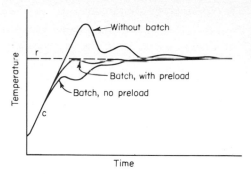

**FIG. 4.5** To be effective for temperature control of a batch reactor, the batch controller must be properly preloaded.

reversed. Many controllers will not encounter a combination of large deviation and wide proportional band, so that preload is unnecessary. Finally, $m_h$ should be set at the point where a limit is naturally encountered in the process. If the control valve is to stroke only over 4 to 8 lb/in$^2$ gage, for example, $m_h$ should be set at 8 lb/in$^2$ gage.

Other methods of avoiding windup in cascade and selective control systems are described in that context in Chap. 6.

### Automanual Transfer

Most modern controllers are capable of "bumpless" transfer between manual and automatic operating modes. In the automatic mode, the manual regulator tracks the controller output, so that at the instant of transfer to manual the output will be held at its last value. In the manual mode, the automatic output is made to track the manual regulator by feedback through the integrating circuit in the manner of Fig. 4.4. A high-gain amplifier replaces the batch unit, driving the automatic output to equal the manual output. Time constant $I$ is bypassed, for faster tracking. Upon transfer to automatic, the output of the controller will start to integrate from the last position maintained in manual; there is no proportional action even with a deviation unless the deviation changes in automatic.

This characteristic allows proportional response to set-point changes to be inhibited by making those changes with the controller in manual. When it is returned to automatic, integral action alone is applied to the set point; the output will then ramp towards its final value without an initial step.

Some controllers are capable of being transferred between auto and manual by a contact closure. This feature is very useful when control action must be coordinated with on-off devices such as pumps. Reference 1 describes a pH control system whose reagent valves are automatically closed when a feed pump is deenergized by a switch acting on low sump level. Because the pH controller is unable to function when the valves are closed, windup will develop unless protection is provided. It is avoided by transferring the controller to manual when the feed pump is stopped and back to automatic when it is restarted.

### The Three-Mode Controller

An ideal three-mode controller consists of a simple combination of the individual modes as they have already been presented

$$m = \frac{100}{P}\left( e + \frac{1}{I}\int e\,dt - D\,\frac{dc}{dt} \right) \tag{4.12}$$

Here derivative action is applied only to the controlled variable, and $e$ is defined as $r - c$. Because the three components of gain at the period of oscillation are out of phase, vector addition is required to obtain controller gain and phase, as shown in Fig. 4.6. The resultant phase and gain of the ideal three-mode controller are

$$\phi_c = \tan^{-1}\left( \frac{2\pi D}{\tau_o} - \frac{\tau_o}{2\pi I} \right) \tag{4.13}$$

$$G_c = \sqrt{1 + (\tan \phi_c)^2} \tag{4.14}$$

A natural interaction between the derivative and integral modes is brought about by the method of construction of most controllers. Figure 4.7 illustrates the functional configuration of controllers with single-stage and two-stage amplification (the amplifiers are identified by triangles). In the single-stage controller, first-order lags providing derivative $D$ and integral $I$ action are connected in negative and positive feedback loops around the amplifier. Derivative gain limitation is produced by instantaneous feedback by the factor $1/K_D$, where $K_D$ is the derivative gain limit of 10 to 20. Observe that the two lags are arranged in series.

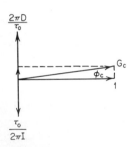

In Fig. 4.7$b$, another stage of amplification is used to apply derivative to the controlled-variable input alone. Derivative and integral action are imposed sequentially, leading to the same type of interaction experienced with the single-stage serially connected controller.

**FIG. 4.6** The resultant gain and phase emanate from a vector summation of the individual modes.

Interaction causes the effective values of the modes to differ from their set values. The effective integrating time is actually the sum of the time constants

$$I_{\text{eff}} = I + D \tag{4.15}$$

and the effective derivative time is the reciprocal of the sum of their reciprocals

$$D_{\text{eff}} = \frac{1}{1/I + 1/D} \tag{4.16}$$

Steady-state gain is also altered by interaction

$$K_c = \frac{100}{P}(1 + D/I) \tag{4.17}$$

These effective values must be used in Eq. (4.13) to determine the true phase and gain of the interacting controller.

Several important observations can be made from the above relationships:

1. The effective value of derivative can never be greater than one-fourth the effective integral time, which is achieved when $D = I$.

2. When $D$ is set above $I$, it changes $I_{\text{eff}}$ more than $D_{\text{eff}}$ and $I$ has more influence over $D_{\text{eff}}$ than $I_{\text{eff}}$.

3. In the range where $D \simeq I$, a change in either $D$ or $I$ affects all three control modes by essentially the same percentage.

The integrated error per unit load change for the interacting controller is

$$\frac{E}{\Delta m} = \frac{I_{\text{eff}}}{K_c} \tag{4.18}$$

If Eqs. (4.15) and (4.17) are substituted into (4.18), we have

$$\frac{E}{\Delta m} = \frac{I + D}{(100/P)(1 + D/I)} = \frac{PI}{100}$$

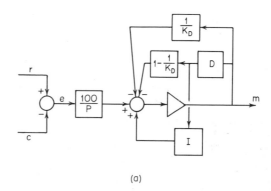

(a)

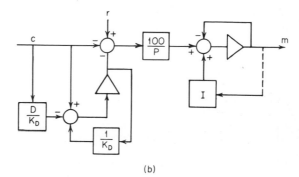

(b)

**FIG. 4.7** Derivative and integrating action are usually applied in series, bringing about interaction between them; (a) single-stage and (b) two-stage controller.

Although interaction may require different settings of $P$ and $I$ to attain stability, $E/\Delta m$ remains the same function of those settings.

The effects of interaction on the step load response of a closed loop is shown in Fig. 4.8. Here derivative time is increased progressively, increasing the gain of the controller and also its effective integral time. Increased gain brings about reduced amplitude and lighter damping; increased integral time appears in the slower approach to set point.

Single-stage controllers requiring batch action or external feedback cannot be connected as in Fig. 4.7$a$. Their derivative and integral time constants must be connected in parallel, as shown in Fig. 4.9. This configuration presents the possibility of positive feedback through $I$ equaling or exceeding negative feedback through $D$ in response speed. Although the effects on the mode settings are the same as in Eq. (4.15) and (4.16), controller gain is radically altered

$$K_c = \frac{100}{P} \frac{1 + D/I}{1 - D/I} \tag{4.19}$$

As indicated, when $D = I$, there is no net feedback and $K_c$ approaches infinity. Furthermore, when $D > I$, $K_c$ becomes negative, so that the action of the controller is reversed. When using one of these controllers, $D$ must be kept substantially below $I$ to assure stability.

### Optimum Controller Settings

Optimum settings for controllers depend not only on the criteria used but also on the type of disturbance encountered. For example, settings which minimize IAE for a step load change will not minimize IAE for a step set-point change or a pulse load change. Optimum settings also vary with the nature of the process. The most common demand seems to be a step change in load, and the most common process is dead time plus capacity. The author has therefore developed an adjustment procedure for this combination. Instead of using IAE as a criterion, however, integrated error is used, in conjunction with a loop gain of 0.5. The results are almost identical with IAE and can be calculated directly from controller settings.

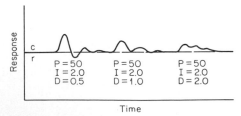

FIG. 4.8 Increasing derivative time reduces error magnitude at the cost of recovery time.

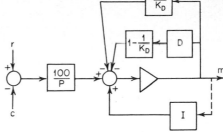

**FIG. 4.9** Single-stage controllers requiring external feedback or batch control of the integral mode have parallel feedback loops.

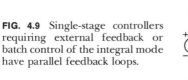

Table 4.3 lists selected combinations of mode settings for PI and both types of PID controllers which produce a loop gain of 0.5 with a dead-time-plus-single-capacity process. The last column gives the integrated error per unit load change calculated from those settings. A controller phase angle of $-28$ to $-29°$ seems to minimize the integrated error for a PI controller. This phase angle is achieved at a period of 5.8 to 5.9 dead times and an integral time constant of about 1.7 dead times. The optimum seems rather broad, there being little penalty encountered anywhere in the range of 27 to 30° phase lag.

The case for the PID controllers is not so clear. The third adjustment gives an additional degree of freedom to be considered. Table 4.3 indicates that a phase lead of 10° gives about the minimum integrated error. That phase angle can be achieved with many combinations of settings, however. For the interacting controller, three combinations of settings are shown which produce 10° phase lead, with minimum integrated error coincident with equal settings.

For the noninteracting controller, equal settings were not chosen in Table 4.3 because of some strange relationships. As it happens, setting $D = I = 0.637\tau_d$

**TABLE 4.3   Optimum Settings for PI and PID Controllers**

| Controller | $\phi_c$, deg | $\tau_o/\tau_d$ | $P$, % | $I/\tau_d$ | $D/\tau_d$ | $PI/100$ |
|---|---|---|---|---|---|---|
| PI | $-27$ | 5.71 | $204K_p\tau_d/\tau_1$ | 1.79 | 0 | $3.643\ K_p\tau_d^2/\tau_1$ |
|  | $-28$ | 5.81 | 209 | 1.74 | 0 | 3.638 |
|  | $-29$ | 5.90 | 215 | 1.69 | 0 | 3.639 |
|  | $-30$ | 6.00 | 220 | 1.65 | 0 | 3.647 |
| Interacting PID | 15 | 3.43 | $226K_p\tau_d/\tau_1$ | 0.71 | 0.71 | $1.607\ K_p\tau_d^2/\tau_1$ |
|  | 10 | 3.60 | $\begin{cases}241 \\ 233 \\ 227\end{cases}$ | 0.66 / 0.68 / 0.70 | 0.71 / 0.68 / 0.67 | 1.590 / 1.589 / 1.590 |
|  | 5 | 3.79 | 242 | 0.66 | 0.66 | 1.594 |
|  | 0 | 4.00 | 255 | 0.64 | 0.64 | 1.621 |
| Noninteracting PID | 15 | 3.43 | $113K_p\tau_d/\tau_1$ | 0.71 | 0.57 | $0.802\ K_p\tau_d^2/\tau_1$ |
|  | 10 | 3.60 | 116 | 0.68 | 0.58 | 0.789 |
|  | 5 | 3.79 | 121 | 0.66 | 0.60 | 0.799 |
|  | 0 | 4.00 | 127 | 0.64 | 0.64 | 0.811 |

**TABLE 4.4    Open-Loop Gain and Phase at Various Periods for PID Controllers**

| Period $\dfrac{\tau_o}{\tau_d}$ | Noninteracting | | Interacting | |
|---|---|---|---|---|
| | $\Pi KG$ | $\Sigma\phi$, deg | $\Pi KG$ | $\Sigma\phi$, deg |
| 4 | 0.500 | −180.0 | 0.500 | −180.0 |
| 6 | 0.979 | −189.8 | 0.813 | −172.6 |
| 8 | 1.807 | −191.3 | 1.250 | −171.9 |
| 10 | 2.916 | −190.5 | 1.818 | −172.4 |
| 20 | 12.29 | −186.2 | 6.516 | −175.4 |
| 40 | 49.88 | −183.2 | 25.30 | −177.6 |

will produce zero phase angle, while $D = I = 0.635\tau_d$ will produce 30° phase lead. This indicates a strong tendency for the loop period to drift when $D$ and $I$ are set close together.

To confirm this tendency the author tested a simulated 10-capacity interacting process controlled by a noninteracting (Foxboro Model 62H-5E) PID controller. Quarter-amplitude damping was achieved at a period of 16 s, using settings of $P = 10$ percent, $I = 4.6$ s, and $D = 2.5$ s. When the proportional band was increased to 100 percent without readjusting $I$ and $D$, the period increased to 55 s although the damping remained the same. A further increase in $P$ to 300 percent caused the period to expand to 110 s, again at the same damping. The natural period of the process was the aforementioned 16 s.

One conclusion that can be drawn from the above calculations and tests is that although a noninteracting controller can provide improved performance, as indicated in Table 4.3, it is sensitive to process parameter variations when $D \rightarrow I$. The interacting controller is not so sensitive because its effective values of $D$ and $I$ cannot be closer than 1 to 4. Hence interacting controllers are safer to use.

A further illustration of the tendency for drifting period is evidenced in the open-loop gain and phase as a function of period. Table 4.4 lists these values for both interacting and noninteracting controllers having $D = I = 2\tau_d/\pi$, which corresponds to the zero controller phase conditions given in Table 4.3. Both controllers are adjusted for a loop gain of 0.5 at $\tau_o = 4\tau_d$, on the given dead-time-plus-capacity process. Observe that for all periods greater than $4\tau_d$, loop phase exceeds $-180°$ for the interacting controller. Consequently the interacting controller will tend to cycle only at a period of $4\tau_d$, yet the noninteracting controller can cycle at longer periods, depending on loop gain. This explains the test results observed earlier. A change in any parameter which would cause loop gain to decrease will shift the point of unity loop gain to a longer period.

Figure 4.10 gives the sensitivity factor of $c$ with respect to $q$ for loops with each of the three controllers described above. The PI controller was adjusted for a phase lag of 30°, and the two PID controllers were set with $D = I = 2\tau_d/\pi$.

The potential superiority of the noninteracting PID controller is apparent, although in most applications that degree of control is too exacting to be safe.

## Procedures for Adjusting Controllers

The following procedures are intended to result in the optimal controller settings of Table 4.3 for PI and PID controllers:

1. Set $D$ to minimum and $I$ to maximum.

2. With the controller in manual, pulse the output enough to initiate an observable response and then transfer the controller to automatic.

3. If no oscillation develops, reduce $P$ in half and repeat step 2.

4. If an oscillation develops, measure its period; this is the natural period $\tau_n = 4\tau_d$.

5.   *a.*  For a *PI* controller, set $I = 0.43\tau_n$ and double $P$; this should increase the period to $\tau_o = 1.45\tau_n$.

     *b.*  For an interacting PID controller, set $D = I = 0.17\tau_n$ and double $P$; this should reduce the period to $\tau_o = 0.9\tau_n$.

     *c.*  For a noninteracting PID controller, set $D = 0.08\tau_n$ and $I = 0.34\tau_n$; this should reduce the period to $\tau_o = 0.9\tau_n$.

6. Pulse the process manually as before and return to automatic. If $\tau_o$ is *less* than it should be (as indicated in step 5), decrease $I$ or $D$ or both to obtain the desired period; if $\tau_o$ is *greater* than desired, increase $I$ or $D$ or both.

7. Readjust $P$ to obtain the desired degree of damping.

Small deviations from the stipulated periods are not considered significant because the optima as given in Table 4.3 are quite broad. When applying batch action to a PI controller, a somewhat longer value of integral time than specified above will allow closer control of set-point overshoot. In general, response to step changes in set point and pulse changes in load will be improved with longer values of integral time than those given above [2].

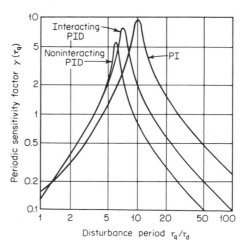

**FIG. 4.10** The sensitivity of a closed loop to periodic disturbances using three types of controllers.

## COMPLEMENTARY FEEDBACK

The question often arises whether proportional, integral, and derivative are really the best control modes for every application. For the easier-to-control processes, their use can be justified. A single-capacity process and some two-capacity processes need only narrow-band proportional action. Derivative is of great value in processes with two or three capacities. But for the more difficult processes, it has been found that integral action is essential.

In processes where dead time is dominant, derivative action has less effect than in processes consisting of capacity alone. This suggests that some other control mode more akin to dead time might be valuable. Derivative and integral are, in actuality, time constants just like the time constants in a process. They bear no resemblance to the dead time that may exist in the plant, however.

### Theory

Several authors [3, 4] have postulated a feedback control system that is modeled after the process. This kind of control action is known as *complementary feedback* because the characteristics of the controller complement the dynamics of the process. A block diagram showing both process and complementary controller appears in Fig. 4.11

The principle employed is that a given error signal $e$ can be made to generate a certain instantaneous restoring force $100e/P$ which will change $c$ exactly enough to cancel the error. At the same time, a complementary signal characterized to match the response of the process is fed back positively to cancel the effect of the negative feedback from the process. This means that the output of the controller will remain at its instantaneous value, which was correct in that it was able to reduce the error exactly to zero.

For this exact sequence of events to occur, the control parameters must have the values

$$P = 100K_p \qquad \mathbf{g}_c = \mathbf{g}_p \tag{4.20}$$

The term $K_p$ is the steady-state gain product of process, valve, and transmitter. The terms $\mathbf{g}_c$ and $\mathbf{g}_p$ are vectors representing the dynamic components in the controller and process, respectively.

It is worthwhile to trace the sequence of events following a set-point change through the block diagram. Initially $e$ is zero because $c = r$, and $m$ is at rest.

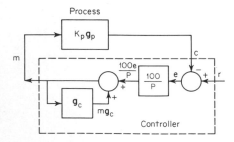

**FIG. 4.11** The complementary controller has a model of the process in its positive-feedback loop.

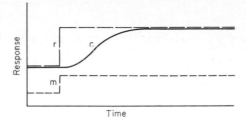

**FIG. 4.12** With complementary feedback $m$ images $r$, producing response which appears to be open-loop.

Following a set-point change $\Delta r$,

$$e = \Delta r$$

and the output jumps instantaneously by

$$\Delta m = 100 \, \frac{e}{P}$$

The output change $\Delta m$ proceeds through the process to cause

$$\Delta c = \Delta m \, K_p \mathbf{g}_p = 100 \, \Delta r \, K_p \, \frac{\mathbf{g}_p}{P}$$

Since $100 K_p/P = 1$, the set-point response is

$$\Delta c = \Delta r \, \mathbf{g}_p$$

Passing through the subtracting junction gives

$$e = \Delta r - \Delta c = \Delta r - \Delta m \, K_p \mathbf{g}_p$$

The signal then appears at the summing point as

$$100 \, \frac{e}{P} = 100 \, \frac{\Delta r}{P} - 100 \, \frac{\Delta m \, K_p \mathbf{g}_p}{P} = 100 \, \frac{\Delta r}{P} - \Delta m \, \mathbf{g}_p$$

Within the controller, complementary feedback is sending $+\Delta m \, \mathbf{g}_p$ to the same summing point, such that $\Delta m$ will retain its original value of $100\Delta r/P$ as $e$ returns to zero.

The controlled variable responds as it would if the loop were open because the output of the controller is constant after the new set point has been inserted. Overshoot is impossible with the proper settings, achieving critical damping. Figure 4.12 shows how a typical process might respond to a step set-point change.

The principal advantage of this type of control scheme is that critical damping can be achieved with a proportional loop gain of unity; in most control loops, the proportional gain exceeds unity. In a single-capacity process, for example, the proportional band may be reduced to zero, placing the proportional loop gain at infinity. Yet there are some processes, notably those dominated by dead time, in which the loop gain must be much less than unity to obtain the desired damping.

Figure 1.19 shows the required proportional band for quarter-amplitude damping for any combination of dead time and capacity. A band of $100K_p$ percent (proportional loop gain of 1.0) is seen to be required for a process having $\tau_d/\tau_1 = 1.2$. But with complementary feedback, the same proportional gain could produce critical damping. Complementary feedback is, by this token, of advantage in controlling the most difficult processes.

### Practical Considerations

To produce the type of response shown in Fig. 4.12, the controller must match the process quite faithfully, both in its steady-state and dynamic components. If the proportional gain of the controller is too low, the output will not move enough in one step to cancel the deviation and additional control action will be required. Approach to set point is then exponential; if the loop gain is one-half, the deviation will be reduced by one-half at intervals represented by the dynamic response of the process. A pure dead-time process will respond to a step change in set point in an exponential series of steps, as shown in Fig. 4.13. A loop gain of 1.5 will produce quarter-amplitude damping, and a loop gain of 2.0 will develop uniform oscillation at a period of $2\tau_d$. These same characteristics will appear with sampled-data control, and are described in more detail under that heading later in the chapter.

If the process is a single lag $\tau_1$ and the controller lag is matched to it, the set-point response will be a lag whose time constant is $\tau_1$ divided by the proportional loop gain. Here a very high loop gain can be tolerated without instability owing to the responsiveness of the process. A PI controller constructed as shown in Fig. 4.3 will be recognized as complementary to a process consisting of a first-order lag.

Matching the dynamic characteristics of the controller to the process is also important. When the proportional loop gain is 1.0,

$$\frac{dc}{dr} = \frac{\mathbf{g}_p}{1 - \mathbf{g}_c + \mathbf{g}_p} \tag{4.21}$$

When $\mathbf{g}_c = \mathbf{g}_p$, the open-loop response of $\mathbf{g}_p$ is achieved to a set-point change. But in the event of a mismatch, the response will be distorted. A dead-time process will erupt with oscillations having a period which is twice the difference between process and controller dead times. Unfortunately processes dominated

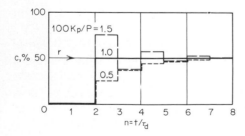

**FIG. 4.13**   A loop gain of 1.0 gives critical damping, and 1.5 gives quarter-amplitude damping; 2.0 produces uniform oscillations.

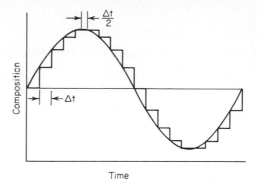

**FIG. 4.14** The act of sampling introduces an effective dead time of half the sample interval.

by dead time, which benefit most by complementary feedback, are also those least tolerant of mismatching by the controller. As a result, a more reliable method of controlling dead-time processes will be found in the sampling controller (below).

## INTERRUPTING THE CONTROL LOOP

In some control loops, feedback of information from the process is available only on an intermittent basis. The on-stream chromatograph is perhaps the most common transmitter of intermittent information, although many less familiar analyzers also have this characteristic. It is certainly true of off-line analyzers, which require a technician to operate, obtain the result, and make the subsequent adjustment to the process. In a loop such as this, only one piece of information is transmitted within a certain space of time known as the *sample interval*. The control loop is open, except at the first instant of each sample interval. This dynamic property differs from anything discussed thus far.

The controller can also be operated on a sampled basis. This characteristic is found in digital control systems where a central processor sequentially performs calculations for many loops. The response of digital control loops can differ substantially from analog loops, which function continuously, and are analyzed in some detail in this section.

### The Dynamics of Sampling

Let the sine wave in Fig. 4.14 represent the true composition of a process stream as a function of time. An analyzer samples that stream periodically, transmitting its analysis to a controller. The composition as seen by the controller changes stepwise at periodic intervals. If there is no dead time in transporting and processing the sample, the controller will see the staircase representation of the composition shown in Fig. 4.14.

Observe that the signal as seen by the controller has virtually the same amplitude as the true composition but is shifted in phase. The phase shift, in fact, is half of what would be produced by an equal amount of dead time

$$\phi_\Delta = -180 \frac{\Delta t}{\tau_o} \tag{4.22}$$

where $\Delta t$ is the sample interval. Effectively, the average age of the information acted on by the controller is half the sample interval.

Normally, there will also be some dead time associated with transporting the sample to and through the analyzer to the detector. A loop containing a sampling element as the only dynamic element or one whose $\Delta t$ exceeds the response of all other elements in the loop will cycle under proportional control with a natural period of $2\Delta t$. Under continuous integral control, however, the period will be $2\Delta t + 4\tau_d$. This can be envisaged by following the sequence of events shown in Fig. 4.15. It shows integral control applied to a process (a), having $\tau_d = \Delta t/2$, and another, (b), having $\tau_d = \Delta t$. In both cases, uniform oscillation results when integral time $I$ is set at $K_p \tau_d$. (In the figure, $K_p$ happens to be 1.0, as indicated by c duplicating m exactly $\tau_d$ later.)

Because the loop is open between samples, the value of c presented to the controller is held at whatever its true value was at the time of sampling. These points are designated $c^*$ in the figure. Integration of the resulting sustained error then produces a ramp in m, duplicated by a later ramp in c. The period of the loop in Fig. 4.15a is $4\Delta t$, while that in 4.15b is $6\Delta t$. Uniform oscillation is to be expected when loop gain is unity, i.e.,

$$G_I K_p = \frac{\tau_o}{2\pi I} K_p = 1.0$$

In the first case, $\tau_o = 4\Delta t = 8\tau_d$, so that

$$G_I K_p = \frac{8\tau_d}{2\pi K_p \tau_d} K_p = 1.27$$

and in the second, $\tau_o = 6\Delta t = 6\tau_d$, giving

$$G_I K_p = \frac{6\tau_d}{2\pi K_p \tau_d} K_p = 0.95$$

The reason loop gain is not exactly 1.0 is the nonsinusoidal nature of the

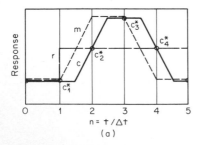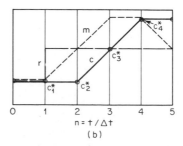

**FIG. 4.15**   In process (a), $\tau_d = \Delta t$, and in (b), $\tau_d = \Delta t$.

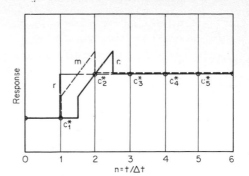

**FIG. 4.16** The proper combination of proportional and integral adjustments can produce critical damping.

oscillation. As $\tau_o >> \Delta t$, oscillations will become more sinusoidal and the effect of sampling can be treated simply as dead time of half its stated value.

### Two-Mode Control

Two-mode control combines the speed of response of proportional action with the elimination of offset brought about by integral action. The proportional mode is just as valuable in a sampled dead-time loop as it was in one without sampling. In fact, proportional action enables any loop whose dead time is less than the sampling interval to be critically damped. Figure 4.16 shows how this is done.

The proportional action produces an instantaneous change in output, which is removed when the error returns to zero. Some overshoot does occur, but it disappears before the next sample. Only one combination of proportional and integral settings will provide this critical damping

$$P = 100K_p \frac{\Delta t}{\tau_d} \qquad I = \tau_d \tag{4.23}$$

If $\tau_d > \Delta t$, critical damping cannot be achieved. As with complementary feedback, reducing $P$ by one-half produces zero damping, by one-third gives quarter-amplitude damping.

Response curves for sampled loops are made of steps. The rate of rise of even a small step is extremely high; therefore derivative control action on a sampled signal produces pulsing of the manipulated variable. This pulsing cannot contribute much to the closed-loop response because sampling prevents the effect of such action from being seen; consequently the manipulated variable is driven severely without cause. Derivative is therefore of little value in the sampled loop.

### A Sampling Controller

Why operate on old information? There is really no value in continuing to drive the controller output when it can have no immediately observable effect. A sampling controller is suggested as being more compatible with the sampled process.

Here is the control strategy. At the start of the sample interval, when the controller sees new information, it is enabled to operate for a very short time. This will be called the control interval $\Delta t_c$. Then the error signal is removed, preventing further integration until the next interval; the action of the controller is similar in effect to a sample-and-hold circuit. By means of this sampling controller, critical damping can be achieved on a sampled dead-time process with integral action alone. Figure 4.17 shows the sequence of events.

A given deviation $e$ is imposed on the controller. Its output changes by an amount

$$\Delta m = e\,\frac{\Delta t_c}{I} \tag{4.24}$$

The process responds within the sample interval by an amount

$$\Delta c = K_p\,\Delta m$$

The change $\Delta c$ will cancel the error $e$ if

$$I = K_p\,\Delta t_c \tag{4.25}$$

This results in critical damping, or what is commonly called *dead-beat response,* the elimination of a deviation in a single action.

If $I = K_p\,\Delta t_c/2$, that is, the loop gain is 2, $\Delta c$ will be $2e$, developing an equal and opposite deviation the next half-cycle, leading to uniform oscillation. Loop gains $<1$ give an exponential response, while those between 1 and 2 produce damped oscillations at a period of $2\Delta t$. Response is essentially as described for complementary-feedback control of a dead-time process in Fig. 4.13.

A sampling integral controller is so effective on dead-time-dominated processes that it is recommended over complementary feedback, to be used even on continuous measurements. It requires only that the sample interval *exceed* the process dead time, rather than the match required by complementary feedback.

An electronic controller can be converted to sampling operation by driving its automanual transfer switch by a repeat-cycle timer. The "off" time (manual) should be set somewhat longer than the process dead time, and the "on" time (automatic) at a much shorter value. The act of integrating produces a certain amount of noise filtering; therefore on noisy signals, the control interval (on

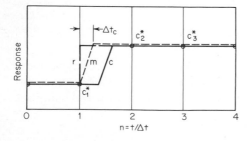

**FIG. 4.17** Performance is improved by operating the controller for only a fraction of the sampling interval.

time) should be long enough to integrate a representative average value of the controlled variable.

Sampling can be added to a pneumatic controller by connecting a repeat-cycle timer to a two-way solenoid valve in the controller's external feedback line (broken line in Fig. 4.3). Integral action is inhibited when the valve is closed and permitted when it is open.

When the controlled variable itself is sampled, as from a chromatographic analyzer, a sampling controller is recommended, even when process response time greatly exceeds $\Delta t$. A sampling PI controller operates according to the function

$$\Delta m = \frac{100}{P}\left(\Delta e + e\,\frac{\Delta t_c}{I}\right) \tag{4.26}$$

Its phase shift is set uniquely by $\Delta t_c/I$ when $\Delta t > \tau_d$

$$\phi_{\Delta c} = -\tan^{-1}\frac{\Delta t_c}{I} \tag{4.27}$$

A sampled integrating process can be controlled with dead-beat response by a sampling PI controller if it contains no dead time. The phase lag required is $-45°$, so that $I = \Delta t_c$; the proportional band required is $100 K_p\,\Delta t/\tau$.

If there is a significant amount of dead time, dead-beat response cannot be achieved and the loop responds essentially as it would with a continuous controller.

Sampling integral action will be obtained if a PI controller capable of bump-less transfer is switched to automatic after a new value of the controlled variable has been received. If PI action is desired, the change in the controlled variable should be accepted with the controller already in automatic.

## DIGITAL CONTROL SYSTEMS

In a digital control system, a central processor makes the required calculations sequentially for a number of control loops. The result of a calculation may be used to drive a control valve directly or to set the set point of an analog controller. The former arrangement is known as *direct digital control* (DDC), and the latter as *set-point control* (SPC). The control algorithms solved by the computer are the same in either case. But the decision whether to use a computer for DDC or SPC is sufficiently important for us to examine the benefits and limitations of each in some detail.

### DDC or SPC

Figure 4.18 shows two ways of implementing DDC. The upper loop is provided with a digital-manual station (HIC)† for manual backup. Should the computer fail, the last valve position is held and the HIC is placed in a manual mode. A

---

†HIC is standard terminology for *hand indicating control station.*

light identifies the station as being in manual to attract the operator's attention. Should the operator transfer the station to manual, a logic signal is sent to the computer to report his action. When the station is again returned to digital control, the computer "initializes" data stored in memory, so that automatic control can proceed bumplessly, starting at the last valve position when in manual.

In the event of computer failure, all loops normally under its control must be tended by the operator. This could be as many as 100 or more, depending on the installation. Some may be sufficiently critical for manual control to be unacceptable. For these, *analog backup* may be chosen, as shown in the lower loop of Fig. 4.18. In the event of computer failure, an analog controller takes over regulation of the loop. However, some consideration must be given to the set point of the analog controller, or an undesirable bump may result upon transfer to analog.

One practice is to transfer to manual upon computer failure and leave the subsequent switch to analog control to the discretion of the operator. Another practice is to cause the analog controller's set point to track its measurement, so that there can be no deviation at the time of transfer. Alternately, the analog set point could track the digital set point, to eliminate the possibility of initiating analog control at an undesirable set point developed during a transient condition.

In the case of analog backup, two controllers have been installed (the computer plus the analog controller) although only one is in use at any given time. Consequently analog backup is prohibitively expensive except for extreme circumstances. To justify that system, the computer must be able to control the process much better than the analog controller. In actual practice, this is not likely to be the case.

If the digital computer is asked simply to duplicate the analog control function, it will not perform it as well, because the act of sampling introduces phase lag into the loop according to Eq. (4.22). If the scan period is much shorter than the loop period, this contribution may be negligible, and therefore

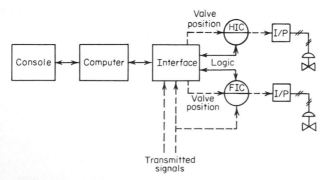

**FIG. 4.18**   Direct digital control can be implemented with manual backup (above) or analog backup (below).

digital control may be justified for slower loops. However, for fast loops like flow and liquid level, even a 1-s sample interval may produce a noticeable deterioration in response. Yet these loops constitute the majority of loops in a typical chemical plant or petroleum refinery. To operate most of the loops at sample intervals of 1 s or less, loads the computer excessively; it is better reserved for complicated tasks involving extensive calculations that can be performed at less frequent intervals.

Flow and liquid level are best regulated by analog controllers. A flow controller can be directed by a computer, using set-point control. The configuration for SPC is identical to the lower loop in Fig. 4.18 *except* that the computer outputs *set point* rather than valve position to the FIC (flow-indicating controller). The flow controller is always controlling; a computer failure simply stops the set point from being updated. The set point may be generated by a feedback algorithm or any one of a variety of calculations based on production scheduling, feedforward control, etc. The operator may place the FIC in manual, in automatic with local set point, or in automatic with computer-driven set point. Whenever the operator returns to this last condition, the computer initializes its stored data to avoid bumping the set point the operator has introduced.

### Digital Control Algorithms

The digital computer performs its control calculations periodically rather than continuously, and therefore its algorithm is a difference equation rather than the familiar differential equation

$$m_n = \frac{100}{P} \left[ e_n + \frac{\Delta t}{I} \sum_0^n e - \frac{D}{\Delta t} (c_n - c_{n-1}) \right] \tag{4.28}$$

where $n$ represents current value and $n - 1$ last value of a signal. While this equation describes the ideal noninteracting position-mode algorithm, it lacks some desirable features, such as external feedback and derivative gain limit.

A preferred algorithm set is the digital simulation of the analog controller given in Fig. 4.7*b*. It is performed in three parts

$$m_n = \frac{100}{P} [e_n - K_D (c_n - y_n)] + b_n \tag{4.29}$$

where $K_D$ is the derivative gain limit, $y_n$ is the lagged value of $c_n$ as given below, and $b_n$ is the lagged value of feedback $f_n$, also as given below

$$y_n = y_{n-1} + \frac{\Delta t(c_n - y_{n-1})}{\Delta t + D/K_D} \tag{4.30}$$

$$b_n = b_{n-1} + \frac{\Delta t(f_n - b_{n-1})}{\Delta t + I} \tag{4.31}$$

These equations are called *positional algorithms* because they develop valve position as an output. At this writing, most digital feedback algorithms are incremental in nature, developing a *change* in valve position to be added to its present value

$$\Delta m_n = \frac{100}{P} \left\{ (e_n - e_{n-1}) \right.$$

$$\left. + \frac{\Delta t}{I} e_n - \frac{D}{\Delta t} [(c_n - c_{n-1}) - (c_{n-1} - c_{n-2})] \right\} \quad (4.32)$$

The incremental algorithm has several disadvantages, including the inability to perform proportional control reliably, the inability to protect the primary of a cascade loop against integral windup, and ineffective implementation of selective control functions. The last two limitations are described in Chap. 6. The limitation of the algorithm in executing proportional control is the lack of a fixed bias term; the bias would be changed each time the algorithm was initialized. Positional algorithms (4.29) can be solved using a fixed value of $b_n$, however.

### Selecting the Sample Interval

From the standpoint of minimizing the phase lag due to sampling, a sample interval that is short relative to the loop response should be chosen. However, intervals that are unduly short may increase the loading on the computer excessively. Furthermore, offset may be produced if the integral gain $\Delta t/I$ appearing in Eqs. (4.28), (4.31), and (4.32) is too low. For example, let $\Delta t = 1$ s and $I = 60$ min. If the roundoff error of the computer is 1 bit in 15, i.e., 1 part in $2^{15}$, the smallest deviation that could be effectively processed is

$$e_{\min} = \frac{3600 \text{ s}}{1 \text{ s} \times 2^{15}} = 0.11 = 11\%$$

If the computer has no means of improving its precision, the ratio $\Delta t/I$ must be selected to reduce the offset to an acceptable minimum.

Selection of the sample interval also affects derivative gain in Eqs. (4.28) and (4.32). If $D/\Delta t$ falls outside the range of 10 to 20, derivative action will be ineffective. The adjustment procedure for a PID controller calls for setting $D = 0.17\tau_n$. If $D/\Delta t$ is to be 10 to 20, then $\Delta t$ must be $0.008\tau_n$ to $0.017\tau_n$. Thus, a loop having a period of 1 min must be sampled every 0.5 to 1 s to provide the necessary derivative gain. The same is true with the filtered derivative algorithm (4.30). However its gain limit offers protection against too frequent sampling which the others do not. If $D/\Delta t$ is too high, algorithms (4.28) and (4.32) will respond excessively to noise, creating spurious disturbances and giving erratic performance.

If $I$ in Eq. (4.31) is set to zero, $b_n = f_n$, producing complementary feedback of dead time equal to $\Delta t$. This can produce dead-beat control of a dead-time process as long as $\Delta t > \tau_d$.

For processes whose controlled variable is sampled, e.g., by an analyzer, the algorithm should be executed just once following each report. This is the only way derivative action can be effective on sampled signals. If $\Delta t$ is shorter than the analysis period, the change in output corresponding to a change in the controlled variable will last for only one sample interval. But if $\Delta t$ is set equal to the period of the analyzer, the same action is sustained over the entire analysis period.

## REFERENCES

1. Shinskey, F. G.: Adaptive pH Controller Monitors Nonlinear Process, *Control Eng.*, February 1974.
2. Sood, M., and H. T. Huddleston: Tuning PID Controllers for Random Disturbances, *Instrum. Technol.*, February 1977.
3. Giloi, W.: Optimized Feedback Control of Dead-time Plants by Complementary Feedback, *Trans IEEE*, May 1964.
4. Smith, O. J. M.: A Controller to Overcome Dead Time, *ISA J.*, February 1959.

## PROBLEMS

**4.1**  Find the optimum combination of proportional and integral settings for a dead-time process from the information given in Table 1.1. Why is it different from the situation described for the PI controller in Table 4.3?

**4.2**  Given a process controlled with a PI controller whose proportional band is 200 percent and whose integral time is 10 min, estimate the maximum error developed by a step load change of 5 percent. What would the error be if the same load change were made gradually over an interval of 30 min? What would the error be if the load change entered as a sine wave of 5 percent amplitude and 2-h period?

**4.3**  Calculate $PI/100$ for values of $I = 0.90\tau_d$ and $D = 0.45\tau_d$ set into an interacting controller, following the example given in Table 4.3. What conclusion can you draw from the result?

**4.4**  Find the optimum settings for two-mode control of a process consisting of a 30-min lag, a 2-min dead time and an analyzer with a 5-min sampling interval. Leave the proportional band in terms of $K_p$.

**4.5**  This chapter describes three different methods for controlling a process dominated by dead time. Select the best method, calculate the optimum values of all the parameters, and estimate the integrated area per unit load change, for a dead time of 2 min and a process gain of 2.5.

# Chapter Five

# Nonlinear Control Elements

Elements with nonlinear properties appear both in processes and their control systems. Up to this point, the simpler nonlinear elements have been discussed, and compensation has been applied to maintain uniform loop gain. In this chapter more severely nonlinear elements are presented, e.g., those having discontinuities and negative resistance. In some cases compensation is not possible, and stability cannot be attained without altering the process being controlled. It is important for the control engineer to be able to recognize these situations, lest he expend substantial time, effort, and resources trying to correct something which cannot be corrected.

Nonlinear control elements can often be introduced into a loop to *improve* performance or to lower costs while achieving adequate performance. The use of on-off valves and single-speed bidirectional motors can reduce systems cost and complexity where their performance is acceptable. But prerequisite to applying or even analyzing the role of these nonlinear elements is an understanding of their effects on loop stability.

## NONLINEAR ELEMENTS IN THE CLOSED LOOP

A linear control loop is identified by its constant loop gain, which applies the same damping to disturbances of all magnitudes. This statement holds true whether the loop consists entirely of linear elements or includes a nonlinear function which was intentionally introduced to compensate another function naturally occurring in the process.

In a nonlinear control loop, gain varies with the amplitude of the oscillation. Loop gain could either increase or decrease as amplitude increases; both cases are illustrated in Fig. 5.1. If loop gain increases with amplitude, small upsets will result in heavy damping. However, the loop gain could cross 1.0 in the event of a sufficiently large disturbance. Then the amplitude can only expand; this is identified as the *point of no return*. Stability can be restored only by intervention; control must be taken away from the controller until the amplitude is reduced below the point of no return.

The opposite characteristic (loop gain varying inversely with amplitude) is far more common. This type of nonlinearity causes small-amplitude excursions to expand and large disturbances to be dampened, both culminating in a constant-amplitude cycle. The amplitude of the cycle is that which produces unit loop gain. The oscillation is then of constant amplitude and is known as a *limit cycle*. It *may* exist between two natural limits such as the extreme positions of a valve but not necessarily.

A limit cycle can be reduced in amplitude by reducing the controller gain, but it cannot disappear. The reduced controller gain simply shifts the point at which loop gain crosses unity to a lower amplitude. The period of the cycle may also change as a result of the adjustment if the phase angle of the nonlinear element varies with amplitude. Limit cycles tend to be nonsinusoidal: clipped sine waves and sawtooth waves are common, the latter being the integral of the former.

## Describing Functions

A *describing function* is a device used to characterize the gain and phase of a nonlinear element in response to a sinusoidal input. If the nonlinear device were to produce a sine wave in response to one, its gain could be expressed simply as the amplitude ratio of input to output. However, the output tends not to be sinusoidal, and therefore amplitude ratio is only an estimate of gain.

In passing through a lag or series of lags, the higher harmonics of a wave

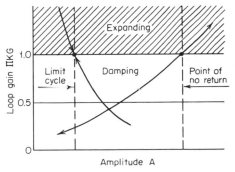

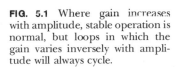

**FIG. 5.1** Where gain increases with amplitude, stable operation is normal, but loops in which the gain varies inversely with amplitude will always cycle.

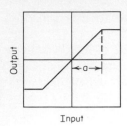 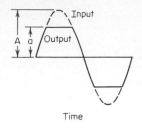

**FIG. 5.2** The limiter produces a clipped sine wave of amplitude $a$ in response to a sine wave of amplitude $A$.

tend to be filtered, leaving the fundamental. The Fourier series for a square wave having a mean value $y_0$, an amplitude $A$, and a period $\tau_o$, is

$$y(t) = y_0 + \frac{4A}{\pi}\left(\sin\frac{2\pi t}{\tau_o} + \frac{1}{3}\sin\frac{6\pi t}{\tau_o} + \cdots\right) \qquad (5.1)$$

Only the fundamental need be considered, owing to the aforementioned filtering, so that the sinusoidal equivalent of a square wave has an amplitude $4/\pi$ times that of the square wave. Similarly, the Fourier series for a triangular wave, which is the integral of a square wave, is

$$y(t) = y_0 + \frac{8A}{\pi^2}\left(\sin\frac{2\pi t}{\tau_o} - \frac{1}{9}\sin\frac{6\pi t}{\tau_o} + \cdots\right) \qquad (5.2)$$

The amplitude of its sinusoidal equivalent is $8/\pi^2$ times the amplitude of the triangular wave. These and similar factors are applied to correct the observed amplitude ratio of nonlinear elements to develop their true gain and phase.

### Limiters

A limiter or saturating element will pass a sine wave undistorted until the amplitude of the wave reaches the value of the limit. Above that point, the output amplitude is fixed, and so the gain of the limiter effectively decreases. Figure 5.2 shows the effect of limiting on a sine wave; the amplitude is reduced, but there is no shift in phase.

The amplitude ratio due to limiting is simply the limit divided by the input amplitude, that is, $a/A$. The gain is not simply $a/A$, however, due to the distortion caused by limiting. Considine [1] gives the describing function for a limiter as

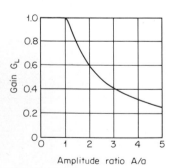

$$G_L = \frac{\alpha}{90°} + \frac{2a}{\pi A}\cos\alpha \qquad (5.3)$$

where $\alpha = \sin^{-1}(a/A)$. The solution to (5.3) is plotted in Fig. 5.3. Note that as $a/A$ approaches zero, the output waveform becomes square and $G_L$ approaches $(4/\pi)a/A$.

**FIG. 5.3** The limiter produces an inverse relationship between gain and amplitude.

The shape of the gain curve classifies the limiter as the type of function capable of producing a limit cycle. In fact, when any loop enters an expanding

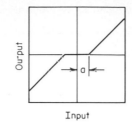

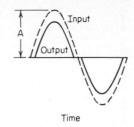

FIG. **5.4** A dead-zone element removes the center of a sine wave.

Input

Time

cycle, the amplitude of the controlled variable can only increase to the point where valve limiting begins. The loop gain then returns to 1.0 as a large-amplitude limit cycle develops.

The author successfully applied a limiter like that shown in Fig. 5.2 to the input signal of a controller in a loop which had the capability of developing an expanding cycle [2]. The limits were set just within the point of no return for the expanding cycle. In effect, the reduced gain of the limiter to higher-amplitude signals offset the rising gain of the rest of the loop, stabilizing the system for all disturbances. Regulation was much more satisfactory than could be obtained simply by widening the controller's proportional band because the gain applied to small disturbances was not affected.

### Dead Zone

A dead-zone element and its output in response to a sine wave appears in Fig. 5.4. It is used in some control systems as an amplitude-sensitive noise filter and to prevent overlapping of sequenced functions. One example of the latter is the sequential delivery of acid and base reagents to control pH of a solution. To avoid concurrent addition, the valves may be adjusted to allow a finite dead zone between one closing and the other opening. Then the valves would respond to a sine wave at the controller output as shown in Fig. 5.4.

The describing function for the dead zone is the complement of the limiter

$$G_z = 1 - \frac{\alpha}{90°} - \frac{2a}{\pi A} \cos \alpha \qquad (5.4)$$

where $\alpha = \sin^{-1}(a/A)$. Gain is plotted against $A/a$ in Fig. 5.5. Although gain increases with amplitude, there is little danger of its promoting an expanding cycle because gain never exceeds 1.0.

There is one cautionary note to observe when working with a loop containing a dead zone. If the process is non-self-regulating or self-regulating but with a large time constant and high steady-state gain (as in pH control), the signal will not remain in the dead zone. Because of the lack of control action there, the signal will drift into one

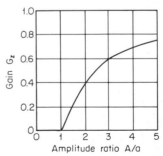

FIG. **5.5** To ensure stability to all upsets, loops containing a dead zone should be adjusted for damping at large amplitudes.

**TABLE 5.1   Loop Gain vs. Amplitude for Proportional Control of an Integrating Process with Dead Band**

| $\dfrac{A}{a}$ | $G_b$ | $-\phi_b$, deg | $-\phi_d$, deg | $\dfrac{\tau_o}{\tau_d}$ | $G_1$ | $\Pi KG$ |
|---|---|---|---|---|---|---|
| $\infty$ | 1.0 | 0 | 90 | 4.00 | $0.64\tau_d/\tau_1$ | 0.50 |
| 10 | 0.96 | 6.9 | 83.1 | 4.33 | 0.69 | 0.52 |
| 4 | 0.84 | 16.5 | 73.5 | 4.90 | 0.78 | 0.51 |
| 2 | 0.59 | 32.5 | 57.5 | 6.26 | 1.00 | 0.46 |
| 1.5 | 0.41 | 44.1 | 45.9 | 7.84 | 1.25 | 0.40 |

live zone or the other, eventually coming to rest at one of the corners, or limit-cycling between them.

## NONLINEAR PHASE-SHIFTING ELEMENTS

Certain nonlinear elements produce a shift in phase as well as attenuation. The phase shift may be a function of amplitude alone or of a combination of amplitude and period of the input signal. These characteristics complicate the analysis of a loop considerably, in that their phase contribution alters loop period, which can affect the gain of linear elements such as integrators and controllers. Two phase-shifting nonlinearities common to control valves, dead band and velocity limiting, are examined in detail.

### Dead Band

Dead *band* is distinguished from dead *zone* in that different paths are taken for increasing and decreasing signals, as shown in Fig. 5.6. (Dead band is also known as *square-loop hysteresis* and was called by that name in the first edition of this book.) It is common to valve motors, caused by friction in the packing and guides. Friction opposes motion in either direction. Therefore on a change in direction of the input signal, motion ceases until a deviation between input and output develops enough force to overcome the friction.

The phase angle is a function of the ratio of dead-band width to input amplitude, and can be approximated as

$$\phi_b \approx -\sin^{-1}\frac{a}{A} \tag{5.5}$$

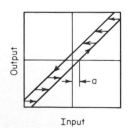

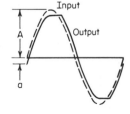

**FIG. 5.6** Dead band produces phase lag as well as attenuation.

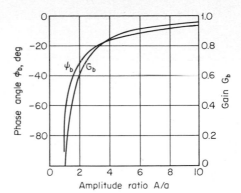

**FIG. 5.7** The increasing phase lag developed by dead band as amplitude diminishes causes the loop period to increase.

The shape of the output wave is similar to that caused by saturation but not identical; note that the leading edge is rounded while the trailing edge is sharp. The exact gain and phase calculated from the describing function in Ref. 1 are plotted against $A/a$ in Fig. 5.7.

The dead-band element contributes an increasing phase lag as amplitude decreases. Therefore the period of oscillation tends to lengthen as the oscillation decays. When the amplitude falls within the dead band, oscillation will stop entirely *if* there are no integrating elements in the loop. But an integrator increases its gain with period, tending to raise loop gain as amplitude diminishes. The gain increase can be offset by the gain reduction contributed by the dead band itself. However, if there are *two* integrators in the loop, e.g., a liquid-level process and a PI controller, loop gain will increase with decreasing amplitude, developing a limit cycle.

Table 5.1 summarizes the loop gain at various amplitudes for an integrating process with dead time and dead band under proportional control. The controller was adjusted for a loop gain of 0.5 without any phase or gain contribution from the dead band. Observe that the loop gain decreases as amplitude falls; the system is stable under all conditions.

Table 5.2 compiles loop gains for the same process under PI control. The controller was adjusted for a phase lag of 30° at an amplitude ratio of 4.0. Note that loop gain increases as amplitude falls. A limit cycle is developed at an amplitude ratio above 3.3 and at a period nearly double that observed at an amplitude ratio of 4.0. The limit cycle will appear as a clipped sine wave in the

**TABLE 5.2  Loop Gain vs. Amplitude for PI Control of an Integrating Process with Dead Band**

| $\dfrac{A}{a}$ | $G_b$ | $-\phi_b$, deg | $-\phi_d$, deg | $-\phi_{Pb}$, deg | $\dfrac{\tau_o}{\tau_d}$ | $\Pi KG$ |
|---|---|---|---|---|---|---|
| 10 | 0.96 | 6.9 | 60.6 | 22.5 | 5.94 | 0.38 |
| 4 | 0.84 | 16.5 | 43.5 | 30.0 | 8.28 | 0.5 |
| 3.5 | 0.81 | 18.5 | 37.9 | 33.6 | 9.50 | 0.57 |
| 3.3 | 0.79 | 19.4 | 22.5 | 48.1 | 16.00 | 1.18 |

record of the manipulated variable, i.e., flow, and as a triangular wave in the record of the controlled variable, i.e., liquid level. Widening the proportional band of the controller will allow a slight reduction in cycle amplitude, accompanied by a commensurate increase in period. Increasing integral time reduces both period and amplitude.

The limit cycle can be eliminated by removing the integral control mode or by closing the loop around the nonlinear element. Valve positioners and cascade flow loops are both effective in the latter role, as described in Chap. 6. Dead band can also be mitigated by superimposing on the control signal a high frequency "dither," exceeding $a$ in amplitude; naturally occurring noise sometimes achieves the same end.

### Velocity Limiting

Most final actuators operate in a limited-velocity mode. There is a speed beyond which they cannot move due to limitations in the rate at which energy can be supplied, regardless of the magnitude of the control signal. This is known as the *stroking speed* of the actuator. Thus the device appears to respond faster to small signals than to large inputs, as shown in Fig. 5.8.

A velocity-limited device can track a sine wave as long as the maximum velocity of the wave does not exceed that limit. Maximum velocity occurs at the midpoint of the wave and has the value

$$\left(\frac{dy}{dt}\right)_{max} = \frac{2\pi A}{\tau_0} \tag{5.6}$$

When maximum velocity substantially exceeds the limit $u$, the output of the limited device cycles in a triangular wave, shifted in phase and attenuated as shown in Fig. 5.9. The gain of the limiter in this mode is

$$G_u = \frac{2\tau_0}{\pi^2 A} \tag{5.7}$$

and its phase angle is

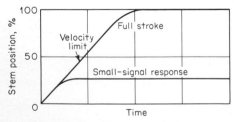

**FIG. 5.8**  Because of its velocity limit, a control valve appears to follow small signals faster than large signals.

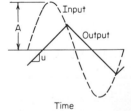

**FIG. 5.9**  When the velocity of the sine wave exceeds the limit of the actuator, a triangular wave is produced.

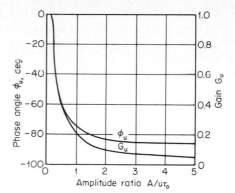

FIG. 5.10 The velocity limiter becomes more linear as amplitude approaches zero.

$$\phi_u = -90 + \sin^{-1} \frac{u\tau_o}{4A} \qquad (5.8)$$

Gain and phase are plotted against $A/u\tau_o$ in Fig. 5.10.

Because the velocity limiter exhibits phase lag increasing with amplitude, large excursions in the controlled variable tend to cause cycles of a longer period than small upsets. If the controller has been adjusted for stability to large-amplitude fluctuations, damping is assured because loop gain and period decrease with falling amplitude. Eventually, the amplitude decays into the linear regime. The principal contribution of velocity limiting is the extension of the period during recovery from large upsets.

## Negative Resistance

Some processes are inherently unstable due to a negative resistance resident in one element. They have no steady state in the negative-resistance range, being uncontrollable either automatically or manually. This characteristic sometimes appears in a combination of steam-jet ejectors drawing a vacuum [3]. Figure 5.11 illustrates such a characteristic.

The ejectors draw a vacuum by pumping air out of a chamber. Absolute pressure within the chamber is controlled by admitting air into it. In the *positive-resistance* regions, i.e., below 10 torr and above 30 torr, an *increase* in airflow into the chamber will cause its pressure to rise, *increasing* the pumping rate of the ejectors until an equilibrium is reached. But when the pressure rises above 10 torr, the pumping rate *falls*, as indicated by the curve. The pressure then rises

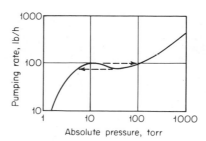

FIG. 5.11 The negative resistance prevents pressure from being controlled either automatically or manually. *(From F. G. Shinskey, "Energy Conservation through Control," Fig. 4.8, Academic Press, Inc., New York, 1978, with permission.)*

without any adjustment to inflow until that pumping rate is again reached at a higher pressure. By the same token, reducing the pressure below 30 torr causes pressure to fall spontaneously to about 7 torr. Any attempt to control in the middle of the range will result in limit cycling between 10 and 30 torr, which is the range over which negative resistance appears. There is no way to stabilize this process because it is inherently unstable without controls. The only solution is to move the negative-resistance region away from the operating range by modifying the process.

A similar characteristic sometimes appears in a boiler subject to an excessive temperature differential. When the temperature difference between a heating fluid and a boiling liquid is below about 40°F (22°C), boiling takes place at nucleation points on the heat-transfer surface. Nucleate boiling is efficient due to the turbulence formed at the heat-transfer surface. If a higher rate of boiling is forced by increasing the temperature of the heating medium, a vapor film begins to form on the heat-transfer surface and the rate of boiling actually begins to fall [4]. This reduces conduction from the heating medium, further raising its temperature and thereby reducing heat transfer even more. Automatic control in this transition boiling range is not possible.

## VARIATIONS OF THE ON-OFF CONTROLLER

A pure on-off control function produces an output of either 0 percent (off) or 100 percent (on), according to the sign of the error. Because a sinusoidal error of any amplitude will produce a square wave of unit amplitude, the device is said to have a variable gain. In order to establish a figure of gain for any device, its output should be expressed in the same terms as its input. So instead of a 100 percent peak-to-peak square wave the output will be thought of as the fundamental sinusoidal component of the square wave. This component has an amplitude of $4/\pi$ times that of the square wave. Then the gain of the on-off controller $G_o$ becomes the output amplitude $2/\pi$ over the input amplitude $A$

$$G_o = \frac{2}{\pi A} \tag{5.9}$$

An on-off controller employed on any process capable of shifting phase beyond 180° will cause a limit cycle. The loop gain $G_o K_p G_p$ under these conditions is 1.0, $K_p G_p$ representing the process gain at its natural period.

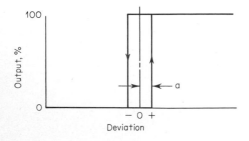

FIG. 5.12 The state of the output will not be changed until the deviation exceeds the width of the dead band.

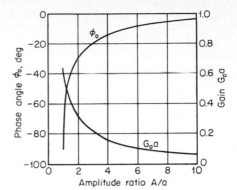

**FIG. 5.13** Dead band in an on-off controller develops a phase lag similar to that in a valve operator.

Therefore the amplitude of the limit cycle, expressed in percent, is

$$A = 200 \frac{K_p G_p}{\pi} \tag{5.10}$$

Strictly speaking, the output of an on-off controller is a rectangular wave rather than square. It would be perfectly square only if the durations of on and off were identical; this can only happen at 50 percent load. At either lower or higher loads, one condition will dominate. After passing through an integrating process, or one with a large time constant, the rectangular wave is converted into a sawtooth form. Additional lags will round its corners.

### Dead Band

A perfect on-off device cannot be used for control because infinitesimal levels of noise would cause constant chattering. Therefore real on-off controllers are limited in gain by either (1) negative feedback, which reduces controller gain to a tolerable level, or (2) positive feedback, which creates a dead band, also known as *differential gap* or *lockup*.

Figure 5.12 illustrates the characteristic of an on-off controller with dead band. Its gain is zero as long as the amplitude of the input signal is less than the dead band. To larger signals, it has the gain of an on-off controller. Phase lag is related to signal amplitude

$$\phi_o = \tan^{-1} \frac{a/A}{\sqrt{1 - (a/A)^2}} \tag{5.11}$$

The phase angle is plotted against amplitude ratio in Fig. 5.13.

When an on-off controller is used to close a loop around an integrating process with dead time, the period and amplitude of the resulting limit cycle depend on dead-band width and the characteristics of the process. Table 5.3 lists various sets of conditions which could result in a limit cycle if the last column, loop gain, were equal to 1.0. To use this table, enter values of $K_p \tau_d/a\tau_1$ into the last column and calculate loop gain. The limit cycle will have an amplitude $A/a$ and a period $\tau_o/\tau_d$ corresponding to unit loop gain.

**TABLE 5.3   Limit-Cycle Amplitude with On-Off Control of an Integrating Process**

| $\dfrac{A}{a}$ | $-\phi_o$, deg | $G_d a$ | $\dfrac{\tau_o}{\tau_d}$ | $\Pi KG$ |
|---|---|---|---|---|
| 1.1 | 65.4 | 0.579 | 14.63 | $1.35 K_p \tau_d / a \tau_1$ |
| 1.2 | 56.4 | 0.531 | 10.71 | 0.91 |
| 1.5 | 41.8 | 0.424 | 7.47 | 0.50 |
| 2.0 | 30.0 | 0.318 | 6.00 | 0.30 |
| 4.0 | 14.5 | 0.159 | 4.77 | 0.12 |
| 10.0 | 3.7 | 0.064 | 4.17 | 0.04 |

The parameter having the most effect on loop gain is dead band. By adjusting dead band, any amplitude may be achieved above the minimum indicated for zero dead band in Eq. (5.10). By setting the last column in Table 5.3 to 1.0, $a$ becomes equal to the indicated numerical value in the column multiplied by $K_p \tau_d / \tau_1$; then amplitude $A$ can also be reported in terms of $K_p \tau_d / \tau_1$ by multiplying the numerical values in the first and last columns. The relationship between amplitude and dead band thus generated is plotted in Fig. 5.14.

The figure indicates that there is relatively little penalty for introducing a moderate amount of dead band into the loop. Because of the direct proportionality between gain and period in an integrating process, the period varies directly with the amplitude. A single curve is used in Fig. 5.14 to represent both, although their values differ by $\pi^2$, that is, 9.87, rather than 10 as shown.

## Proportional-Time Control

In systems where on-off control produces a limit cycle that is both too long and too high, certain modifications may be applied. Proportional-time control is a technique by which the on-off output is modulated with a signal of fixed period but variable "on" time. The percentage of each period during which the controller output is maximum is proportional to deviation. Thus the average

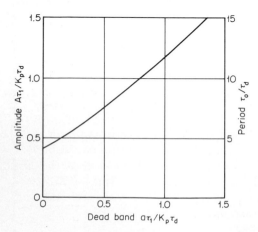

**FIG. 5.14** Amplitude increases uniformly with dead band.

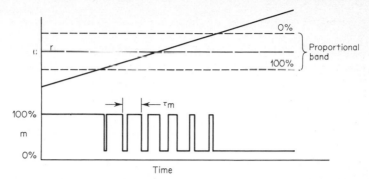

**FIG. 5.15** Percentage "on" time varies as the measurement passes through the proportional band.

value of output is the same as it would be with a proportional controller. Figure 5.15 shows the relationship between deviation and controller output.

Because this type of controller naturally oscillates, the loop is forced to limit-cycle at the period of modulation $\tau_m$. This period should be selected so that the gain of the process is low enough for its limit cycle to be of negligible amplitude. When this is assured, the loop approaches proportional control with a linear final operator. The criteria for adjusting the proportional band are essentially the same as with a linear loop, and offset is encountered for the same reason.

### A Constant-Speed Motor as the Final Operator

Up to this point only two-state on-off control has been presented, where the final operator would be a solenoid valve or an electric heating element. These are either on or off. But very often a constant-speed reversible motor is used to drive a valve or to position a lever. This type of operator has three states, drive upward (opening), stop, and drive downward (closing). So the controller must be similarly arranged. The simplest controller for this function consists of two on-off devices whose inflection points are separated by a dead zone. Within this zone, the motor would be stopped. In practice, each on-off device also contains a small dead band. The input-output relation is pictured in Fig. 5.16.

A finite dead zone must exist, first to ensure that the on-off switches do not overlap, for this would energize both windings of the motor simultaneously and could result in damage. But beyond this, it provides a state of rest for the loop that did not exist with previous on-off configurations. Thus the control system is not bound to limit-cycle. Gibson [5] gives the describing function for the three-state controller, from which the following phase and gain characteristics were derived:

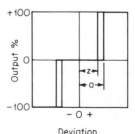

**FIG. 5.16** A three-state control device can drive a motor up or down or leave it at rest in a dead zone.

$$\phi_3 = -\tan^{-1}\frac{a - z}{A(\cos \alpha + \cos \beta)} \tag{5.12}$$

$$G_3 = \frac{2}{\pi A} \sqrt{(\cos \alpha + \cos \beta)^2 + [(a - z)^2/A^2]} \tag{5.13}$$

where $\alpha = \sin^{-1}(a/A)$ and $\beta = \sin^{-1}(z/A)$. Gain and phase are plotted against amplitude ratio $A/a$ for three values of dead zone $z/a$ in Fig. 5.17.

Because gain for an on-off device varies inversely with amplitude $A$ rather than amplitude ratio $A/a$, it is shown plotted as $G_3 a$. Observe that $G_3 a$ approaches $4/\pi$ at $A/a = 1$, rather than $2/\pi$, as might be expected from Eq. (5.9). This doubling is caused by the bidirectional capability of the output, as indicated in Fig. 5.16.

While the dead zone provides stability, it also allows offset to exist up to $\pm a$. The fact that the control loop contains an integrator (the valve motor) precludes the use of this type of control system on non-self-regulating processes. If the process is self-regulating but has a time constant that is long relative to that of the valve motor, a very wide dead zone may be necessary to avoid limit cycling. Then the resulting offset may be unacceptable, and a proportional-time controller would be needed to reduce loop gain. The integral time of the control loop would be the time constant of the valve multiplied by the proportional band of the controller.

Another solution to the problem is to add a PID controller to the loop as shown in Fig. 5.18. Then the controller gain can be adjusted for stability, and the integral mode eliminates offset. Response time tends to be poor, however, due to the double integration of valve and controller. A preferred arrangement uses feedback from valve position or flow to close the loop through the three-state controller, as shown by the broken line in Fig. 5.18. Then the valve becomes a proportional device; this arrangement can then be used with non-self-regulating processes.

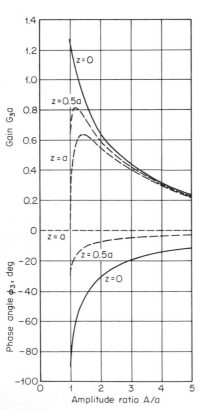

**FIG. 5.17** The presence of the dead zone reduces both gain and phase lag at any amplitude.

### THE DUAL-MODE CONCEPT

What would you like a controller to do, forgetting for the moment what it is reasonable to expect? The most severe demand would certainly be to follow a step change in set point perfectly. This could be demanded of the controller but not of the process, because it requires infinite process

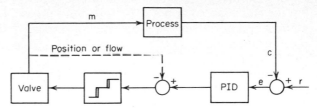

**FIG. 5.18**  Feedback of valve position or flow converts the valve from an integrator to a lag, improving the response of the control loop.

gain. The speed at which a variable can change is limited by the maximum rate at which energy can be delivered to the process. A valve may only open fully, not infinitely. Therefore it can only be asked that the controller not interfere with the maximum speed of the process. To duplicate the remainder of the step input, the control loop must be stable to the point that no overshoot or oscillation is observable. Nor should there be any offset. Finally, the controller ought to be insensitive to input noise, which is usually present in some form. To summarize, the ultimate controller should be capable of achieving the following loop-response characteristics:

1. Maximum speed
2. Critical damping
3. No offset
4. Insensitivity to noise

Any control system that can satisfy the above demands will also satisfy any minimum-integral-error criterion, regardless of what function of the error may be used and regardless also of the nature of the input signals. The character of the process determines the complexity of the controller which is to accomplish the goals listed above. If the process is a pure single capacity, an on-off controller will provide maximum speed, critical damping, and no offset. An on-off controller is sensitive to noise, however. Significantly, this simplest control device is capable of achieving the ideal closed-loop response on the simplest process. As the process complexity increases, on-off control is no longer optimum, and combinations of less severe linear or nonlinear elements must be used to provide stability. Obviously the nature of the process determines the design of the controller which will elicit the best loop performance.

With regard to difficult processes, control functions which approach the demands of the four points of performance listed above need to be set forth:

1. Maximum speed implies that the controller be saturated for any measurable deviation. Such a demand limits the selection to an on-off controller. But if the tolerance may be widened somewhat, the controller need only saturate in response to a large signal. (How large the signal must be will vary with the difficulty of the process.)

2. Critical damping can be achieved by both low gain and derivative action,

but the latter amplifies noise. Critical damping implies an asymptotic approach to the set point. To accommodate maximum speed, the zone of critical damping must be restricted to a narrow band about the set point. Therefore this criterion and its solution apply specifically to small-signal response.

3. Zero offset requires a controller with infinite gain in the steady state. An integrator is sufficient to satisfy this criterion.

4. Low noise response can be obtained through low-gain or low-pass filtering, but low-pass filtering degrades the speed of response of the loop. The only condition, then, which tends to reconcile this requirement with the others, is the application of low gain to small signals.

Of particular significance is the combination of high gain to large signals and low gain to small signals. The exact combination of parameters that will be most effective for a specific application may not be obtainable in a single controller. It may then be necessary to use two controllers, intelligently programmed to take the best advantage of their individual features. The combination of two controllers operating sequentially in the same loop has been called a *dual-mode system.*

**Selecting the Two Controllers**

The only place where stability is assured with the nonlinear devices presented thus far is in a dead zone. But in the dead zone there is no control whatever, which results in offset. Therefore the region about the set point should not have zero gain, but its gain should be low compared with on-off control.

A sensible approach is to use a linear controller in this region, with gain adjustable by the proportional band. Because the gain is expected to be low, integral action is also required. Derivative is also recognized as valuable in promoting damping and rapid response to load change. Consequently the most logical controller to use in the small-signal region, for most applications, is a conventional PID controller.

On-off action provides the highest available gain for large-signal response. So the choice of on-off to operate sequentially with the linear controller is obvious. But the transition from one to another is not obvious. And the boundary between what is considered a small signal and what is considered large is not at the moment clear.

It would seem that a nonlinear control mode would not be very effective in combating load changes. To begin with, only a relatively severe load change would ordinarily cause an error large enough to enter the large-signal zone. But all load changes are acted upon by the linear controller, because the errors which they induce all originate and subside in the small-signal zone.

The same is not true of set-point changes. A set-point change may easily be introduced fast enough to pass directly into the large-signal zone. The conclusions drawn earlier concerning desirable controller characteristics were based on particular demands of set-point response. For these reasons, it may be ventured that nonlinear controllers may enhance set-point response, while linear controllers are more effective against changing load. This is further evidence in support of the selection of a linear controller for small errors and a nonlinear for large.

### Optimal Switching

The performance that a control system is capable of giving is relative to the amount of intelligence built into it. It is possible to design a programmed system that will outperform any conventional feedback system in response to command, i.e., set-point, inputs. But the program must embody more of the characteristics of the process than conventional controllers do. It has been found necessary to use an *antiwindup* feature in a controller with integral action in order to prevent overshoot on large set-point changes. It has been further noted that the controlled variable will respond best if the integral mode is preloaded to the steady-state conditions that are expected to prevail. In a sense, this is programmed control.

An optimum program is one which will place the controlled variable at the designated set point in the minimum time (or with minimum energy or minimum cost).

On-off control will drive the process at maximum speed but will cause overshoot with more than a single capacity in the loop. Overshoot means that some oscillation, however damped, is present, requiring a specified settling time after the set point is crossed. This is scarcely consistent with minimum time.

To explore what is required for minimum-time control, consider the application to a dead-time-plus-integrating process. In Fig. 5.19 the tracks of both the intermediate variable, i.e., the output of the integrator, and the controlled variable are plotted. Minimum time requires that $m$ be switched from 100 percent to equal the load $q$ before the set point is reached.

The optimum control program must include the expected load at the designated set point, for two reasons: (1) if the manipulated variable fails to match the load after the new set point is reached, the process will be in an unsteady state, and (2) the rate of rise to the new set point, hence the error $e_l$ at the time of switching, is load-dependent.

It may be recalled from Chap. 1 that the rate of rise of the intermediate variable, and also of the controlled variable, is

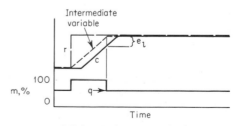

**FIG. 5.19** Minimum-time control of a dead-time-plus-integrating process.

$$\frac{dc}{dt} = \frac{m - q}{\tau_1}$$

where $\tau_1$ is the time constant of the integrating element. The controlled variable is delayed by the dead time $\tau_d$ behind the intermediate variable. It therefore lags in magnitude by

$$\tau_d \frac{dc}{dt} = \frac{\tau_d}{\tau_1} (m - q)$$

Overshoot will be prevented if $m$ is switched from 100 percent to $q$ when the intermediate variable reaches the set point. The deviation $e_l$ at that time is

$$e_l = \frac{\tau_d}{\tau_1} (100\% - q) \tag{5.14}$$

The parameters $e_l$ and $q$ must be manually set into the control system.

The control loop will not oscillate, for if the load is well matched, the switching point will be crossed only once. The loop is actually open from then on. As described above, the system will only respond to increasing set points. With an optimum program for decreasing set points, the on-off controller will switch between 0 percent and $q$ at the point $e_h$

$$e_h = \frac{\tau_d}{\tau_1} q \tag{5.15}$$

A control system designed for bidirectional optimal switching requires two on-off operators, as depicted in Fig. 5.20. The distance between the two switching points, expressed in percent, is the dead zone $z$, analogous to the proportional band of a linear controller

$$z = e_l + e_h = 100 \frac{\tau_d}{\tau_1} \tag{5.16}$$

If the same program is applied to a two-capacity process, the controlled variable will be more heavily damped than necessary. Therefore this program provides the minimum-time switching only for dead-time-plus-integrating processes.

Referring back to Fig. 2.2, notice that the overshoot for a two-capacity

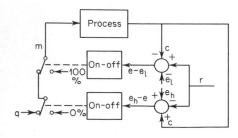

FIG. 5.20 Bidirectional programming requires two on-off controllers with separately adjustable switching points.

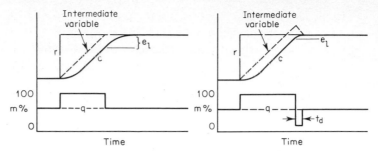

**FIG. 5.21** The minimum-time program (right) actually causes the intermedi-ate variable to overshoot.

process under on-off control was less than with dead time because of the reduced gain of the second capacity at that period. It is possible to take advantage of that gain reduction to save time. Figure 5.21 compares the response of the previous program to that of the optimum program for a two-capacity process.

The program consists of switching from 100 percent output at $e_l$ to 0 percent output for a specified time $t_d$, after which the steady-state value is selected. Values of $e_l$ and $t_d$ necessary for minimum-time control can be found by solving the equations

$$t_d = -\tau_2 \ln \frac{q}{100} \quad \text{and} \quad e_l = \frac{100\tau_2 - q(\tau_2 + t_d)}{\tau_1} \tag{5.17}$$

For the particular case where $q = 50$ percent

$$t_d = 0.693\tau_2 \quad \text{and} \quad e_l = \frac{15.35\tau_2}{\tau_1}$$

If the program designed for a dead-time-plus-integrating process had been used, $e_l$ would have been $50\tau_2/\tau_1$. With the minimum-time program, 100 percent output is retained $0.693\tau_2$ longer, which corresponds to the time required to dissipate that additional energy.

In order to accommodate this more complicated program, a delay timer must be added to the system. The arrangement of the loop for an increasing set-point change is shown in Fig. 5.22.

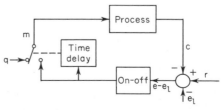

**FIG. 5.22** The delay timer is actuated by the on-off controller.

The on-off controller deenergizes when $e_l$ is reached, sending 0 percent output to the process while the timer is operating.

Again, processes do not fall into such neat classifications as two-capacity or single-capacity plus dead-time. The bulk of difficult processes lies between these limits. But the same control function described by Fig. 5.22 and Eq. (5.17) can be adjusted to accommodate dead time in addition to two capacities. Equation (5.18) indicates the required settings for optimal switching

$$t_d = -\tau_2 \ln \frac{q}{100} \qquad e_l = \frac{100(\tau_2 + \tau_d) - q(\tau_2 + \tau_d + t_d)}{\tau_1} \tag{5.18}$$

This is one control function whose exact settings can be determined numerically for a process with three dynamic elements. But however difficult the process, settings for the switching parameters can be found which will provide absolute optimum set-point response using this system.

### Adding a Linear Controller

Optimal switching programs were developed originally for positioning systems and vehicle control. The final state of these processes is generally quiescent, i.e., zero velocity, where no control is required over the steady state. But in fluid processes, control is needed to provide mass and energy balance in the steady state. As a result, programmed control, whose final state is open-loop, is incomplete.

The loop may be closed simply by adding a linear feedback controller to operate in conjunction with the programmed mode. When the programmed action is completed following a set-point change, the linear controller is switched into the loop. In effect, two controllers constitute the system, one for the steady state, one for the unsteady state.

The output of the control system should match the load, just as was done without the linear controller. If it does not, an error will develop after transfer is made to the linear controller. The addition of a linear controller should not be construed as license to discount the settings of the program. Instead, the program should be designed just as if there were no linear controller. Every effort should be made to place the controlled variable exactly on the set point with zero velocity when transfer is made. In this way, the controller will have no work to do and hence will not disturb the process. To be sure, the linear controller will compensate as well as it can for inaccuracies in the program, and this is beneficial.

The linear controller must, above all, be preloaded to the anticipated process conditions at the new set point. The arrangement of the system for increasing set-point changes is shown in Fig. 5.23.

The sequence of events is as follows:

1. While $e > e_l$, the on-off operator is energized, sending 100 percent output to the process. The preload setting $q$ is sent to the PID controller.

2. When $e = e_l$, the on-off output drops to zero and the two timers are started.

3. At the end of the first delay, the output signal is transferred from the on-off controller to the feedback for the PID controller, which remains at the preload value.

4. At the end of the second delay, the PID controller is placed in automatic and its external feedback loop is closed. By this time, the deviation should be nearly zero and so should its rate of change.

The dual-mode system gives the best set-point response attainable. Optimal switching, by definition, is unmatched in the unsteady state, while the linear controller provides the regulation necessary in the steady state. But any control system is only as good as the intelligence with which it is supplied. In the event of maladjustments in the three parameters $e_l$, $q$, and $t_{d1}$, the track of the controlled variable will be imperfect. The value of $e_l$ will vary directly with the difficulty of the process. As the process difficulty decreases, the controlled variable is less a function of load, and hence has more tolerance for inaccuracies in the control parameters. But the degree of performance *improvement* provided by dual-mode control also varies directly with process difficulty.

The dual-mode system needs seven adjustments, which fall into two independent groups. Settings of proportional, integral, and derivative only pertain to the steady state, while the program settings are in effect elsewhere. Consequently, adjusting the dual-mode system is no more difficult than adjusting two separate controllers. Rules for setting the program parameters are self-evident:

1. Maladjustment of $e_l$ causes overshoot or undershoot.

2. Excess $t_{d1}$ turns the controlled variable downward after the set point is reached.

3. An incorrect preload setting introduces a bump after the first time-delay interval.

The effects of these maladjustments are graphically demonstrated in Fig. 5.24. The second delay is not critical.

Recall the specifications which were set forth at the beginning of the section on dual-mode control. Maximum speed has been provided by the on-off controller. The programmed switching critically damps the loop as the set point is approached. Offset is eliminated by integral action in the linear controller.

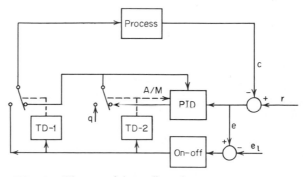

**FIG. 5.23** The second timer allows the process to come to rest before initiating PID control.

Finally, noise of magnitude less than $e_l$ will not actuate the on-off operator and therefore will be no more of a problem than in a linear system. Although complicated and costly, dual-mode control cannot be matched for performance.

## NONLINEAR PID CONTROLLERS

It has been demonstrated that a loop whose gain varies inversely with amplitude is prone to limit-cycle. Any controller with similar characteristics can promote limit cycling in an otherwise linear loop. Since on-off controllers are in this category, any nonlinear device that is purposely inserted into a loop for the sake of engendering stability must have the opposite characteristic, i.e., gain increasing with amplitude. The only stabilizing nonlinear devices discussed up to this point have this property; it was manifested as a dead zone in the three-state controller and as the linear mode in the dual-mode system.

It is not difficult to visualize a desirable combination of properties for a general-purpose nonlinear controller. In fact, the characteristics outlined for a dual-mode system apply: the controller should have high gain to large signals, low gain to small signals, and integrating action. The variation of gain with error amplitude can be accomplished continuously or piecemeal.

### A Continuous Nonlinear Controller

It is possible to create a controller with a continuous nonlinear function whose gain increases with amplitude. In contrast to the three-state controller, its gain in the region of zero error would be greater than zero, with integrating action to avoid offset. But its change in gain with amplitude should be less severe than that of a dual-mode system. Thus it would be more tolerant of inaccuracy in the control parameters.

The continuous nonlinear controller could be mathematically described by the expression

$$m = \frac{100}{P} f|e| \left( e + \frac{1}{I} \int e\, dt - D \frac{dc}{dt} \right) \tag{5.19}$$

In this way its gain varies with the absolute magnitude of the error. A suitable linear function can be used.

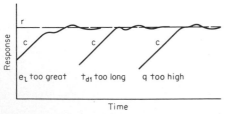

FIG. 5.24 Maladjustments in the program parameters are easy to diagnose.

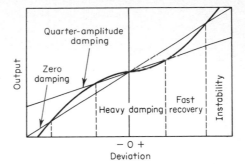

**FIG. 5.25** The proportional characteristic of a continuous nonlinear controller displays variable damping.

$$f|e| = k + \frac{(1-k)|e|}{100} \tag{5.20}$$

where $k$ is an adjustable parameter representing linearity and $e$ is expressed in percent. If $k = 1.0$, the controller is linear. But as $k$ approaches zero, the control function becomes square-law, taking the shape of the parabolic sections shown in Fig. 5.25. It is not desirable for $k$ to equal zero, since this would render the controller essentially insensitive to small signals, and offset would result. A value of $k$ in the vicinity of 0.1 would make the minimum gain of the controller $10/P$.

A characteristic of this sort produces varying degrees of damping in the closed loop. If a linear controller were used to regulate a given linear process, a certain proportional gain could be found which would produce uniform oscillations. A straight line representing this gain, labeled "zero damping," is superimposed on the curve in Fig. 5.25. If the proportional gain of the linear controller were halved, the closed loop would exhibit quarter-amplitude damping. The controller gain representing quarter-amplitude damping is also indicated.

The nonlinear characteristic crosses both these contours of constant damping. Between the intersections are three distinct stability regions. In the region surrounding zero deviation, damping heavier than quarter-amplitude persists, while adjacent to it on both sides are regions of lighter damping and consequently faster recovery. There is still another region on each side where damping is less than zero—representing instability. Should a deviation arise large enough to fall into this last area, it will be *amplified* with each succeeding cycle.

Damped oscillations in a linear loop theoretically go on forever. But with a nonlinear characteristic of the kind shown, damped oscillations cannot persist beyond one or two cycles. On the other hand, a large signal causes more corrective action than a linear controller, appropriately damped, could provide. A sufficiently larger deviation could promote instability, however, so the proportional band of the nonlinear controller must be adjusted for the largest anticipated deviation.

As with other nonlinear controllers, set-point response exceeds what is obtainable with linear modes. This is because set-point changes are normally

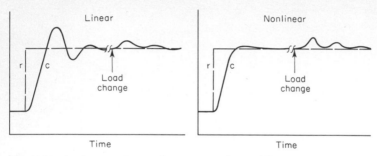

**FIG. 5.26** A three-mode nonlinear controller exhibits better set-point response but poorer load response than its linear counterpart.

greater and more rapid than load disturbances, taking advantage of the region of higher gain. Load disturbances make their appearance as a slow departure of the controlled variable from the set point. Since a linear controller has more gain in the region close about the set point, it will generally respond more effectively to small load changes. A comparison of the responses of linear and nonlinear three-mode controllers is shown in Fig. 5.26.

A flow measurement is always accompanied by noise. This noise is attenuated somewhat by the wide proportional band of the controller and passed on to the valve. If the noise is of any magnitude, the valve may be stroked sufficiently to introduce actual changes in flow. The nonlinear function is an efficient noise filter, in that it rejects small-amplitude signals. The result is smoother valve motion and a more stable loop. Figure 5.27 shows comparative records for linear and nonlinear control of a noisy flow loop. The nonlinear controller has proved quite effective on pulsating flows too, where the disturbance is periodic rather than random.

Level measurements are often noisy because of splashing and turbulence. In addition, the surface of a liquid tends to resonate hydraulically, producing a periodic signal superimposed on the average level. Since the liquid-level process cannot respond fast enough for a change in valve position to dampen these fluctuations, they ought to be disregarded by the controller. A nonlinear controller does just this, sending a smooth signal to the valve.

It was pointed out in Chap. 3 that many tanks with level controls are intended

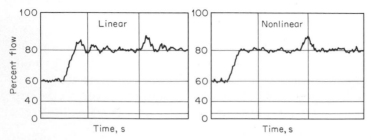

**FIG. 5.27** The nonlinear PI controller is superior in all respects on a noisy flow loop.

as surge vessels. In these applications, tight control is inadmissible because it frustrates the purpose of the vessel. A wide proportional band with integral action was suggested for control. But the nonlinear controller is, in fact, ideal for this application for two reasons: (1) minor fluctuations in liquid level will not be passed on to the valve, providing smooth delivery of flow, and (2) major upsets will be met by vigorous corrective action, ensuring that the upper and lower limits of the vessel will not be violated.

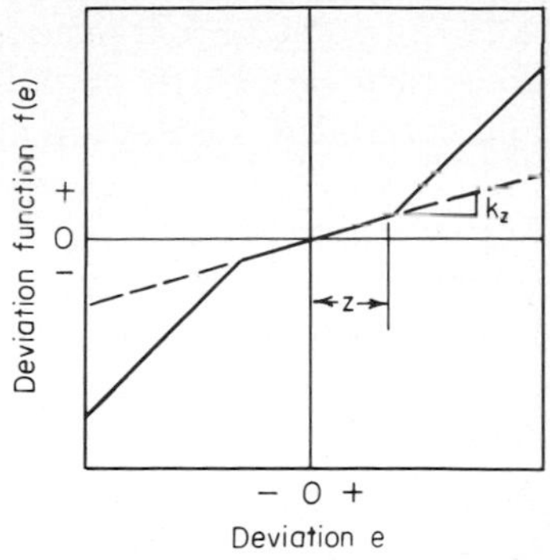

**FIG. 5.28** To effectively control pH, a zone of low-gain $k_z$ with adjustable width $z$ is needed.

### Three-Zone Nonlinear Controllers

Figure 2.17 identified the principal problem associated with pH control; i.e., the titration curve for the process tends to be severely nonlinear. A nonlinear controller is usually needed to stabilize this loop, because of the shape of the process curve and its extreme sensitivity in the region about the set point. However, the type of nonlinear function described in Eq. (5.20) and Fig. 5.25 is not sufficiently severe or adjustable to satisfy most pH curves. Instead, the segmental nonlinear function shown in Fig. 5.28 has been found more useful.

There are two adjustable parameters in this function, gain $k_z$ and width $z$ of the low-gain zone. Both are necessary to achieve effective pH control. It is essential that $k_z$ *not* be zero. In fact, if it is too low to provide the needed regulation, pH will limit-cycle between the two high-gain zones, tracing a triangular wave whose period is much longer than the natural period of the loop. By contrast, too high a value of $k_z$ will produce a smaller cycle closer to the natural period.

Occasionally, but not often, the set point must be positioned away from the center of the titration curve, requiring an unsymmetrical function. In that case, a simple solution to the problem is to remove one high-gain region altogether, extending the low-gain region as shown by the broken lines in Fig. 5.28.

An alternate form of piecewise nonlinear controller is shown in Fig. 5.29.

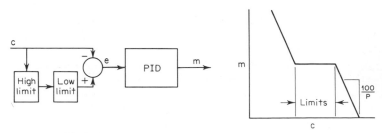

**FIG. 5.29** This discontinuous nonlinear controller is made from a conventional controller and a high-low limiter.

The set point to a conventional PID controller is generated by passing the controlled variable through high and low limits. Between these limits, set point and controlled variable are identical, so that the controller has no gain. Beyond each limit, a deviation is developed which will bring the controlled variable back to that limit. Consequently, the controlled variable tends to hang at one or the other limit; there is no inducement to remain in the dead zone. This controller has probably been unsatisfactory in more applications that it has been satisfactory, principally because it provides no regulation whatever within the dead zone.

## REFERENCES

1. Considine, D. M.: "Process Instruments and Controls Handbook," pp. 11–87, McGraw-Hill, New York, 1957.
2. Shinskey, F. G.: Limit Cycles and Expanding Cycles, *Instrum. Control Syst.*, November 1974.
3. Shinskey, F. G.: Controlling Unstable Processes, pt. I: The Steam Jet, *Instrum. Control Syst.*, December 1974.
4. Adiutori, E. F.: "The New Heat Transfer," pp. 7–24, Ventuno, Cincinnati, 1974.
5. Gibson, J. E.: "Nonlinear Automatic Control," p. 360, McGraw-Hill, New York, 1963.

## PROBLEMS

**5.1** A linear process is found to be undamped under proportional control with a band of 20 percent. What will happen if the band is reduced to 10 percent; to 5 percent?

**5.2** A thermal process with a 10-s dead time and a 5-min lag is to be cooled with refrigerant supplied from a solenoid valve. If the valve is left on, the temperature falls to 0°F; when it is off, the temperature rises to 60°F. Estimate the period and amplitude of the limit cycle if the on-off controller were perfect.

**5.3** The on-off controller used for the process in Prob. 5.2 actually has a dead band of 2°F. Estimate the period and amplitude of the limit cycle, taking the dead band into account.

**5.4** A lever is driven by a bidirectional constant-speed motor to a position determined by a three-state controller. The motor has a speed of 10 percent of full stroke per second and an inertial time constant of 1.0 s. Dead band in the controller is 2 percent of full stroke. How wide does the dead zone have to be to prevent limit cycling? What would be the period of the cycling?

**5.5** A batch chemical reactor is to be brought up to operating temperature with a dual-mode system. Full controller output supplies heat through a hot-water valve, while zero output opens a cold-water valve fully; at 50 percent output, both valves are closed. While full heating is applied, the temperature of the batch rises at 1°F/min; the time constant of the jacket is estimated at 3 min, and the total dead time of the system is 2 min. The normal load is equivalent to 30 percent of controller output. Estimate the required values for the three adjustments in the optimal switching program.

**5.6** A given linear process is undamped with a proportional band setting of 50 percent for a linear controller. If a continuous nonlinear controller used with a linearity setting of $k = 0.2$, how narrow can the proportional band be set and still tolerate an error of 20 percent?

# Part Three

# Multiple-Loop Systems

# Chapter Six

# Improved Control through Multiple Loops

This chapter deals with situations where a single variable is manipulated to satisfy the specification of a certain combination of controlled variables. In any system with a single manipulated variable, only one controlled variable is capable of independent specification. To put it in other words, there can be only one independent set point at any given time. This, however, does not exclude the incorporation of several controlled variables as long as their combination contains but one degree of freedom.

Thus we encounter the cascade control system, where the final element is manipulated through an intermediate or secondary controlled variable whose value is dependent on the primary. Multiple-output control systems manipulate several actuators, but they all have basically the same effect on the primary variable. Selective control embodies the logical assignment of the final element to whichever controlled variable (of several) is in danger of violating its specified limits. Finally, adaption is the act of automatically modifying a controller to satisfy a combination of functions of a controlled variable.

## CASCADE CONTROL

The output of one controller can be used to manipulate the set point of another. The two controllers are then said to be *cascaded*, one upon the other. Each controller will have its own measurement input, but only the primary controller can have an independent set point and only the secondary controller has an output to the process. The manipulated variable, the secondary controller, and its measurement constitute a closed loop within the primary loop. Figure 6.1 shows the configuration.

The principal advantages of cascade control are these:

1. Disturbances arising within the secondary loop are corrected by the secondary controller before they can influence the primary variable.

2. Phase lag existing in the secondary part of the process is reduced measurably by the secondary loop. This improves the speed of response of the primary loop.

3. Gain variations in the secondary part of the process are overcome within its own loop.

4. The secondary loop permits an exact manipulation of the flow of mass or energy by the primary controller.

Cascade control is of great value where high performance is mandatory in the face of random disturbances or where the secondary part of the process contains an undue amount of phase shift. For example, a secondary loop should be closed around an integrating element whenever practicable, to overcome its inherent 90° lag. On the other hand, flow is used as the secondary variable whenever disturbances in line pressure must be prevented from affecting the prime variable.

It must be recognized, however, that cascade control cannot be employed unless a suitable intermediate variable can be measured. Many processes are so arranged that they cannot readily be broken apart in this way.

### Properties of the Inner Loop

The secondary, or inner, loop confronts the primary controller as a new type of dynamic element. The inner loop can be represented as a single block, the diagram of Fig. 6.1 being resolved into the simpler configuration shown in Fig. 6.2.

Heretofore the dynamic properties of a closed loop were of little concern. The controller was simply adjusted for a damping which satisfied certain

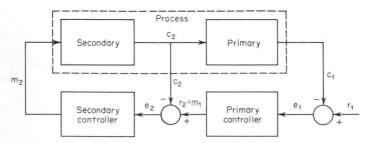

**FIG. 6.1** Cascade control resolves the process into two parts, each within a closed loop.

transient-response specifications. Moreover there was only one period of oscillation to be considered.

But each loop has its own natural period, and, as may be expected, the period of the primary loop is to a great extent determined by that of the secondary.

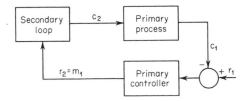

**FIG. 6.2** The primary controller sees a closed loop as a part of the process.

Consequently the gain and phase of the secondary loop, whose natural period will be designated $\tau_{n2}$, must be known for any value of the primary period $\tau_{o1}$, since the latter is dependent on the former. The dynamic properties of the open secondary loop can be converted into its closed-loop characteristics by solving for the response of $c_2$ with respect to $r_2$.

The response of the secondary loop to changes in set point can be evaluated by using the familiar closed-loop relationships developed earlier

$$\frac{dc_2}{dr_2} = \frac{1}{1 + 1/K_c\mathbf{g}_cK_p\mathbf{g}_p} \tag{6.1}$$

where $K_c$ and $K_p$ are steady-state controller and process gains and $\mathbf{g}_c$ and $\mathbf{g}_p$ their dynamic-gain vectors. This response will be identified as $\mathbf{g}_{o2}$, that is, the dynamic gain of the secondary loop to inputs from the primary controller; its steady-state gain is unity.

If the secondary loop consists of an integrating element, dead time, and a proportional controller, $\mathbf{g}_{o2}$ varies with $\tau_{o1}$ as

$$\mathbf{g}_{o2} = \frac{1}{1 + 2\tau_{n2}/\tau_{o1}, \angle (90° + 90°\tau_{n2}/\tau_{o1})} \tag{6.2}$$

where 2 represents the reciprocal of the gain of the secondary loop at its natural period of $\tau_{n2}$ and $\tau_{o1}$ is the period of the primary loop. The resulting gain and phase representing vector $\mathbf{g}_{o2}$ are plotted against the ratio of the loop periods in Fig. 6.3.

The phase angle of the closed secondary loop will always be less than that of the secondary process, so that cascading naturally improves the response of the primary loop. But the gain of the closed secondary loop may be more or less than the secondary process, depending on the period of the input. At very high ratios $\tau_{o1}/\tau_{n2}$, the gain of the secondary process would tend to exceed unity; if

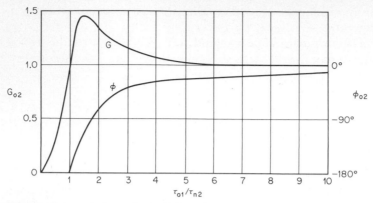

**FIG. 6.3** Gain and phase of a dead-time-plus-capacity process under proportional control.

the natural period of the primary loop is relatively long, the primary-controller gain can be increased by cascading. But as $\tau_{o1}$ decreases, $G_{o2}$ rises, whereas the gain of the secondary process tends to fall. Therefore as $\tau_{o1}$ approaches $\tau_{n2}$, the gain of the primary controller may have to be reduced if the secondary loop is closed. In this case, cascading can actually reduce performance.

It is, of course, impossible for $\tau_{o1}$ to equal $\tau_{n2}$ because there is always phase shift within the primary part of the process. The region to be avoided is that where the gain is high, i.e., periods from 1 to $3\tau_{n2}$. If a PI controller is used to close the secondary loop, the gain peak is increased fourfold. Figure 6.4 gives the closed-loop gain and phase angle for a dead-time-plus-capacity process under PI control, with the controller adjusted for 30° phase lag at the loop period. Note that $G_{o2}$ and $\phi_{o2}$ are 1, $\angle -180°$ at $1.5\tau_{n2}$; this is the period of the secondary loop as established by the aforementioned phase lag of the controller.

Whether integral action is really necessary depends on the proportional band of the secondary controller. Offset in the secondary loop may be of no consequence to the primary loop, because there are no specifications on the secondary variable. However, if the proportional band of the secondary controller must be very wide for stability, as in a flow loop, secondary offset could be so great that $c_2$ could not be driven full scale by the output of the primary controller. Additionally, load changes arising within the secondary loop would not be entirely counteracted without the integral mode.

### Protection against Windup

While the secondary controller can be protected against windup with a batch unit, this will not help the primary controller. Should the secondary controller be unable to respond to the primary for any reason, the primary controller will tend to integrate to the point of saturation. For massive load or set-point changes that would cause the output of the primary controller to reach a limit through *proportional* action alone, a batch unit can protect it against windup. But if the secondary controller were to reach saturation gradually, the primary loop would be open without the primary output reaching any limit. Then protection against windup should be based on secondary deviation.

A system for providing this protection is shown in Fig. 6.5. Normally, the primary output, which is the secondary *set point*, would be fed back to its integral mode *internally*, but Fig. 6.5 shows the secondary *controlled variable* fed back *externally* to the primary controller. If there is no secondary deviation, the primary controller integrates normally. But if a secondary deviation develops, due to a constraint, the positive feedback loop is opened; $c_2$ becomes an independent variable. Then the primary controller loses its integral mode, responding as a proportional controller

$$m_1 = r_2 = \frac{100}{P_1} e_1 + c_2 \tag{6.3}$$

A steady-state secondary deviation will be developed proportional to the primary deviation

$$e_2 = \frac{100}{P_1} e_1 \tag{6.4}$$

When the secondary deviation is again permitted to return to zero, the primary deviation will too; but if the secondary controller does *not* have integral action, a primary *offset* will develop, as Eq. (6.4) demonstrates.

Another feature of this arrangement is that the secondary loop is enclosed in the positive-feedback loop of the primary controller. This necessarily improves primary-loop performance because it adds complementary feedback, providing a better match of the primary controller to the process.

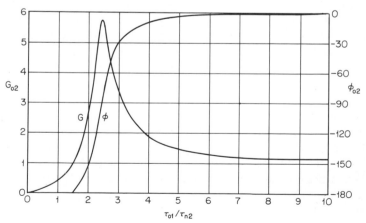

**FIG. 6.4** Addition of integral mode to the secondary loop increases its closed-loop gain peak by a factor of 4.

## Valve Position as the Secondary Loop

A positioner is used to close the loop around a valve motor. It drives the motor until a mechanical measurement of stem position is balanced against its input signal. In this way, forces that would act to impede the motion of the stem are

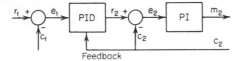

**FIG. 6.5** The primary controller can be protected from windup by using $c_2$ rather than $r_2$ as feedback to its integral mode.

overcome. A positioner is essentially a high-gain proportional controller closing the loop around what is essentially single capacity plus dead band and velocity limit.

The forces affecting the motion of a valve stem are principally friction and pressure drop. It has been pointed out that friction is the cause of dead band. And dead band can cause limit cycling in the presence of two integrating elements, such as a liquid-level loop with proportional-plus-integral control. But a positioner is stable in the presence of dead band and will succeed in eliminating from the primary loop that source of phase shift.

High pressure drop across the seat of a control valve acts against the area of the seat in opposing the force of the valve motor. It can easily be sufficient to keep the valve from closing tightly. A positioner will place as much pressure on the valve motor as is available if the stem is very far from its directed position and so is able to overcome this difficulty.

Figure 6.6 gives the sinusoidal response of a Foxboro P50 pneumatic valve actuator, with and without a positioner; the amplitude of the input wave was ±5 percent. At long periods, that is, >2 s, the actuator alone will show a residual attenuation and phase lag due to dead band. The valve positioner forces the gain closer to unity and the phase lag toward zero in this range. At shorter periods, the time constant of the actuator and its velocity limit bring both gain and phase downward. The positioner is effective here in providing additional amplification and air delivery. An additional factor not shown is the role of the positioner in decoupling the capacity of the actuator from its transmission line. If the line were included in the response test, the actuator would have an even lower gain and greater phase lag than indicated.

A valve positioner is helpful in every kind of loop except flow or liquid pressure. These processes are self-regulating and therefore stable in the presence of dead band. Furthermore, the addition of a positioner is seen to increase the gain of the valve by a factor of 2 or more in the dynamic range of a 0.5- to 2.0-s period, where those loops cycle. This requires the doubling of an already wide proportional band if the same stability is to be achieved as without a positioner. And a wider proportional band reduces the capability of the loop to respond to set point and load changes.

### The Cascade Flow Loop

Cascade flow loops are used most often to provide consistent delivery of material to or from the process in response to the demands of the primary controller. They overcome variable pressure drop, valve friction, and nonlinear valve characteristics.

But if the measurement is in the differential form, its nonlinearity becomes part of the primary loop because flow is being delivered to the process while flow squared is set by the primary controller. The nonlinearity of the differential meter was recognized earlier as a problem in a flow loop. But using a differential meter in the secondary loop of a difficult thermal or composition process is asking for trouble. Suppose the process is linear with respect to flow. The output of the primary controller manipulates differential pressure, however, which varies as flow squared

$$h = kF^2 \qquad \frac{dF}{dh} = \frac{1}{2kF} \tag{6.5}$$

Loop gain now varies inversely with flow (which is much worse than varying directly with flow because it can approach infinity). And since many processes are started up or operated for extended periods at low flow, the problem is serious. If the primary controller is not placed in manual, the loop will limit-cycle around zero flow. The best solution is to insert a square-root extractor in the flow-measurement line to linearize the secondary loop.

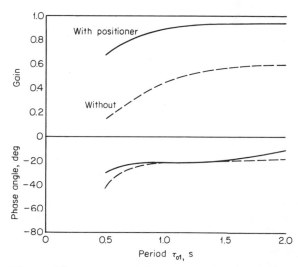

**FIG. 6.6** The response of a Foxboro P50 valve actuator improves markedly in gain but only slightly in phase angle when a positioner is added.

The problem of variable dead time was discussed in Chap. 2. This problem was resolved by using an equal-percentage control valve. If a flow loop were placed around the valve, however, its characteristic would be lost. Furthermore, if the flow loop were of the differential type, its nonlinearity would be in the wrong direction, making the primary-loop gain vary inversely as the square of flow. These factors deserve careful consideration before deciding on a cascade flow loop.

In Chap. 2 the problem of a variable secondary lag was also presented, in association with the liquid-level process of Fig. 2.9. In that example, period and loop gain both *increased* with flow, requiring a valve characteristic of the quick-opening type. The characteristic which will exactly compensate the gain variation of that process *is* the cascade differential-flow loop, as indicated by Eq. (6.5).

Figure 6.4 illustrates the difficulties that may be encountered if the period of the primary loop is too close to that of a secondary loop under PI control, which is always used in a flow loop. Instability can be expected if $\tau_{o1}$ is less than about $4\tau_{n2}$. Therefore one flow controller should not be used to set another in cascade, nor should a gas or liquid-pressure controller set flow in cascade. Undesirable interaction may also be encountered in a level loop. If the situation cannot be avoided, one or both of the controllers must be detuned. Increasing the integral time of the secondary controller will reduce its closed-loop gain and period of oscillation, moving the peak of Fig. 6.4 toward that of the proportional loop of Fig. 6.3.

**Temperature as the Inner Loop**

Perhaps the third most common cascade configuration is that of a temperature controller setting another temperature controller. The stirred-tank reactor described in Fig. 3.7 can be controlled effectively by setting coolant *outlet* temperature in cascade. This is preferred to the selection of coolant *inlet* temperature, in that it moves the dynamic response of the jacket from the primary to the secondary loop. The period of the primary loop is reduced typically from 40 to 20 min, allowing a proportional band reduction from perhaps 30 to 15 percent. Furthermore, disturbances in heat evolution affect coolant outlet temperature, bringing about secondary-loop correction with less upset to reactor temperature.

Equation (3.25) shows that reactor temperature responds linearly to coolant outlet temperature. The principal nonlinearity in the system is the relationship between outlet temperature and flow, as given in Eq. (3.22). Compensating it with an equal-percentage valve and enclosing it within the secondary loop keeps the primary loop free of this source of variability.

Because coolant outlet temperature can be controlled stably with 10 to 20 percent proportional band, its controller does not need an integral mode. This reduces the resonant peak of the secondary loop and increases its speed of response—important factors, particularly in the control of batch reactors.

## MULTIPLE-OUTPUT CONTROL SYSTEMS

Very often there arises a need to control a single variable by coordinating the manipulation of several parallel variables, all having essentially the same effect on the process. The situation is presented whenever two or more parallel units serve a common user. One example would be several boilers discharging into a common steam header, whose pressure must be controlled by manipulating their individual firing rates. Another would involve parallel compressors drawing suction from, or discharging into, a common header whose pressure is to be controlled.

The division of load among the multiple manipulated variables should be performed in such a way that:

1.  Loop gain is constant, regardless of the number of units in automatic.
2.  The operator may be permitted to distribute the load at his own discretion.
3.  Units can be brought on and off line with minimal disturbance to the common controlled variable.
4.  Units can enter or leave constraints with minimal disturbance to the common controlled variable.

These objectives can be achieved either through careful coordination of the manipulated variables on a programmed, calculated basis or by a fast secondary feedback loop. The choice depends on the accuracy with which the effects of manipulations can be predicted and the speed of response attainable.

### Cascade Control of Total Output

This system is based on the assumption that the effect of each individual manipulated variable upon the common controlled variable is reasonably uniform across its operating range. If this is the case, the effect of all the output signals acting together is the weighted sum of the individual signals. Figure 6.7

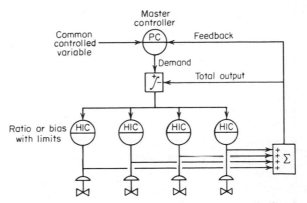

**FIG. 6.7**  Any independent action taken on one manipulated variable is fed back immediately to reposition the others.

describes the system wherein the weighted sum of the output signals is fed back to a high-speed integrator, which also receives a command from the master controller. Because there are only instruments in the feedback loop, the integrator may have a very short time constant (1 s for pneumatic instruments and less for electronic).

The integrator drives all manipulatable outputs to force the weighted sum of *all* outputs to equal that demanded by the master controller. The operator is free to adjust any individual output through the HIC (hand indicating control) station. These stations may include a ratio or a bias function, or possibly both. A ratio station multiplies its input signal by a manually set ratio to produce the output, whereas a bias station adds or subtracts a manually set bias. Adjustable output limits are normally included, along with the capability to position the output manually.

When an adjustment is made, the resulting change in output affects the weighted sum, causing the integrator to readjust its command to all outputs until the original sum is reestablished. The high speed of response of this loop should prevent the upset from having any discernible effect on the common controlled variable. Similarly, any action taken in manual affects the total output and is corrected without delay by the integrator. And when one unit encounters a limit, the integrator will drive the others to make up for that insufficiency.

Units that are shut down, in manual, or operating at limits are unable to respond to control action. This does reduce loop gain, but only in the fast loop closed by the integrator. Its time constant must then be adjusted for the highest-gain condition, i.e., when all units are on line; then all other conditions will result in slower response. But if the speed of response of this loop is two orders of magnitude faster than that of the master loop, the performance of the latter should not be materially degraded. Certainly this arrangement is preferred over allowing the gain of the master loop to vary.

To avoid windup of the master controller when *all* units are in manual or limited, the total output signal is fed back to its integral mode.

Linearity and reproducibility of the multiple-output control system can be improved by feeding back total *flow*, as shown in Fig. 6.8. The flow-control loop will be somewhat slower than the secondary loop of Fig. 6.7, precluding its use on rapidly responding processes such as compressors. This arrangement is quite adequate for steam-pressure control of a multiple boiler installation, however.

While the total-flow loop keeps the valve characteristics and dead band out of the primary loop, they may still affect the way the system is balanced. Ratio or bias adjustments and limits are applied to output signals as before. These controls would be more accurate if they could be applied to individual flow controllers, and this is sometimes done. But then cascading is doubled, with a total-flow controller setting individual flow controllers. To avoid instability, the total-flow loop must have a very low gain at the natural period of the individual flow loops.

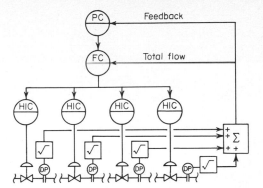

**FIG. 6.8** Nonlinearities associated with valves are removed from the master loop by feeding back total flow.

The performance of the system can be analyzed by referring to Fig. 6.4, which describes the *entire* process enclosed in the total-flow loop. Because the process gain is so high, proportional action will be ineffective. Then if the total-flow controller is an integrator, its period of oscillation will be that at which the individual flow loops generate 90° phase lag. Figure 6.4 places this point exactly where the closed-loop gain peaks, at about 5.7. For quarter-amplitude damping to be achieved, the integral time of the total-flow controller would have to be

$$\tau_I = \frac{\tau_{o1}}{2\pi}\frac{5.7}{0.5} = \frac{2.5\tau_{n2}}{2\pi}\frac{5.7}{0.5} = 4.54\tau_{n2}$$

Without the individual flow controllers, the total-flow controller could be PI, with an integral time of about $0.41\tau_{n2}$, as estimated from Table 4.3. The addition of the individual flow loops therefore reduces the speed of the total flow loop by a factor of 10. While there are some applications that would not be affected adversely by this loss in response, many would.

### Combining Modulating and On-Off Devices

To lower equipment cost and to save energy, many installations use several constant-rate devices in parallel with a single modulating unit. An example would be a bank of fan coolers, with only one equipped with a variable-speed driver. The variable-speed fan can be used for control of the process temperature, but only over a narrow load range. When the load reaches the maximum- or minimum-speed capability of the fan, others must be turned on or off to retain control. However, this sequencing must be properly coordinated to minimize upsets to the process.

Figure 6.9 shows a system for accomplishing such coordination. The output of the master controller represents demand for output from the entire system. On-off controllers are arranged to energize drivers at specified levels of demand, with the balance of the load delivered by the modulating unit. Consider $n$ units sharing the load. When the load is below $100/n$ percent, only the modulating unit would be operating. When the master demand increased above $100/n$ percent, the first on-off device would be started. A relay energized

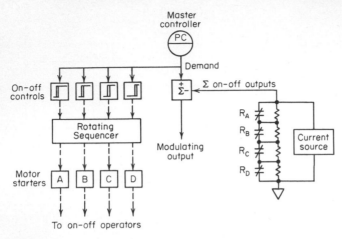

**FIG. 6.9** When a unit is started or stopped, feedback to the subtractor automatically readjusts the output to the modulating unit.

by the starter would feed back a signal to the subtractor, resetting the modulating unit back to low speed. If the system is correctly calibrated, this action should not upset the process.

A rotating sequencer is shown, intended to distribute the load among the on-off units. It causes the first unit that was energized to be the first deenergized; without this feature it would be the last deenergized and would carry more than its share of duty.

### Keeping Valves Open

In many systems with parallel units, there will be one more manipulated variable than needed for control. An example of this situation is shown in Fig. 6.10, where a variable-speed pump is supplying feed to two parallel units, each having its own flow control loop. Ideally, the pump should operate at the minimum speed needed to satisfy both users, although this minimum will vary considerably with load.

The system shown in Fig. 6.10 selects the highest of the flow controller outputs to control pump speed. The valve-position controller (VPC) is an integrating device which attempts to keep the most-open valve nearly wide open but still in the controllable range. This minimizes the power required to pass the flow demanded by the process.

To avoid upsetting the unselected flow loops, the valve-position controller must change its output slowly. But this does not have to be a tight loop; its primary function is to save energy in the steady state.

Valve-position control introduces a nonlinearity which should be considered: differential pressure across the valves must vary with flow because valve position is being held constant in the steady state. Liquid flow through a linear valve varies as

$$F = mC_v \sqrt{\frac{\Delta p}{\rho}} \qquad (6.6)$$

where $m$ = fractional stem position
$\quad C_v$ = rated valve capacity
$\quad \Delta p$ = differential pressure, lb/in²
$\quad \rho$ = fluid density, specific gravity units
Valve gain is

$$K_v = \frac{dF}{dm} = C_v \sqrt{\frac{\Delta p}{\rho}} \qquad (6.7)$$

Substituting (6.6) into (6.7) shows gain varying directly with flow when position is controlled

$$K_v = \frac{F}{m} \qquad (6.8)$$

This essentially produces an equal-percentage characteristic from a linear valve—not a desirable characteristic for flow control but an example of how performance must sometimes be sacrificed to save energy.

## SELECTIVE CONTROL LOOPS

Frequently a situation is encountered where two or more variables must not be allowed to pass specified limits for reasons of economy, efficiency, or safety. If the number of controlled variables exceeds the number of manipulated variables, whichever ones are in most need must logically be selected for control. (This is the case of the squeaky wheel getting the grease.) Signal selector units are available for this type of service. They are employed in four basic areas of application:
1. Protection of equipment
2. Auctioneering

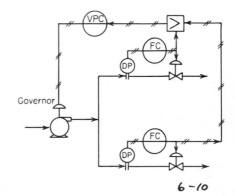

FIG. 6.10   The valve-position controller sets pump speed to keep the most-open valve in the control range.

6-10

3. Redundant instrumentation
4. Variable structuring

As an example of how equipment might be protected by a selective control system, consider a compressor whose discharge is ordinarily on flow control, except that discharge pressure must not be allowed to exceed a given limit. During conditions of low load, the pressure controller must be allowed to assume control, thereby reducing flow. When the demand for gas is high, the flow controller will take over to see that its set point is not exceeded.

Figure 6.11 shows how the controller that has the lower output is selected to manipulate motor speed. Decreasing motor speed will reduce both flow and pressure, and so the use of a low selector guards against an excess of either. The record shows how pressure is allowed to drift below its set point during conditions of high load, while flow is controlled. Conversely, when the load is low enough to raise the discharge pressure to its set point, flow is reduced.

*Auctioneering* is a term used to describe the selection of the highest of a battery of inputs. An example is the control of the highest temperature in a fixed-bed reactor. The possibility exists that the location of the highest temperature may shift with catalyst degeneration, flow, etc. Temperatures all along the reactor would then be compared and the highest used for control, as shown in Fig. 6.12.

To protect against instrument failure's placing the plant in a hazardous condition, key instruments can be duplicated. Often, duplicated devices are used only for record or alarm purposes. But where closed-loop control is involved, automatic selection of the controlled variable must be provided.

Analyzers are generally less reliable than other instruments. Figure 6.13 illustrates a system that would allow control to be maintained in the event of downscale failure of either analyzer. An upscale failure would be allowed, which would shut down the reactor, but this is a safe condition.

If *neither* an accident nor an unwarranted shutdown is considered an acceptable consequence of an analyzer failure, three analyzers must be provided. Their

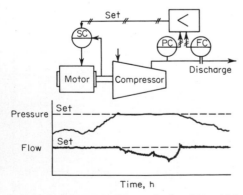

FIG. 6.11   Motor speed is manipulated by whichever controller has the lower output.

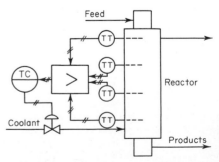

FIG. 6.12   A high selector is used to permit control of the peak reactor temperature.

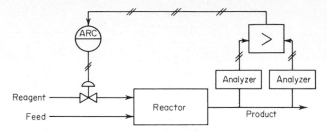

**FIG. 6.13** The high selector prevents a failure of either analyzer from damaging the reactor with excess reagent.

outputs are then compared in a *median* selector, which rejects the lowest and the highest signals. The plant is then protected against an individual failure in either direction.

### Protection against Windup

When one controller is selected from two or more, the others are in an open-loop condition. If these controllers have integral action, which is most often the case, they need to be protected against windup. This is accomplished by using the output of the selector as a common feedback to all controllers. In this way, the selected controller will have its own output fed back and therefore will have integral action. But the others in the system will have a feedback signal which is not their own output, forcing them to respond like proportional controllers. The controller arrangement is shown in Fig. 6.14 for the pressure-flow system that appeared in Fig. 6.11.

Automatic transfer from one controller to another takes place at the instant when the outputs are equal. This fact, coupled with the common feedback signal, means that transfer is bumpless.

Assume that the flow controller is presently manipulating compressor speed, in that its output is the lower of the two. Its output $m_F$ is then

$$m_F = \frac{100}{P_F}\left(e_F + \frac{1}{I_F}\int e_F \, dt\right) \qquad (6.9)$$

But because the output $m_p$ of the pressure controller is greater,

$$m_p = \frac{100}{P_p} e_p + m_F \qquad (6.10)$$

Transfer from flow to pressure control will take place when $m_p = m_F$. This requires that $e_p = 0$ but does not require the same of $e_F$. To generalize, a controller can be selected only when its error crosses zero. For controllers with derivative on the input, $e + D \, de/dt$ would have to be zero. This allows transfer before zero error is reached, avoiding the overshoot that would otherwise be encountered.

A single automanual transfer station at the output of the selector is sufficient

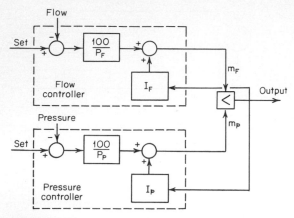

**FIG. 6.14**  The output of the selector is used as a common feedback signal to prevent windup of the unselected controller.

unless one or more of the controllers must remain in automatic to protect the plant.

Occasionally a situation will arise when none of the controllers in a selective loop assumes control. Some processes are prone to this, although in most it cannot happen. In the process shown in Fig. 6.11 neither controller would take over if suction pressure were to fall low enough. In this case, discharge pressure and flow would both be below their set points, and the motor would go to maximum speed. Both controllers would then saturate. When suction pressure was restored, discharge pressure or flow or both could overshoot their set points. If this were a common occurrence, a batch unit would have to be inserted in the common feedback line at the output of the selector.

It should be mentioned that a controller whose output passes through a high or low limiter is also capable of windup when these limits are exceeded. Integral feedback may be taken downstream of the limiters if the problem is serious.

If the selector drives the set point of a secondary loop in cascade, feedback of the secondary measurement to all primary controllers (as shown in Fig. 6.5 for one primary) will protect against windup in the presence of a secondary deviation.

Unselected controllers can also be protected from windup by setting their internal high and low limits from the selected output. The high limit must be biased slightly above the selected output, and the low limit slightly below, to avoid interference with operation of the selected controller. Unselected controllers will then wind up only to the extent of this bias, which may result in a small overshoot during transfer from one controller to another.

### Selection Using Digital Algorithms

The effectiveness of selective control based on controller output signals is limited to positional algorithms. Selection of incremental or velocity signals is *not* equivalent and can lead to serious problems which may not easily be recognized. This constitutes a significant limitation in applying digital control because most computers feature incremental rather than positional algorithms.

The first problem is associated with the sample intervals of the controllers whose outputs are compared. If the selector samples more often than one of the controllers, it will see a $\Delta m$ of zero at times when that control algorithm is not being processed. This is actually false information and causes the system to respond differently to increasing and decreasing signals. In the case of a low selector, the system will drive downscale at the shortest sample interval of any of the associated controllers and upscale at the longest sample interval. Roles are reversed for a high selector.

A second problem is associated with noise response. Incremental algorithm (4.32) shows that output $\Delta m$ responds to a change in deviation from the last sample, that is, $e_n - e_{n-1}$. Noise on the input signal will cause $e_n$ to differ from $e_{n-1}$, even in the steady state, developing a proportional $\Delta m$. The change is even more pronounced with derivative action. Ordinarily the output to the valve simply reflects this noise level. But a selector will pass $\Delta m$ of one sign and reject the next $\Delta m$ of the opposite sign. As a result, the changes induced by noise are rectified by the selector, causing the valve to be driven in the direction preferred by the selector.

According to Eq. (4.32), the system will reach an equilibrium when $\Delta m$ induced by noise through proportional and derivative action is offset by that caused by integral action. After $\Delta m$ has been set to zero, (4.32) can be rearranged to estimate the offset resulting from noise rectification

$$ e = \frac{I}{\Delta t} \, \Delta c \left( 1 + \frac{D}{\Delta t} \right) \tag{6.11} $$

where $\Delta c$ represents the peak-to-peak noise level in the controlled variable. Offset $e$ can be reduced by reducing both $I$ and $D$, by increasing sample interval $\Delta t$, and by filtering the measurement signal to reduce $\Delta c$.

### Variable Structuring

It is occasionally necessary to transfer a controller from one valve to another when a constraint is reached. A case in point is the condenser control system in Fig. 6.15. In its normal operating mode, all the vapor is condensible. The pressure in the system is then controlled by restricting the flow of reflux, thereby maintaining a liquid level within the condenser. A falling pressure indicates that vapor is being condensed faster than it is being generated; the pressure controller reacts by raising the liquid level, reducing the heat-transfer surface exposed to condensation. During this time, liquid level in the reflux drum is high, causing the level controller to produce a high output, which is not selected. The two inputs to the subtractor are equal, producing an output of zero to the vent valve.

Should noncondensible gas accumulate in the condenser, pressure will rise.

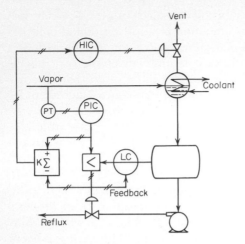

**FIG. 6.15** When the level controller is selected to manipulate reflux, pressure control is transferred to the vent valve.

The pressure controller will increase reflux flow but may not be able to lower pressure enough even when the condenser is empty. The liquid-level controller would then have to take over manipulation of reflux, to avoid emptying the drum and cavitating the pump. With an empty condenser, pressure can be controlled only by venting noncondensible gas.

The system in Fig. 6.15 transfers pressure control to the vent valve smoothly when the level controller is selected for reflux manipulation. At this point, the two inputs to the subtractor begin to differ, developing a signal which opens the vent valve. The output from the pressure controller then manipulates the vent valve instead of the reflux valve. The pressure controller should be tuned while manipulating reflux. Readjustment for vent-valve manipulation is through gain $K$ of the subtractor, which is common to both inputs

$$m_v = K(m_p - m_L) \tag{6.12}$$

where $m_v$ = output signal to vent valve

$m_p$ = output of pressure controller

$m_L$ = output of selector

The level controller needs external feedback to avoid windup, but the pressure controller does not; its loop is always closed, either through one valve or the other.

Other examples of variably structured systems are described in Ref. 1.

## ADAPTIVE CONTROL SYSTEMS

An adaptive control system is one whose parameters are automatically adjusted to compensate for corresponding variations in the properties of the process. The system is, in a word, "adapted" to the needs of the process. Naturally there must be some criterion on which to base an adaptive program. To specify a value for the controlled variable, i.e., the set point, is not enough; adaption is not required to meet this specification. Some *objective function* of the controlled variable must be specified in addition. It is this function which determines the particular form of adaption required.

The objective function for a given process may be the damping of the controlled variable. In essence, there are then two loops, one operating on the controlled variable, the other on its damping. Because damping identifies the dynamic loop gain, this system is designated a *dynamic* adaptive system.

It is also possible to stipulate an objective function of the steady-state gain of the process. A control system designed to this specification is then *steady-state-adaptive*.

There is, in practice, so little resemblance between these two systems that their classification under one single title adaptive has led to much confusion.

A second distinction is to be made, not on the objective function but on the mechanism through which adaption is introduced. If enough is known about a process to enable parameter adjustments to be related to the variables which cause its properties to change, adaption can be *programmed*. However, if it is necessary to base parameter manipulation upon the measured value of the objective function, adaption is effected by means of a feedback loop. This is known as a *self-adaptive system*.

### Dynamic Adaptive Systems

The prime function of dynamic adaptive systems is to give a control loop a consistent degree of stability. Dynamic loop gain is then the objective function of the controlled variable being regulated; its value is to be specified.

The property of the process most susceptible to change is gain. In some cases the steady-state gain changes, which is usually termed a nonlinearity. Other processes exhibit a variable period, which reflects upon their dynamic gain. But by whichever mechanism loop stability is affected, it can always be restored by suitable adjustment of controller gain. (This assumes that the desired degree of damping could be achieved in the first place, which rules out limit cycling.)

Many cases of variable process gain have already been cited. In general, an attempt is made to compensate for these conditions by the introduction of selected nonlinear functions into the control system. For example, the characteristic of a control valve is customarily chosen with this purpose in mind. But compensation in this way can fall short for several reasons:

1. The source of the gain variation lies outside the loop, and hence is not identified by controller input or output.

2. The required compensation is a combined function of several variables.

3. The gain of the process varies with time.

Perhaps the most readily assimilated example of a process which could benefit by adaptive control is that of a single capacity plus variable dead time. Dead time could vary inversely with flow, in the manner shown in the heat-exchanger example given in Chap. 2. Response of the process was presented in Fig. 2.8. In this example, an equal-percentage valve was used to provide gain compensation for changes in flow. This method worked but it in turn made loop gain dependent on the magnitude of the controlled variable. The trade-off was inevitable because the variable which affected the process gain, i.e., flow, is outside the loop.

Exact compensation can be obtained by programming the settings of the controller as functions of flow. Because the period of the loop varies directly with dead time, derivative and integral time ought to vary inversely with flow. And since process dynamic gain varies inversely with flow, the proportional band should too. Knowing this, it is possible to write a flow-adapted control algorithm

$$m = \frac{100f}{P}\left( e + \frac{f}{I}\int e\,dt + \frac{D}{f}\frac{de}{dt}\right) \tag{6.13}$$

The adaptive term $f$ is the fractional flow through the process, and $P$, $I$, and $D$ are the optimum settings at full-scale flow.

Implementation of Eq. (6.13) requires an adaptable controller. Most digital algorithms are adaptable, in that settings of $P$, $I$, and $D$, are name variables stored in memory. They can be calculated through a simple program as a function of measured flow.

The Foxboro SPEC 200 electronic controller is also capable of adaptation, as described in Ref. 2. Derivative and integral time constants are generated by transistors gated by oscillators whose frequency is proportional to an applied voltage. The two time constants then vary inversely with voltage, which can be made proportional to measured flow. Gain adaptation can be achieved through a similar circuit or by using a multiplier, as described under Feedforward Control in the next chapter.

### Dynamic Self-adaptive Controllers

A great deal of research effort has been spent in several industries in the quest for a self-adaptive controller. The application goes beyond compensating for variable loop gain because a device which could adapt itself could also relieve the operator of the task of adjustment altogether. Thus the performance of a

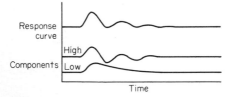

**FIG. 6.16** If the loop is properly damped, high- and low-frequency components will exist in a certain ratio.

critical loop could be made independent of the skill of the operator. (Although, instead, it is made doubly dependent on the skill of the control designer.)

Again, the purpose of this adaptive loop is to regulate system damping. If the normal state of the system is steady, no measurement of damping is available. If

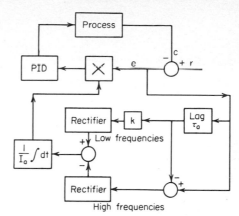

**FIG. 6.17** The adaptive controller reduces loop gain in the presence of oscillation and raises it under drifting conditions.

the self-adaptive function is to work, some means of perturbing the state of the process must be decided upon. Either a periodic disturbance may be introduced as a test signal, or the system must wait for disturbances to occur naturally. Each of these methods has certain disadvantages. The first is generally inadmissible in that it effectively robs the process of its rightful steady state.

Since the second method does not test the process, the current value of loop gain is unknown until a disturbance identifies it. Identification must then be carried out and parameter adjustment made carefully to prevent overcorrection. Identification consists principally of factoring the response curve into high- and low-frequency components whose ratio represents the dynamic gain of the closed loop. The load-response curve shown in Fig. 6.16 is so separated.

Having thus identified the damping of the loop, the task of adjusting it remains. This must be recognized as a feedback operation—manipulating a parameter on the basis of a measurement made on the controlled variable. The arrangement of the loop is shown in Fig. 6.17.

Separation of the deviation into two frequency bands is arranged so that oscillations near the natural period of the loop pass through the high-frequency channel. The signal is rectified so that adaptation is applied equally to positive and negative deviations. The integrator responds in proportion to the magnitude of the deviation by reducing the gain of the controller. At the same time, low-frequency components of deviation are adjusted in amplitude by gain $k$, rectified, and sent to the integrator with the opposite sign. Offset, drift, and sluggishness then cause the integrator to raise the gain of the controller to bring the deviation to zero.

This concept was applied to control the pH of an effluent from a chemical plant [3] to compensate for unmeasurable variations in the titration curve of the waste being neutralized. Because the process was nonlinear as well as time-variant, the output of the adaptive unit adjusted the width of the low-gain zone in a nonlinear pH controller [4]. Frequency discrimination was accomplished by a single first-order lag of the time constant $\tau_a$. To place 70 percent of the natural cycle of the control loop in the high-frequency channel, $\tau_a$ must be adjusted for a gain of 0.3 to signals of $\tau_n$

$$G_a = \frac{1}{\sqrt{1 + (2\pi\tau_a/\tau_n)^2}} = 0.3 \tag{6.14}$$

Solving for $\tau_a$ gives

$$\tau_a = \frac{\sqrt{10}}{2\pi}\,\tau_n = 0.5\tau_n$$

If the natural period should change, $\tau_a$ must also change. In the pH control system, ultrasonic cleaning was applied to minimize electrode coating, which would raise $\tau_n$.

It was found that a $k$ of 1.0 promoted *burst cycling*, i.e., periodic bursts of cycles signaling instability in the adaptive loop. Reducing $k$ to 0.2 promoted a very slow cycle, which was far more damaging than the natural cycle the adaptive controller was designed to eliminate. Settings in the 0.5 to 0.7 range brought stable performance to the pH loop; other processes may require different settings. It was also found that the integral time of the adapter can be safely set at or somewhat below that of the process controller.

Pessen [5] describes an adaptive PID controller in which derivative and integral time constants were set relative to the observed period of the loop. The intent is to maintain optimum ratios of $D$ and $I$ to $\tau_o$, as described in Table 4.3. This function also forms a feedback loop because $D$ and $I$ do affect the period.

Self-adaptive control ought to be reserved for processes where the need exists and there is no satisfactory alternative. If the cause of the parameter change is known and measurable, programmed adaptation should be used because it can keep the loop constantly in proper adjustment. By contrast, poor performance must appear *before* the self-adaptive controller takes any action. Usually two or three cycles are developed before loop gain can be corrected. In cases where loop gain is constantly changing, the self-adaptive controller will always lag behind the needs of the process. If the gain of the process changes with load, as in a heat exchanger, the self-adaptive controller will always adjust to the *last* load change, whereas adaptation is needed for the *next* one.

### The Steady-State Adaptive Problem

Where the dynamic adaptive system controlled the dynamic gain of a loop, its counterpart seeks a constant steady-state process gain. This implies, of course, that the steady-state process gain is variable and that one particular value is most desirable.

Consider the example of combustion control system whose fuel-air ratio is to be set for highest efficiency. Excess fuel or air will both reduce efficiency. The true controlled variable is efficiency, while the true manipulated variable is the fuel-air ratio. The desired steady-state gain in this instance is

$$\frac{dc}{dm} = 0 \tag{6.15}$$

The system is to be operated at the point where either an increase or decrease in ratio decreases efficiency. This is a special case of steady-state adaption known as *optimizing*. A gain other than zero may reasonably be stipulated, however.

Where the value of the manipulated variable which satisfies the objective function is known relative to conditions prevailing within the process, the adaption can easily be programmed. As an example, the optimum fuel-air ratio may be known for various conditions of airflow and temperature. The control system may then be designed to adapt the ratio to airflow and temperature much in the way that the controller settings were changed as a function of flow in the example of dynamic adaption.

If a reasonably accurate mathematical model of the process can be obtained in a simple form, it can be differentiated to solve for the adaptive control program. In the following expressions, let $K_r$ represent the desired gain of the process. Consider the example of a variable-gain process affected by a load term $q$

$$c = am - qm^2 \tag{6.16}$$

(Note that if $c$ were directly proportional to $m$, the process gain would be constant and there would be zero degrees of freedom.) Differentiation solves for process gain, which is set equal to $K_r$

$$\frac{dc}{dm} = a - 2qm = K_r \tag{6.17}$$

Next, Eq. (6.17) is solved for $m$, which is the output of the control system

$$m = \frac{a - K_r}{2q} \tag{6.18}$$

Figure 6.18 shows how such a control system would be arranged.

Because the system described above has no feedback loop, it does not rightly belong in this chapter. Therefore further discussion of this class of system, which is growing in importance, will be relegated to Chap. 7.

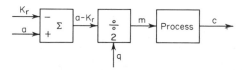

**FIG. 6.18** This steady-state adaptive system does not have a feedback loop.

## A Continuous Self-optimizing Controller

Like the self-tuning controller, the self-optimizing controller requires no prior knowledge of plant conditions but instead conducts its own search. Its goal is to keep the manipulated variable at the point where process steady-state gain $K_p$

satisfies a specification $K_r$. But before this can be done, the controller must first test the process for its gain, either continuously or intermittently.

Since process gain $K_p$ cannot be measured directly, it must be inferred by correlating changes in input and output. Recognize, however, that output $c$ and input $m$ are displaced in time by the process dynamic elements. A best estimate of steady-state gain then requires dynamic compensation of the manipulated variable signal by a model of the process. The controlled variable responds to manipulation as

$$dc = K_p \mathbf{g}_p \, dm \qquad (6.19)$$

Differentiation using a time constant $\tau$ yields

$$\tau \frac{dc}{dt} = K_p \mathbf{g}_p \tau \frac{dm}{dt} \qquad (6.20)$$

Division by the dynamically compensated derivative of $m$, as shown in Fig. 6.19, then gives

$$\frac{K_p \mathbf{g}_p \tau \, dm/dt}{\mathbf{g}_p \tau \, dm/dt} = K_p \qquad (6.21)$$

Calculated process gain $K_p$ is then compared with desired gain $K_r$ to develop a deviation for control action.

Control can be achieved at any desired slope $K_r$ if that slope appears but once over the operating range of the process. For a process characteristic curve that passes through a *maximum*, gain $K_p$ decreases with increasing $m$, so that the controller output should increase with increasing $K_p$. This is the action described in Fig. 6.19. Should the characteristic curve pass through a *minimum*, the signs at the summing junction must be reversed.

There are some peculiarities in this system which need attention: (1) Calculation of process gain is based on differentiation, which is an inexact operation. Any noise on the input signals tends to be amplified, so that the dynamic gain of

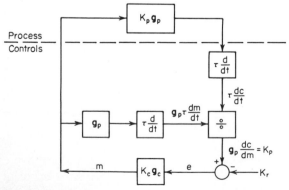

**FIG. 6.19** A model of the process dynamic response is needed to phase numerator and denominator properly.

the differentiators must be limited. This consists in applying a first-order lag, as described for the derivative control mode in Chap. 4. (2) The two differentiators must have the same time constant $\tau$; otherwise their ratio would appear in the quotient of the divider. (3) Failure of the dynamic compensator to match the process model can cause errors in the gain calculation in the unsteady state.

In a conventional control loop, dynamic errors would not ordinarily be bothersome, but this system has a nebulous steady state. If neither $c$ nor $m$ is changing, $K_p$ is indeterminant, so that there is no assurance of the objective's having been achieved. The control system itself will not tolerate a steady state, however, in that the gain of the divider rises to its maximum value under this condition. As a result, small perturbations in the controlled variable are greatly amplified, developing significant deviations from set point, which tend to keep the loop in constant motion.

If control is to be achieved at either the maximum or minimum of a curve, the decreasing process gain in the vicinity of the control point may partially offset the rising gain of the divider. However, many processes do not have a rounded maximum or minimum. A plot of flame temperature vs. fuel-air ratio, for example, consists essentially of two straight lines forming a sharp vertex. The conductivity of certain solutions is observed to pass through a minimum corresponding to phase or composition changes. These minima tend to be even sharper than the vertex described above, having the appearance of a crevasse, rather than the wide valley usually associated with double-valued functions. The extreme gain in the region about the control point develops a limit cycle that cannot be dampened. Adjusting the gain $K_c$ of the controller will ultimately determine its amplitude.

Because of the low accuracy of differentiators, particularly to slowly changing signals, this system is limited in application to fast processes such as component blending and fuel-air-ratio control.

## A Sampling Optimizer

To overcome the equipment-limitation problem on slow processes, the optimizing search may be carried out discretely, using sampled data. This amounts to supplanting the differentials in the previous example with differences

$$\frac{\Delta c}{\Delta m} = \frac{\Delta c / \Delta t}{\Delta m / \Delta t} \tag{6.22}$$

The sample interval $\Delta t$ must be long enough to let the process return to equilibrium after each change in controller output. When $\Delta t$ has expired, the most recent $\Delta c / \Delta m$ is calculated and compared with $K_r$

$$e_n = \frac{\Delta c_n}{\Delta m_n} - K_r$$

Next, the output of the controller is stepped proportional to the error signal by a gain $K_c$

$$\Delta m_{n+1} = K_c e_n$$

The effective integral time is related inversely to gain and directly to sample interval $\Delta t$

$$I = \frac{\Delta t}{K_c} \tag{6.23}$$

The sampling optimizer is not affected by the same equipment limitations as its continuous counterpart, but its other characteristics are similar. The sampling is of some advantage on processes dominated by dead time but introduces the same uncertainty factor as other sampling controllers encountered. Notice the similarity of the control programs to that of the incremental digital algorithms. Naturally a digital computer can readily be programmed with this optimizing function. Sampling, however, still does not permit its use on processes without self-regulation.

## SUMMARY

In the earlier chapters the point was made that the characteristics of a process determine how well it can be controlled. Furthermore, the settings of the control parameters were shown to depend directly on these properties. But this chapter demonstrates that control improvement is possible, and additional specifications can be satisfied as well, by using more information from the process. This necessitates, however, a deeper understanding of the process than the earlier work. The trend will continue, culminating in application work that is exclusively process-oriented.

The treatise on adaptive control just concluded should verify the value of such an orientation. Although it is possible to design a controller to adapt itself to the process characteristics with no foreknowledge, a controller already equipped with this knowledge is faster, more accurate, and more reliable. These conclusions will be expanded in Chap. 7, where the role of feedback will be largely eclipsed.

## REFERENCES

1. Shinskey, F. G.: Process Control Systems with Variable Structure, *Control Eng.*, August 1974.
2. The Foxboro Company: Signal Distribution Module Model 2AX+DFC, *Spec. Instruc.* 1-00706.
3. Shinskey, F. G.: Adaptive pH Controller Monitors Nonlinear Process, *Control Eng.*, February 1974.
4. Shinskey, F. G.: Adaptive Nonlinear Control System, U.S. Patent 3,794,817, Feb. 26, 1974.
5. Pessen, D. W.: Investigation of a Self-adaptive Three-Mode Controller, *ISA Pap.* 30.3.63.

## PROBLEMS

**6.1**  The reactor in Example 3.5 is to be controlled by setting coolant exit temperature in cascade. Estimate the natural period of the primary loop and the optimum PID settings.

**6.2**  Repeat Prob. 6.1, moving the secondary controller to coolant inlet temperature.

**6.3**  The flow of liquid from a storage tank is to be controlled at a fixed rate as long as the level in the tank is within certain limits. Design a control system to control flow while honoring high and low limits of level.

**6.4**  The heat exchanger described in Figs. 2.7 and 2.8 is under interacting PID control, optimally adjusted as in Table 4.3, at 50 percent flow. At what flow does the loop gain reach 1.0?

**6.5**  Repeat Prob. 6.4, compensating steady-state gain variations through use of an equal-percentage valve. What conclusions can be drawn regarding the relative merits of gain compensation and adaptation of all three modes?

# Chapter Seven

# Feedforward Control

It has been shown that the nature of the process largely determines how well it can be controlled: the proportional band, integral and derivative times, and the period of cycling are all functions of the process. Processes which cannot be controlled well because of their difficult nature are very susceptible to disturbances from load or set-point changes. When a difficult process is expected to respond well to either of these disturbances, feedback control may no longer be satisfactory for three reasons: (1) The nature of feedback implies that there must be a measurable error to generate a restoring force, hence perfect control is unobtainable. In the steady state, the controller output will be proportional to the load. When the load changes, the controller output must change. In going from one output to another, a controller must integrate because in each steady state, proportional and derivative offer no contribution. Consequently the net change in output has been shown to be a function of the integrated error

$$\Delta m = \frac{100}{PI} \int e \, dt$$

Any combination of wide band and long integral time (characteristic of difficult processes) results in a severe integrated error per unit load change

$$\frac{\int e \, dt}{\Delta m} = \frac{PI}{100}$$

This explains why difficult processes are sensitive to disturbances. (2) The feedback controller does not know what its output should be for any given set of conditions, and so it changes its output until measurement and set point are in agreement; it solves the control problem by trial and error, which is characteristic of the oscillatory response of a feedback loop. This is the most primitive method of problem solving. (3) Any feedback loop has a characteristic natural period. Should disturbances occur at intervals less than about three periods, it is evident that no steady state will ever be reached.

There is a way of solving the control problem directly, called *feedforward control.* The principal factors affecting the process are measured and, along with the set point, are used in computing the correct output to meet current conditions. Whenever a disturbance occurs, corrective action starts immediately, to cancel the disturbance before it affects the controlled variable. Feedforward is theoretically capable of perfect control, notwithstanding the difficulty of the process, its performance being limited only by the accuracy of the measurements and computations.

Figure 7.1 is a simplified diagram illustrating the arrangement of the feedforward control system as it has been described. Its essential feature is the forward flow of information. The controlled variable is *not* used by the system, because this would constitute feedback; this point is important because it shows how it is possible to control a variable without having a continuous measurement of it available. A set point is essential, however, because any control system needs a "command" to give it direction.

Although a single controlled variable is indicated in the figure, any number can be accommodated in one feedforward system. Three forward loops are shown, to suggest that all the components of load which significantly affect a controlled variable can be used in solving for the manipulated variable. Although their configuration differs from the commonly recognized feedback loop, these loops are truly closed. Therefore feedforward control should not be construed as merely an elaborate form of programmed or open-loop control.

## THE CONTROL SYSTEM AS A MODEL OF THE PROCESS

In practice, the feedforward control system continually *balances* the *material* or *energy* delivered to the process against the demands of the load. Consequently the computations made by the control system are material and energy balances

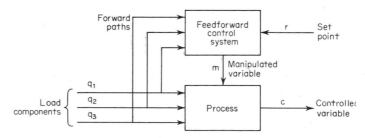

**FIG. 7.1**  The control system embodies a forward flow of information.

on the process, and the manipulated variables must therefore be accurately regulated flow rates. An example is the balancing of firing rate vs. thermal power that is being withdrawn as steam from a boiler. Some material and energy are inevitably stored within the process, the content of which will change in passing from one state to another. This change in storage means a momentary release or absorption of energy or material, which can produce a transient in the controlled variable unless it is accounted for in the calculations.

To be complete, then, the control computer should be programmed to maintain the process balance in the steady state and also in transient intervals between steady states. It must consist of both steady-state and dynamic components, like the process: it is, in effect, a model of the process. If the steady-state calculations are correct, the controlled variable will be at the set point as long as the load is steady, whatever its current value. If the calculations are in error, an offset will result, which may change with load. If no dynamic calculations are made, or if they are incorrect, the measurement will deviate from the set point while the load is changing, and for some time thereafter, while new energy levels are being established in the process. If both the steady-state and dynamic calculations are perfect, the process will be continually in balance and no deviation will be measurable at any time. This is the ultimate goal.

In the procedure followed in the design of a feedforward system, the process model is reversed. The manipulated variables are solved for in terms of load components and controlled variables, but where the controlled variables appear in the equations set points are used instead. It is the intent of a feedforward system to force the process to respond as it was designed, to follow the set points as directed without regard to load upsets.

### Systems for Liquid Level and Pressure

In Chap. 3, a distinction was made between variables which are integrals of flow and those which are properties of a flowing stream. This distinction takes on added significance now, being reflected in the configuration of the feedforward system. Load is a flow term, of which liquid level and pressure are integrals. Therefore feedforward calculations for liquid level and pressure are generally linear. But where a property of the flowing stream, such as temperature or composition, is to be controlled, the system will be found nonlinear in appearance.

In general, liquid-level and pressure processes appear mathematically as

$$\tau \frac{dc}{dt} = mK_m\mathbf{g}_m - qK_q\mathbf{g}_q \tag{7.1}$$

The terms $K_m$, $\mathbf{g}_m$, $K_q$, and $\mathbf{g}_q$ represent the steady-state and dynamic-gain terms for manipulated variable and load. The feedforward control system is to be designed to solve for $m$, substituting $r$ for $c$

$$m = \frac{\tau \, dr/dt + qK_q\mathbf{g}_q}{K_m\mathbf{g}_m}$$

Since $dr/dt$ is normally zero,

$$m = q \, \frac{K_q \mathbf{g}_q}{K_m \mathbf{g}_m} \tag{7.2}$$

Feedforward is commonly applied to level control in a drum boiler. Because of the low time constant of the drum, level control is sensitive to rapid load changes. In addition, constant turbulence prevents the use of a narrow proportional band because this would cause unacceptable variations in feedwater flow. The feedforward system simply manipulates feedwater flow to equal the rate of steam being withdrawn, since this represents the load on drum level. The system is shown in Fig. 7.2.

If the two flowmeters have identical scales, which is to be expected, the ratio $K_q/K_m$ of Eq. (7.2) is 1.0. Furthermore, the dynamic elements $\mathbf{g}_q$ and $\mathbf{g}_m$ are virtually nonexistent. The control system then simply solves the equation

$$W_F = W_s + m_L - 0.5$$

The terms $W_F$ and $W_s$ are mass flows of feedwater and steam, respectively; $m_L$ is the output of the level controller, whose normal value is 0.5.

It must be remembered that liquid-level processes such as this are non-self-regulating. The controlled variable will consequently drift unless feedback is applied. Since integral feedback may not be used alone, because instability would result, a PI controller is always used. In the steady state, feedwater flow will always equal steam flow, so the output of the level controller will seek the bias applied to the computation. If the controller is to be operated at about 50 percent output, that bias must be 0.5, as indicated in the formula. The controller does not have to integrate its output to the entire extent of the load change with a forward loop in service but need only trim out the change in error of the computation during that interval.

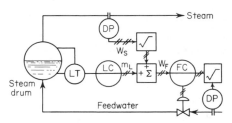

**FIG. 7.2** Feedwater flow is set equal to steam flow in a drum boiler.

This feedforward system has two principal advantages: (1) feedwater flow does not change faster or farther than steam flow, and (2) control of liquid level does not hinge upon tight settings of the feedback controller.

Because this feedforward system, like many, is based on a material balance, accurate manipulation of feedwater flow is paramount. In general, the output of a feedforward system is the set point for a cascade flow loop and does not go directly to a valve. Valve position is not a sufficiently accurate representation of flow.

## Systems for Temperature and Composition

Temperature and composition are both properties of a flowing stream. Heat and material balances involve multiplication of these variables by flow, producing a characteristic nonlinear process model. Feedforward systems for control of these variables are similarly characterized by multiplication and division. The general form of process model for these applications is

$$c = K_p \frac{m\mathbf{g}_m}{q\mathbf{g}_q} \tag{7.3}$$

A single coefficient $K_p$ is sufficient to identify the steady-state gain.

The feedforward equation to control this general process is simply the solution for $m$, replacing $c$ with $r$

$$m = \frac{rq\mathbf{g}_q}{K_p\mathbf{g}_m} \tag{7.4}$$

Notice that the manipulated variable is affected equally by the load and set point, which are multiplied. In level and pressure processes, the set point is added and contributes little to the forward loop.

Because temperature and composition measurements are both subject to dead-time and multiple lags, they are relatively difficult to control. As a result, it is perfectly reasonable to expect that feedforward can be more readily justified in these applications. But along with the need, there likewise exists the problem of defining these processes well enough to use computing control. In addition, nonlinear operations and dynamic characterization are required. Yet multipliers and dividers did not come into common use in control systems until about 1960. It is easy to understand, therefore, why level control was perhaps the first but hardly the most significant application of the feedforward principle.

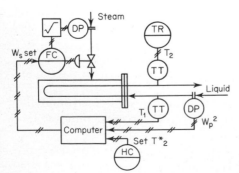

**FIG. 7.3**  The feedforward control system calculates the correct steam flow to match the heat load.

## Application to a Heat Exchanger

The most easily understood demonstration of feedforward is in the control of a heat exchanger. The computation is a heat balance, where the correct supply of heat is calculated to match the measured load. The process is pictured in Fig.

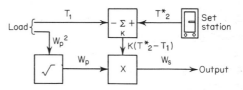

**FIG. 7.4**  Three computing elements and a set station provide the steady-state heat balance.

7.3. Steam flow $W_s$ is to be manipulated to heat a variable flow of process fluid $W_p$ from inlet temperature $T_1$ to the desired outlet temperature $T^*_2$.

The steady-state heat balance is readily derived:

$$Q = W_s H_s = W_p C_p (T^*_2 - T_1)$$

where $Q$ = heat-transfer rate
$H_s$ = latent heat of the steam
$C_p$ = heat capacity of the liquid
Solving for the manipulated variable, we get

$$W_s = W_p K (T^*_2 - T_1)$$

The coefficient $K$ combines $C_p/H_s$ with the scaling factors of the two flowmeters and is included as an adjustable constant in the computer; $T^*_2$ is the set point; $W_p$ and $T_1$ are load variables. Witness the multiplication of flow by temperature.

In the control computer shown in Fig. 7.4, the coefficient $K$ is introduced as the gain of the summing amplifier. The measurement of liquid flow is linearized before multiplying; then steam flow must also be linearized, to be compatible with its set point.

Steam flow is begun automatically by increasing both the liquid flow and the set point, since it is proportional to their product. If the exit temperature fails to reach the set point, it indicates that the ratio of steam flow to liquid flow is incorrect. In practice, this ratio is easily corrected by adjusting $K$ until the offset is eliminated. This is the principal calibrating adjustment for the system; it sets the gain of the forward loop. If the system is perfectly accurate, exit temperature will respond to a change in liquid flow as shown in Fig. 7.5.

Two failings of the steady-state control calculation should be noted: (1) each load change is followed by a period of dynamic imbalance, which makes its appearance as a transient temperature error, and (2) the possibility of offset

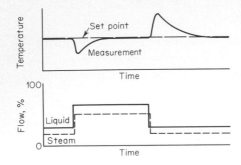

**FIG. 7.5** If the steady-state calculation is correct, temperature will eventually return to the set point following a flow change.

exists at load conditions other than that at which the system was originally calibrated.

On the other hand, the performance of the system exhibits a high level of intelligence. It is inherently stable and possesses strong tendencies toward self-regulation. Should liquid flow be lost for any reason, steam flow will be automatically discontinued. Feedback control systems ordinarily react the other way upon loss of flow, because the measurement of exit temperature is no longer affected by heat input.

The importance of basing control calculations on mass and energy balancing cannot be stressed too highly. First, they are the easiest equations to write for a process, and they ordinarily contain a minimum of unknown variables. Second, they are not subject to change with time. It was not necessary, for example, to know the heat-transfer area or coefficient or the temperature gradient across the heat-exchanger tubes in order to write their control equation. And should the heat-transfer coefficient change, as it surely will with velocity, or fouling, etc., control is unaffected. It may be necessary for the steam valve to open wider to raise the shell pressure in the event of a reduction in heat-transfer coefficient, but steam flow consistent with the heat-balance equation will be maintained nonetheless.

Some unknown factors do exist, however. No allowance was made for losses. If they are significant, and particularly if they change, an offset in exit temperature will result. Steam enthalpy could also vary, as well as the calibration of the steam flowmeter, should upstream pressure change. But for the most part, these factors are readily accountable, whereas heat- and mass-transfer coefficients may not be.

Response to a set-point change will be exponential, appearing as if the loop were open. Moving the set point causes steam flow to move directly to the correct value, i.e., the response that was sought with complementary feedback (see Fig. 4.12).

## RATIO CONTROL SYSTEMS

Ratio control systems are feedforward systems wherein one variable is controlled in ratio to another to satisfy some higher-level objective. For example, additives are controlled in ratio to the principal component in a blend, to maintain the composition of the blend constant. The primary controlled variable is then composition, which is a function of the ratio of the components. However, it cannot always be measured, in which case it cannot be controlled by feedback.

In a ratio control system, then, the true controlled variable is the ratio $R$ of two measured variables $X$ and $Y$

$$R = \frac{X}{Y} \qquad (7.5)$$

Control is usually effected by manipulating a valve influencing one of the variables, while the other is uncontrolled, or "wild." The obvious way to implement the ratio control function is by computing $X/Y$, as shown in Fig. 7.6. But this is not the best way.

Figure 7.6 has a divider within the closed loop, regardless of which variable is affected by the output of the controller. If $X$ is manipulated, loop gain changes with the wild variable $Y$; if $Y$ is manipulated, the loop becomes nonlinear in that the gain changes with controller output

$$\frac{dR}{dX} = \frac{1}{Y} \qquad \frac{dR}{dY} = -\frac{X}{Y^2} = -\frac{R}{Y} \qquad (7.6)$$

All these problems can be overcome by moving the calculation out of the closed loop. Ratio control is then brought about in the set-point circuit, making $r = RY$ if $X$ is controlled or $r = X/R$ if $Y$ is controlled. Figure 7.7 shows the set-point calculation. In this configuration, one of the variables becomes controlled and the other serves to generate a set point. The wild variable is multiplied by the adjustable coefficient $R$ in a device called a *ratio station*.

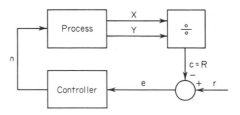

**FIG. 7.6**   Using the ratio of two variables as the input to a controller is not recommended.

## Ratio Flow Control

Most ratio stations have a gain range of about 10:1. Since the normal ratio setting for most applications is in the vicinity of 1.0, this is selected as the midscale position. (This assumes that the flowmeters for the various streams are sized in proper proportion, so that under normal operation they will all read about the same percentage flow.) Thus the gain of a typical ratio station would be adjustable from 0.3 to 3.0.

But most flowmeters are of the differential type, in which case flow squared is the transmitted variable. Then

$$X^2 = R^2Y^2$$

The gain of the ratio station is the square of the ratio setting; in other words,

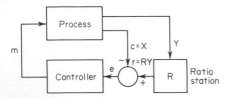

**FIG. 7.7** In the recommended system, the ratio calculation is outside the loop.

the ratio setting is the square root of the gain. The ratio scale for differential meters is compared with a linear scale below:

| Gain $R^2$ | 0.3 | 1.0 | 3.0 |
|---|---|---|---|
| Ratio $R$ | 0.6 | 1.0 | 1.7 |

The ratio stations discussed thus far are nothing more than amplifiers with adjustable gain. The ratio setting was introduced manually. If the ratio setting is to be introduced as a control signal, a different device must be used. The device is, in fact, defined as a multiplier, because its output is the product of two analog signals. The gain of a multiplier is directly proportional to each of its inputs. If a midscale ratio of 1.0 is desired, the gain range of the multiplier must be 0 to 2.0. The multiplier is then said to have an element factor of 2.0. Linear and differential flowmeter ratio scales for the multiplier are compared below:

| Linear flow ratio | 0 | 0.5 | 1.0 | 1.5 | 2.0 |
|---|---|---|---|---|---|
| Differential flow ratio | 0 | 0.7 | 1.0 | 1.2 | 1.4 |

The use of a multiplier permits setting of the ratio by the output of a primary controller. This is desirable where the primary variable is a function of the ratio of the flowing streams, apart from their total flow. A typical application is the blending of ingredients to form a mixture of controlled composition. Figure 7.8 shows the arrangement of the control loops for a two-component system. Notice that the signals are in the differential form. Although the gain of the cascade flow loop varies inversely with its flow, the gain of the multiplier varies with the wild flow squared. Thus loop gain is

$$\frac{dX}{dR} = \frac{Y^2}{2X} = \frac{X}{2R}$$

Direct variation of gain with flow is much preferable to inverse variation.

### Digital Blending Systems

The use of turbine and positive displacement meters has generated a new type of flow ratio system. These devices are inherently linear and are of a fairly wide range, but, of most significance, they do not produce an analog output. Each rotation of their moving elements produces a discrete number of pulses representative of a particular volume of fluid that has passed through. The pulse rate or frequency is proportional to the flow rate, and the total number of pulses transmitted over a given length of time is a measure of the volume of fluid delivered.

To take advantage of the discrete nature of the measurements, a special type of control system has been developed. In the digital blending system, the volume delivered through each meter is continuously compared with the volume desired for that stream. An error between the two generates a control signal that manipulates the valve in that stream. A master set station transmits a pulse rate that paces the entire system, demanding a certain total flow rate. This pulse rate is multiplied by the ratio setting for each flow loop, thus producing a rate set point for each loop. There the similarity to an analog system ends.

The object of the controller is to maintain the set volume delivery. This allows a temporary reduction of flow in any loop, to be made up to the correct total once its cause has been removed. The basic control mechanism is a digital up-down counter. Set-point pulses cause an increase in the register, while measurement pulses bring about a decrease. The difference between the number of pulses from the two sources is stored and converted into an analog signal, which can be used to drive a valve. The valve would then be driven proportionally to

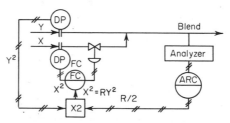

**FIG. 7.8** The ratio of the two flows is automatically set in cascade to maintain composition control.

the volume error or integrated flow error. Thus the control mode is integral. Let the volume error required for 100 percent change in output be identified as $V$ and the maximum flow rate as $F$. The percent output to the valve in terms of percent demanded and measured flow is then

$$m = \frac{1}{V} \int F(r - c) \, dt \qquad (7.7)$$

Then the time constant of the integrator is

$$I_1 = \frac{V}{F} \qquad (7.8)$$

To be sure, the valve position changes in discrete steps, as does the output of the up-down counter. The steps are about 1 percent, and the valve will attempt

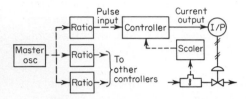

**FIG. 7.9**   The output of the controller is the only analog signal in a typical digital blending system.

to follow each pulse registered by the counter. But because most valves are incapable of following the pulse rate of the counter, which may be as high as 100 pulses/s, delivery is reasonably smooth.

Although integral control is capable of reducing a flow-rate error to zero, integral offset can exist. Examine the flow delivery following startup, as shown in Fig. 7.10. The set flow rate is already at its proper value, but it takes time to bring the measured flow rate to the same value. Eventually there will be no rate error, but in the meantime a significant volume error appears as the area between the curves.

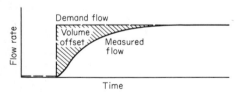

**FIG. 7.10**   An integral control system is subject to volume offset.

The integrated error $E$ required to generate a given controller output is

$$E = \int e \, dt = I_1 m$$

The actual volumetric offset is the percent integrated error $E$ times the maximum flow rate

$$EF = mV \qquad (7.9)$$

Each valve position, hence each flow rate, has a related volume offset.

On the face of each digital control station are two counters. One registers the total demand, while the other registers the total delivered flow. Volume offset appears as a difference of perhaps 5 or 10 between the two counters. In the steady state, the two count rates are identical, and so the volume offset is constant. In a continuous process, volume offset is of little significance, repre-

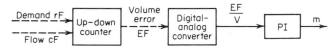

**FIG. 7.11**  A second integration is required to eliminate the volume error.

senting only a few counts out of increasing thousands. But in small batches of only a few hundred counts, it can be cause for concern.

The obvious solution is to add a second integral mode. But double integral by itself is unstable, in that it produces 180° phase lag at all periods. But if the first integral, i.e., the volume error, is acted upon by a proportional-plus-integral controller, the system can be stable. Such an arrangement is functionally described in Fig. 7.11.

Mathematically, the up-down counter totals the difference between the demand and the measured rates, appropriately scaled

$$EF = \int F(r - c)\, dt$$

The D-A converter fixes the range $V$ of volume error $EF$ over which the control valve is driven. Substituting Eq. (7.8) and adding proportional-plus-integral ($I_2$) action gives

$$m = \frac{100}{P}\left(\frac{E}{I_1} + \frac{1}{I_1 I_2}\int E\, dt\right) \tag{7.10}$$

The expanded control equation is

$$m = \frac{100}{PI_1}\left[\int (r - c)\, dt + \frac{1}{I_2}\int\int (r - c)\, dt^2\right] \tag{7.11}$$

Figure 7.12 shows how the second integral works to eliminate volume offset.

Flow loops with conventional controllers, i.e., proportional-plus-integral,

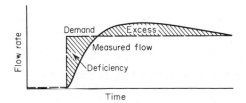

**FIG. 7.12**  The second integral cancels volume deficiency with an equal excess but takes longer to reach a steady state.

have their natural period in the region of 1 to 10 s. The phase contribution of a properly adjusted controller at the natural period is normally 30° or less. But the blending controller contributes about 100° lag, generally doubling the period of a loop. Thus blending loops will oscillate from about a 5- to a 20-s period.

Presence of any valve dead band in combination with the two integrations, will cause a limit cycle, as explained in Chap. 5. Although a valve positioner is not ordinarily recommended for flow, the period of a blending loop is usually far enough removed from that of the position loop for one to be safely used.

## APPLYING DYNAMIC COMPENSATION

The transient deviation of the controlled variable depicted in Fig. 7.5 was attributed to a dynamic imbalance in the process. This characteristic can be assimilated from a number of different aspects.

If the load on the process is defined as the rate of heat transfer, increasing load calls for a greater temperature gradient across the heat-transfer surface. Since the purpose of the control system is to regulate liquid temperature, steam temperature must increase with load. But the steam in the shell of the exchanger is saturated, so that temperature can be increased only by increasing pressure, which is determined by the quantity of steam in the shell. Before the rate of heat transfer can increase, the shell must contain more steam than it did before.

In short, to raise the *rate* of energy transfer, the energy *level* of the process must first be raised. If no attempt is made to add an extra amount of steam to overtly raise the energy level, it will be raised inherently by a temporary reduction in energy withdrawal. This is why exit temperature falls on a load increase.

Conversely, on a load decrease, the energy level of the process must be reduced by a temporary reduction in steam flow beyond what is required for the steady-state balance. Otherwise energy will be released as a transient increase in liquid temperature.

The dynamic response can also be envisioned simply on the basis of the velocity difference between the two inputs of the process, although this is less representative of what actually takes place. The load change appears to arrive at the exit-temperature bulb ahead of a simultaneous steam-flow change. To correct this situation, steam flow must be made to lead liquid flow.

The technique of correcting this transient imbalance is called *dynamic compensation.*

### Determining the Needs of the Process

Capacity and dead time can exist on both the manipulated and the load inputs to the process. There may also be some dynamic elements common to both, such as the lags in the exit-temperature bulb for the heat exchanger. The relative locations of these elements appear as shown in Fig. 7.13.

A feedback controller must contend with $g_m \times g_p$, which are in series in its closed loop. But the feedforward controller need only be concerned with the ratio $g_q/g_m$ in order to make the corrective action arrive at the divider at the same time as the load. Recall the appearance of this ratio in both Eqs. (7.2) and (7.4). In some difficult processes, the manipulated variable enters at the same location as the load, e.g., in a dilution process where all streams enter at the top

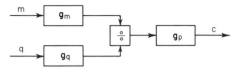

FIG. 7.13 Compensation is needed when the dynamic elements of the two inputs differ.

of a vessel. In this case, even though $g_p$ may be quite complex, $g_m$ and $g_q$ could be nonexistent, making dynamic compensation unnecessary.

Perhaps the easiest way to appreciate the need for dynamic compensation is to consider a process in which $g_q$ and $g_m$ are dead time alone. Let $\tau_q$ and $\tau_m$ represent their respective values. The response of the controlled variable as a function of time is

$$c(t) = K_p \frac{m(t - \tau_m)}{q(t - \tau_q)}$$

The division makes the process fundamentally nonlinear, which complicates dynamic analysis. To allow inspection of the transient response of the process, analysis must be made on an incremental basis, by differentiating both sides of the equation:

$$dc(t) = K_p \left[ \frac{dm(t - \tau_m)}{q} - \frac{m\, dq(t - \tau_q)}{q^2} \right] \tag{7.12}$$

If only a steady-state control calculation is made,

$$m = \frac{rq}{K_p} \tag{7.13}$$

Differentiating gives

$$dm = \frac{1}{K_p}(r\, dq + q\, dr) \tag{7.14}$$

Substituting for $m$ and $dm$ in Eq. (7.12) yields the closed-loop response

$$dc(t) = dr(t - \tau_m) + dq \frac{r}{q}(\tau_q - \tau_m) \tag{7.15}$$

Equation (7.15) shows that the set-point response is delayed by $\tau_m$ and that a

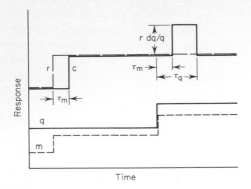

FIG. 7.14  Lack of dynamic compensation produces a transient equal to the difference in dead times.

load change will induce a transient of duration $\tau_q - \tau_m$ and magnitude $r\, dq/q$. Both responses appear in Fig. 7.14.

Of the two, load response is the more important because set-point changes are ordinarily less frequent. Ideally, the load signal should be delayed by $\tau_q$ before it is multiplied, and then advanced by $\tau_m$. It is impossible to create a time advance, however. So dynamic compensation is best introduced in this application by delaying the feedforward signal by an amount $\tau_q - \tau_m$. If $\tau_m > \tau_q$, compensation is impossible.

It has been pointed out that dynamic compensation generally takes the form $\mathbf{g}_q/\mathbf{g}_m$. It may be recalled, however, that the ratio of two vector quantities like these resolves into the ratio of their magnitudes and the difference between their phase angles. Since dead-time elements have unity gain, their ratio is also unity; their only contribution is phase lag. This is why the ratio $\mathbf{g}_q/\mathbf{g}_m$ appears as the difference $\tau_q - \tau_m$ between the dead times.

The complete forward loop, including dynamic compensation, appears in Fig. 7.15. Note the complete cancellation of all elements in the load path by the elements in the forward loop.

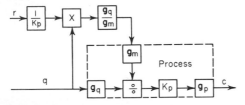

FIG. 7.15  Observe how faithfully the forward loop images the properties of the process.

Because forward loops exhibit absolutely no oscillatory tendencies, to talk of gain and phase is rather inconsequential. Step responses will be used throughout, since they constitute the most severe test of system performance. The response of systems under feedforward control, both with and without dynamic compensation, differs markedly from that experienced with feedback control.

For this reason, it is not surprising that dynamic elements in the forward loop bear little resemblance to the conventional modes of feedback controllers.

Although dead time serves as a useful demonstration of why dynamic compensation is necessary, it rarely appears alone in a process. In fact, multiple lags are most commonly encountered in actual applications. Fortunately, there is usually one dominant lag on each side of the process, which acts as the principal element to be compensated. The response of a process wherein $\mathbf{g}_m$ and $\mathbf{g}_q$ are first-order lags of time constants $\tau_m$ and $\tau_q$, respectively, can be found by substituting their individual response terms into Eq. (7.15). Thus $t - \tau_m$ becomes $1 - e^{-t/\tau_m}$, and $\tau_q - \tau_m$ is replaced with $e^{-t/\tau_q} - e^{-t/\tau_m}$

$$dc(t) = dr(1 - e^{-t/\tau_m}) + dq\,\frac{r}{q}\,(e^{-t/\tau_q} - e^{-t/\tau_m}) \tag{7.16}$$

Figure 7.16 gives both set-point and load-response curves described by this equation, for the case where $\tau_q > \tau_m$. Compare it with the heat-exchanger response, Fig. 7.5, where $\tau_m > \tau_q$.

A qualitative appraisal of the requirement for dynamic compensation can be obtained from a comparison of open-loop response curves. Because an increase in the manipulated variable acts in opposition to the load, their individual step-response curves will diverge. One or the other response will have to be inverted so that the two curves can be superimposed, as is done in Fig. 7.17. The response of such a process under uncompensated feedforward control appears as the difference between these two curves.

If the curves do not cross, the uncompensated forward-loop response will lie wholly on one side of the set point, as in Figs. 7.5 and 7.16. Which side of the set point depends on whether the difference $\mathbf{g}_m - \mathbf{g}_q$ is positive or negative. If the curves cross, the uncompensated forward-loop transient will cross the set point.

### The Lead-Lag Unit

For the bulk of processes to which feedforward control can be applied, the dynamic elements $\mathbf{g}_q$ and $\mathbf{g}_m$ are similar in nature and value. Although dead

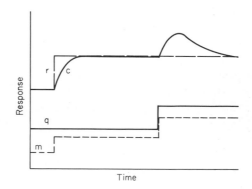

**FIG. 7.16** Lack of dynamic compensation shows up principally as a load-response transient.

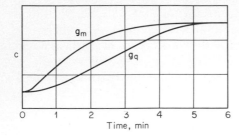

**FIG. 7.17** Comparison of the open-loop response of $c$ to $m$ with the reverse response of $c$ to $q$ shows that $m$ must be made to lag $q$ for this process.

time may be encountered in both, their values are usually close enough to provide nearly complete cancellation. So in most cases, only the dominant lags need to be considered. In addition, the presence of the common element $\mathbf{g}_p$ provides enough attenuation to make exact dynamic compensation unnecessary. Fortunately this allows one dynamic compensator to be used almost universally, the lead-lag unit.

A lead was defined earlier as the inverse of a lag; the lead term to be used here represents $1/\mathbf{g}_m$ and the lag represents $\mathbf{g}_q$. The output $m(t)$ of a lead-lag unit follows a step input $m$ as

$$m(t) = m\left(1 + \frac{\tau_1 - \tau_2}{\tau_2}\, e^{-t/\tau_2}\right) \tag{7.17}$$

where $\tau_1$ is the lead time and $\tau_2$ the lag time; either may be greater, allowing an overshoot or an undershoot, as Fig. 7.18 demonstrates.

The step-response curve reveals an instantaneous gain of $\tau_1/\tau_2$, and recovery to 63 percent of the steady-state value is effected in time $\tau_2$.

Oddly enough, the most stringent specification on a lead-lag unit is steady-state accuracy. If it cannot accurately repeat its input in the steady state, the lead-lag unit degrades the performance of the forward loop. Consequently, linearity, repeatability, and freedom from hysteresis are mandatory, more so than for a conventional controller. In addition, lead and lag times need to be adjustable to match the time constants of most processes.

Exact compensation may be impractical for very slow processes, particularly those with electronic components, because of impedance limitations. Pneumatic devices have a greater potential range in this respect, because extremely large-capacity tanks can be used without danger of leakage. References 1 and 2 describe several arrangements for obtaining lead-lag functions using standard pneumatic control components.

Digitally, the lead-lag function can be realized with a simple iterative procedure. To demonstrate this procedure, $x$ will represent the input, $y$ the input lagged by $\tau_2$, and $z$ will be $y$ led by $\tau_1$. The differential equations are

$$x = y + \tau_2 \frac{dy}{dt} \qquad z = y + \tau_1 \frac{dy}{dt}$$

Rearranging, we get

$$z = x + (\tau_1 - \tau_2)\frac{dy}{dt} \qquad \frac{dy}{dt} = \frac{1}{\tau_2}(x - y)$$

Digital computers are sampling devices, repeating their calculations at some regular interval $\Delta t$. Therefore the differentials above must be rewritten as difference equations. First the current value of $z$ can be calculated at each interval, from current values $x$ and $y$

$$z_n = x_n + \frac{\tau_1 - \tau_2}{\tau_2}(x_n - y_n) \qquad (7.18)$$

But $y$ must be incremented before the next calculation can be made

$$y_{n+1} = y_n + \frac{\Delta t}{\tau_2}(x_n - y_n) \qquad (7.19)$$

Unfortunately, (7.19) tends to produce a steady-state error when $\Delta t/\tau_2$ becomes a low number. This error can be eliminated, although with some loss in dynamic accuracy, by resolving the differential equation into another form

$$y_n = x_n - \frac{\tau_2}{\tau_2 + \Delta t}(x_n - y_{n-1}) \qquad (7.20)$$

A lead-lag function with dominant lag can also be generated using a PI controller in a closed loop. The input signal is entered as set point, while the output is fed back as the controlled variable. In this configuration, the output responds to the input with a lead equal to the integral time and a lag augmented by the proportional band

$$\tau_1 = I \qquad \tau_2 = I\left(1 + \frac{P}{100}\right) \qquad (7.21)$$

Exceptionally long time constants can be achieved by operating the controller for only a fraction of real time, using a repeat-cycle timer to transfer it between automatic and manual. It can also be used to hold its last output by transfer to

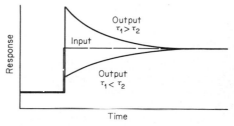

**FIG. 7.18** The lead-lag unit can be made to overshoot or undershoot a step input.

manual on a signal failure. It is especially useful on feedforward input signals from analyzers which are prone to failure and require frequent maintenance.

### Adjusting the Dynamic Terms

The lead-lag function enables the delivery of more (or less) energy or mass to the process to alter its potential during a load change. The integrated area between its input and output should match the area of the transient in the uncompensated response curve. If this is done, the net area of the response will then be zero.

The integrated area between input and output of the lead-lag unit can be found from Eq. (7.17). First the difference between input and output should be normalized by dividing by the input magnitude

$$\frac{m(t) - m}{m} = \frac{\tau_1 - \tau_2}{\tau_2} e^{-t/\tau_2}$$

Integrating over the limits of zero to infinity yields

$$\int_0^\infty \frac{m(t) - m}{m} \, dt = \tau_1 - \tau_2 \tag{7.22}$$

The normalized integrated area of the uncompensated loop response of Eq. (7.16) and Fig. 7.16 is similar

$$\int_0^\infty (e^{-t/\tau_q} - e^{-t/\tau_m}) \, dt = \tau_m - \tau_q \tag{7.23}$$

This is further proof that $\tau_1$ should equal $\tau_m$ and $\tau_2$ should equal $\tau_q$.

Area alone is an insufficient index of proper compensation. A lead of 10 min and a lag of 9 min would produce the same area as a lead of 2 min and a lag of 1 min, but that area would be distributed differently. The location of the transient peak of the uncompensated response can be of help in estimating the actual values of $\tau_1$ and $\tau_2$. Let $\tau_1$ and $\tau_2$ be substituted for $\tau_m$ and $\tau_q$ in Eq. (7.16). By differentiating and then equating to zero, the time $t_p$ of the maximum (or minimum) can be found

$$t_p = \frac{1}{1/\tau_2 - 1/\tau_1} \ln \frac{\tau_1}{\tau_2} \tag{7.24}$$

A plot of this relationship is given in Fig. 7.19.

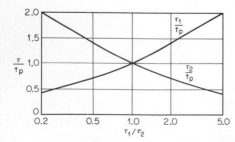

FIG. 7.19 The location of the peak in the uncompensated response transient can be used to infer the required compensation.

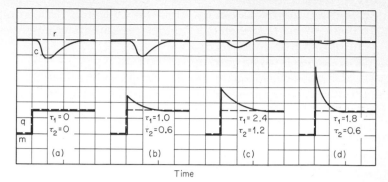

**FIG. 7.20** Curve (*a*) is the uncompensated loop response; curves (*b*) and (*c*) reveal incorrect compensation; curve (*d*) is nearly perfect.

Equation (7.24), like (7.23), is a single relationship containing two unknowns. Yet some definite conclusions can be drawn which will be useful in making preliminary adjustments: (1) If $\tau_1$ must exceed $\tau_2$, based upon the *direction* of the uncompensated transient, $\tau_2$ can safely be set in the vicinity of $0.7t_p$. If $\tau_1$ must be less than $\tau_2$, $\tau_2$ should be about $1.5t_p$. (2) Initially, $\tau_1$ can be set at about $2\tau_2$ in the former case or $0.5\tau_2$ in the latter.

Once these preliminary adjustments have been introduced, the load response should be repeated, from which finer adjustments can be made. Figure 7.20 compares the load response, having varying degrees of compensation, with the uncompensated response of a typical process.

Notice that curve (*b*) in Fig. 7.20 lacks area compensation. Curve (*c*), on the other hand, shows adequate compensation with respect to area, in that it is distributed equally about the set point. In this case, the difference between $\tau_1$ and $\tau_2$ is correct, but their individual values are not. Once the correct area compensation has been found, $\tau_1$ and $\tau_2$ should *both* be adjusted in the same direction, to maintain their difference. In the example shown in Fig. 7.20, $\tau_1$ and $\tau_2$ need to be reduced; this will make their ratio increase, which will move the centroid of the lead-lag area to the left. Curve (*d*) is the result of such an adjustment: it crosses the set point at approximately time $t_p$.

Perfect compensation is unattainable. For one thing, any process sufficiently problematic to warrant feedforward control can be expected to display some dead time in addition to capacity. This is true of the heat exchanger. But compensation for dead time is, at best, approximate. Second, the dynamics of most processes are subject to change. This is also true for the heat exchanger, whose dead time varies with the rate of flow through the tubes.

It might be possible to construct an extremely complete dynamic model of a process, but any compensator with more than three terms to adjust would be unreasonably difficult to cope with. Furthermore, the purpose of dynamic compensation is to minimize an error which is already transient, so perfection is not really warranted. In most cases, a simple lead-lag function will be perfectly adequate and will be able to reduce the absolute area of the response curve by tenfold or more, distributed uniformly. Figure 7.21 compares the load response of the heat exchanger under dynamically compensated feedforward control with that under feedback control.

For processes whose uncompensated load-response curves cross the set point, lead-lag may be inadequate. For these applications, an additional lag is useful in canceling the first part of the curve, while the lead-lag function compensates the balance.

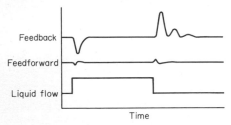

**FIG. 7.21** Feedforward control is capable of reducing both the area and the duration of the load response transients.

## ADDING FEEDBACK

The only serious failing of the feedforward technique is its dependency on accuracy. To provide perfect control, a system must model the plant exactly; otherwise whatever error may exist in positioning the manipulated variable causes offset. Errors may arise from several sources:

1. Inaccuracy in the measurement of load and manipulated variables
2. Errors in the computing components
3. Failure of the computing system to represent the characteristics of the plant adequately
4. Exclusion of significant load components from the feedforward system

The first and second items alone limit the accuracy of practical systems to the vicinity of 1 to 2 percent [1].

Some processes, such as the heat exchanger described earlier, are easy to model. But this is not always the case, particularly when mass- and heat-transfer coefficients must be used over a wide range of operating conditions. Therefore item 3 may be of considerable importance in the more complicated processes.

The feedforward system cannot be all-inclusive. Some load components are so slight, or invariant, or ill defined that their inclusion is not warranted. If payout and ease of operation are important, which is to be expected, the control system had best remain simple. To this extent, certain terms such as heat losses and ambient-temperature effects are usually neglected. Yet their variation can induce a measurable offset.

If offset is intolerable, some means must be provided for recalibration while the system is operating. In general, this can be done most directly by readjusting the set point, which is already scaled in terms of the controlled variable. Other adjustments could be made, such as the coefficient $K$ in the heat-exchanger control system, but with less predictable results.

## The Role of Feedback

Regardless which term is trimmed to remove offset, the procedure amounts to manual feedback. Automatic feedback is perfectly capable of effecting the same result if the controlled variable can be measured with sufficient reliability. (This qualification is significant in that feedforward control is occasionally used because a feedback measurement is unavailable.)

Proportional feedback trim is insufficient to eliminate offset, for the same reason that it was insufficient in a conventional loop. The presence of feedforward control components within the feedback loop induces no substantial change in the mode settings required of the feedback controller. The process is just as *difficult* to control as it was without feedforward; only the *amount* of work required of the feedback controller has been markedly reduced.

Integration, then, is necessary if offset is to be eliminated altogether. Whether proportional and derivative are useful modes depends on the nature of the process. If rapid load changes outside the forward loop may be encountered, proportional and derivative action could be advantageous. If the process is fundamentally non-self-regulating, as in level control, proportional action is essential. Finally, if the process is fairly easy to control because of the absence of dead time, derivative may be useful in improving the dynamic load response, but this is unusual.

In general, since feedforward control is warranted only on the most demanding and most difficult applications, integral is the only useful feedback mode. Experiments conducted on a heat exchanger, which is not particularly difficult to control, indicated that proportional and derivative feedback modes responded too slowly to contribute anything to the load response of the system, either with or without dynamic compensation. On the other hand, feedback can be detrimental by promoting oscillation in an otherwise stable system. To this extent, the settings of the feedback controller, regardless of what modes have been selected, ought to be relaxed.

## Where to Introduce Feedback

This is not always an easy question to resolve. The feedback controller may be asked to perform a number of different services. In the heat-exchanger application it can be useful in correcting for heat loss, in which employment it should add an increment of heat to the process at all loads; this would amount to a zero adjustment. Or its principal function might be to correct for variable steam enthalpy, in which case it should apply a span adjustment by setting the coefficient $K$. In another process, linearity could be the largest unknown factor. But a single feedback controller can hardly be called upon to do all these things.

When no one source of offset is outstanding, the argument for readjusting the set point of the feedforward system is irresistable. Figure 7.22 shows how the feedback controller would fit into this arrangement.

Feedback is added simply by replacing the manual set station with a feedback controller; no additional computing elements are needed. Startup may proceed in orderly fashion in manual, simply by setting the output of the feedback controller equal to its set point. No guesswork is involved, since this is the known operating region for the system. When in automatic, there is ultimately only one set-point signal because the controller output is now a variable.

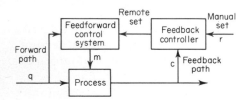

**FIG. 7.22**   In general, the feedback controller should reposition the set point of the feedforward system.

Another feature of this configuration is that it displays the inherent inaccuracies of the forward loop. The difference between the set point and the output of the feedback controller is the offset which would have appeared if feedback had not been used. A plot of controller output vs. load could conceivably identify the principal sources of error in the computing system. Bear in mind, however, that the amount of offset $dc$ resulting from an error in the manipulated variable $dm$ varies inversely with load

$$dc = K_p \frac{dm}{q}$$

Bristol [3] assembled an adaptive system in which the feedback controller alternately adjusted the gain and bias to calibrate a feedforward system on-line. Gain was selected for adjustment on increasing loads, and bias on decreasing loads. Eventually the system would converge on a solution.

Although feedforward systems are designed primarily for regulation, some rules regarding set-point changes are noteworthy. To avoid the usual dynamic problems associated with feedback loops, the feedback controller should be placed in manual during set-point changes. In the case where this is a frequent occurrence, set-point changes can be introduced in automatic if sent directly to the forward loop; this requires addition of the set point to the output of the controller, as shown in Fig. 7.23.

It is impossible, of course, for the process to respond instantaneously to a step in set point. Since the controlled variable will lag behind the set point, a positive error will develop before the new set point is reached. The feedback controller, being in automatic, will integrate the error, changing its output to a new but incorrect value; it must then bring its output back to the previous state by generating a negative error, equal in area to the earlier positive error. The effect is the same as that shown in Fig. 7.12, produced by the blending control system.

This situation can be remedied simply by inserting a lag in the set point to the feedback controller (but not to the multiplier), as Fig. 7.23 illustrates. The lag should be adjusted to prevent the overshoot that would be realized without it.

The feedback controller will always equalize the integrated error promoted by any disturbance entering a forward loop. For this reason, it should remain in manual while dynamic compensation is being adjusted.

Another important consideration when adding feedback is the location of the dynamic compensator. Although lead-lag can be beneficial to the response of a feedback loop, it interferes with manual operation. When the output of the controller is changed manually, the operator likes to see that action reproduced exactly by the manipulated variable. With a lag or lead-lag between the controller and the manipulated variable, several minutes—possibly even an hour—could elapse before the effect of the adjustment is complete. Therefore it is mandatory to arrange the system so that dynamic compensation is out of the feedback loop.

In theory, each forward loop should have its own dynamic compensator, but ordinarily only flow inputs can change fast enough to warrant dynamic compensation. Observe the location of the lead-lag unit in the system with two forward loops and one feedback loop, which is shown in Fig. 7.24.

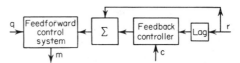

**FIG. 7.23**   If set-point response is important, the set point should go directly into the forward loop.

### Mutual Adaptation

An adaptive control system was defined as having the ability to change its parameters in accordance with the changing character of the process. A feedforward control system, by itself, can only generate an output relative to known and measurable inputs as prescribed by a fixed program. Some factors relating to the process may be unknown and variable. For optimum performance, the feedforward system ought to be supplied with information regarding these unknowns. A feedback controller, on the other hand, is geared to solve for unknowns. So the inclusion of a feedback signal in a forward loop actually adapts the forward loop to unmeasured changes in the process.

Remarkably enough, the feedforward system also adapts the feedback loop to variations in process gain. Figure 7.21 shows the load response of the cited heat exchanger under feedback control. With increasing load, the transient is overdamped. With decreasing load, the transient is greater and underdamped, indicating that the process gain changes inversely with liquid flow. (This characteristic was discussed under Variable Dynamic Gain in Chap. 2, and again under Dynamic Adaptive Systems in Chap. 6.) Since the process gain varies inversely with flow, the controller gain ought to vary directly with flow. The complete control system for the heat exchanger, Fig. 7.24, illustrates how this is brought about.

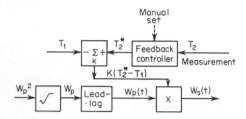

**FIG. 7.24**  The complete control system for the heat exchanger includes feedback and dynamic compensation.

The feedback loop sees $T_2$ as its input and $W_s$ as its output. But within the loop $T_1$ is subtracted from, and $W_p$ multiplied by, the controller output. Subtraction is a linear operation, so gain is not changed therein; but multiplication is nonlinear, causing feedback gain to vary directly with flow $W_p$. Correct loop-gain adaptation cannot be achieved if the feedback is introduced in *any* other place. If the output of the feedback controller were to set $K$, feedback gain would vary both with $W_p$ and with $T^*_2 - T_1$. But process gain does not vary with $T^*_2 - T_1$. This is another argument in favor of using feedback to trim the feedforward set point.

This mutual adaptation is a further indication how perfectly feedforward and feedback complement each other. Feedforward is fast, intelligent, and responsive but also inaccurate; feedback is slow but accurate and is capable of regulation in the face of unknown load conditions.

## ECONOMIC CONSIDERATIONS

Feedback control can be enforced on the heat exchanger using only three elements—transmitter, controller, and valve. Adding feedforward control requires another temperature transmitter, two flow transmitters, two square-root extractors, a steam-flow controller, a summer, a multiplier, and a lead-lag unit—nine items. Such an expense must be justified.

Although the heat exchanger serves as a convenient demonstration of feedforward control, only in rare instances would such refined control be justified for this type of application. Management likes to see investments such as advanced control systems pay for themselves in less than 2 years. If improved temperature control had no value, there would be no payout. But if it pro-longed the life of the exchanger, or saved steam, or reduced maintenance, it would have a measurable worth.

Economic incentive for improved control is most likely to be found where consistent quality of a valuable product is important. Many unit operations are conducted at reduced capacity or low recovery, or use excessive amounts of utilities to ensure that quality will surpass specifications even with poor control. Naturally, the rate of payout from these sources will vary directly with the production rate of the plant. Large plants therefore encounter reduced risks.

But the payout of a control system is as much a factor of cost as of savings. A control system can be too perfect; i.e., its cost can be out of proportion to the job to be done. In the control system for the heat exchanger, for example, both flow and inlet temperature were accepted as load changes. If inlet temperature was nearly constant or subject only to slow variations, it could be eliminated from the system as long as feedback trim was available. This would reduce cost by saving a transmitter and a summer.

There are three principal areas where feedforward control can produce results unobtainable with any other technique:

1. A difficult process subject to frequent disturbances may never settle out under feedback control. Load changes occurring at intervals of less than three cycles are sufficient to develop this situation. This is not uncommon on some towers, furnaces, and coupled processing units which may cycle at periods exceeding an hour.

2. In many plants, the variable of interest cannot be measured continuously, accurately, or quickly enough for adequate feedback control. Often a secondary variable is used as an inferential measure of the first, simply because it is the best available. In a case like this, however, product quality may suffer both from poor control of the inferential variable and from its indifferent relationship to the primary variable.

3. More interest is developing in the control of economic variables: cost, debits, yield, etc. These variables are not directly measured, and often not computed, to allow the closing of the economic loop. But even if they were available, conventional feedback control could not be used, because the intent is to maximize or minimize their value rather than to control at a given set point.

## Optimizing Programs

Feedforward control systems are not limited solely to regulatory duties. In fact, the controlled variable may be easily programmed with respect to any measured term, simply by making the appropriate substitution in the process equation. If, for some reason, we wanted to vary the temperature of the liquid leaving the heat exchanger relative to its flow, this could readily be done. To accomplish it the computing system would become somewhat different but no more difficult to implement than for a constant set point. If feedback trim were used, however, the set point to the feedback controller would have to be similarly programmed, which might increase the complexity of the system significantly.

An important observation is that *any* control program can be followed, even one that would cause the process gain $dc/dm$ to change sign. Hence feedforward is the logical means of optimizing (steady-state adaptive) control. This has already been demonstrated in Chap. 6.

The first step in optimizing is the definition of the sources of loss within the plant. The process will generate valuable product somewhat proportional to the rate of material pumped into it, with the result that the profit can be subject to wide variation. If the losses are maintained at a minimum, however, the highest profit will always result and values and rates of raw materials can be ignored.

Losses, or debits, are not difficult to define. They consist principally of utility costs, streams of unrecoverable materials, and lowered market value brought about by contaminants in the product. Debits peculiar to a process may vary with feed composition, catalyst activity, atmospheric conditions, and demands elsewhere within the plant, but feed rate generally exercises the greatest influence upon them. The purpose of the computer in this domain is to manipulate those variables which can best offset the influence of the above uncontrolled variables on the plant economy.

Complicated as all this seems to be, often a fairly simple relation can be derived in an effort to optimize part of the plant or to partially optimize the whole plant. As an example, consider a simplified absorption tower where a gas stream $F$ containing $z$ percent of a valuable material is absorbed by a liquid stream $L$, which is to be manipulated to maintain minimum-cost operation. Assume that gas-exit composition $y$ varies as

$$y = kz\frac{F}{L}$$

where $k$ is the absorption-rate coefficient.

The principal debits $l$ associated with such an operation might be losses of valuable material in the exit gas and costs of processing the absorbent

$$l = v_1 Fy + v_2 L$$

where $v_1$ is the product value and $v_2$ is the processing cost. The debit equation can be rewritten on the basis of independent variables alone

$$l = v_1 kz\frac{F^2}{L} + v_2 L$$

The minimum point on a curve of debits vs. $L$ would occur where the slope $dl/dL$ is zero

$$\frac{dl}{dL} = -v_1 kz\frac{F^2}{L^2} + v_2 = 0$$

This defines the locus of optimum $L$

$$L_{opt}^2 = v_1 kz \frac{F^2}{v_2}$$

Having solved the minimum-cost equation, it is only necessary to build a computer which will program $L^2$ to follow variations in $F^2$ and $z$ in accordance with current figures of $v_1$ and $v_2$. (The manipulated variable has been left in the form $L^2$ because flow rate is most commonly measured by a differential meter.) Adjustable coefficients should be available for perfecting the model in actual operation. Note that if no solution exists to the derivative, the process exhibits no minimum. This would be the case if $v_2$ were zero.

Often such rigorous and simple expressions cannot be established. Shotgun patterns of data sometimes must be analyzed, involving all combinations of wild and manipulated variables until some measure of relationship can be envisioned. This usually culminates in contour plots like Fig. 7.25. Contours of debit are shown as a function of a wild and a manipulated variable. If such a plot can be made, a line can be drawn across it representing the locus of minimum debit. A model of this line can then be made to program the manipulated variable as a function of the wild variable. Again, adjustable coefficients may be incorporated to perfect the model inasmuch as some doubt always accompanies relationships derived from real plant data.

Normally the operating conditions of any plant are surrounded by constraints and limitations. It is not surprising to learn that the optimum conditions for many plants lie outside equipment limitations. Many applications would not result in enough remuneration to pay for the computer or the engineering involved. These two facts severely restrict the number of processes that could benefit by optimizing control.

The total debits which can be expected to be recovered from a given operation with a given computer divided by the installed cost of the computer is the payout in percent per year. A simple analog like the one just discussed cannot be expected to recover all the existing debits in a process, though 50 to 75 percent recovery should not be difficult to realize.

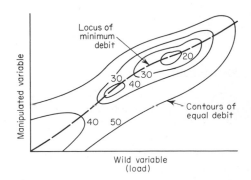

**FIG. 7.25**  The locus of minimum cost can be drawn across a contour plot such as this.

A more complex system designed to a more exact model might be able to recoup an additional 10 to 20 percent but at perhaps twice the installed cost. It is easy to see that the simplest optimizing computer will nearly always result in the greatest rate of payout.

## SUMMARY

There is absolutely no question that feedforward is the most powerful technique that has ever been brought to bear in the regulation of difficult processes. Where certain unusual feedback modes like complementary feedback, sampling, and nonlinear functions were able to reduce the integrated error per unit load change perhaps twofold, feedforward is capable of a hundredfold improvement [1]. A feedforward system would only have to be accurate to ±10 percent to provide tenfold improvement, in conjunction with feedback.

But nothing of great value is ever gained without cost. The cost in this instance is the process knowledge which must go into the design of every system. This precludes the mass production of fully adjustable feedforward systems. The only feedforward systems that will look alike are those being applied to like processes.

Although the feedforward loop carries most of the load, feedback is still important in its execution. The ultimate regulatory system is a triple cascade, the primary feedback controller trimming the set point of the feedforward loop, which in turn manipulates a cascade flow loop. The symmetry of such a configuration is striking.

The basic tenents of the technique were covered in this chapter with examples chosen for illustration rather than practical value. But in each of the chapters that follow, feedforward will be applied to those processes which would most benefit from superior load regulation. Optimizing systems, where practicable, will also be given due consideration.

## REFERENCES

1. Shinskey, F. G.: Feedforward Control Applied, *ISA J.,* November 1963.
2. Samson, J. E.: Improvements in or Relating to Automatic Force Balance Apparatus, Br. Patent 860,485, Feb. 8, 1961.
3. Bristol, E. H.: A Simple Adaptive System for Industrial Control, *Instrum. Technol.,* June 1967.

## PROBLEMS

**7.1**  In the heat-exchanger control system, equal flow rates of steam and liquid in percent of full scale raise the liquid temperature 100°F. Under these conditions, what offset would be induced by a 2 percent error in steam flow at 50 percent and at 25 percent flow?

**7.2**  Write the equation for the feedforward system which controls exit-gas composition $y$ as it leaves the absorber described under the section on optimizing programs. How does this differ from the optimizing program solved in the example?

**7.3** Estimate the peak location of the uncompensated response curve which would result from feedforward control of the process whose dynamic characteristics are given in Fig. 7.17. What settings of lead and lag will be necessary for dynamic compensation?

**7.4** Given a process whose dynamics consist of first-order lags, $\tau_m = 3$ min and $\tau_q = 1$ min. For conditions of $r$ and $q$, both at 50 percent, what is the integrated area per unit load change with an uncompensated forward loop? What is the integrated area with lead-lag compensation if the lead time is 2.5 min and the lag time is 1 min?

**7.5** Control of a given dead-time-plus-integrating process can be improved (*a*) by adding derivative to the existing two-mode controller (interacting) or (*b*) by investing in a noninteracting controller or (*c*) by adding a simple forward loop. In the last case, the feedforward system is likely to be accurate only to ±10 percent. What is the improvement in integrated area per unit load change for each of the three cases? What would be the effect of variable dead time in each case?

# Chapter Eight

# Interaction and Decoupling

Until now, discussion has been confined to control systems with a single manipulated variable. Furthermore, only one controlled variable has been allowed to be independently specified. But any process capable of manufacturing or refining a product cannot do so within a single control loop. In fact each unit operation requires control over at least two variables, product rate and quality.

Whenever two control loops are to be placed in operation on a single process, the question arises: Which valve should be manipulated from which measurement? In some cases the answer will be obvious. But when it is not, some basis must be available to permit the correct decision to be made. In every case these variables interact with each other to some extent, which naturally interferes with their individual control.

The most effective arrangement of control loops cannot be determined without an appreciation for the needs of the process. In this chapter, a procedure is developed for determining the relative response of each controlled variable in a process to each manipulated variable, to guide the designer in configuring his system. The method additionally characterizes the magnitude and type of interaction to be expected and its effect on control-loop performance. Finally, techniques are described for decoupling those variables which interact with particular severity, along with a method for evaluating their effectiveness.

## RELATIVE GAIN

The first step in applying controls to a process with a single controlled and manipulated variable is to evaluate its open-loop gain. The same procedure is followed with multivariable processes, although the dimensionality of the problem expands. In a process with two pairs of controlled and manipulated variables, there are four open-loop gains to be considered. Although only two will be overtly closed within control loops, the choice which two they will be must be made. While in some cases the choice may be obvious, in other situations it will not, and the designer has a 50-50 chance of making a poor selection. In a process with three pairs of variables, the designer's odds at guessing right are reduced to 1 in 6 possible configurations. To eliminate the need for guesswork and provide the designer with a measure of multivariable system performance, the *relative-gain method* was developed.

### Two Open-Loop Gains

Relative gain is a measure of the influence a selected manipulated variable has over a particular controlled variable relative to that of other manipulated variables acting on the process. It is not enough to calculate its open-loop gain with all other manipulated variables fixed. If there is interaction in the process, each manipulated variable will affect more than one controlled variable. Then the response of a particular controlled variable to a selected manipulated variable will depend on what the other manipulated variables are doing.

Bristol [1] suggested that each open-loop gain should be evaluated first with all other loops open, i.e., all other manipulated variables constant, and then reevaluated with all other loops closed, i.e., with all other controlled variables constant. If a particular open-loop gain does not change between these two conditions, it is not affected by the action of the other loops and therefore does not interact with them. In this instance, its *relative gain* would be unity. By this token, the relative gain of a controlled variable $i$ to a manipulated variable $j$ is defined as

$$\lambda_{ij} = \frac{(\partial c_i / \partial m_j)_m}{(\partial c_i / \partial m_j)_c} \tag{8.1}$$

If $c_i$ does *not* respond to $m_j$ when all other manipulated variables are constant, $\lambda_{ij}$ is *zero* and $m_j$ should *not* be used to control $c_i$. If some interaction exists, changing $m_j$ will affect other controlled variables as well as $c_i$. Then if they are to be held constant, the other manipulated variables will have to change during the second evaluation of open-loop gain. The result will be a *difference* between the two open-loop gains, causing $\lambda_{ij}$ to be neither 0 nor 1.

Another possibility would have the denominator in Eq. (8.1) go to zero. Then the presence of other closed loops would prevent $c_i$ from responding to $m_j$. This case is signaled by $\lambda_{ij}$ approaching infinity, although any number greater than perhaps 5 would indicate potential loss of control.

Because the process is typically represented by steady-state and dynamic-gain terms, its relative gains are to be expected to contain both components as well. In most cases, however, the steady-state components will be found to carry the greater significance as well as being easier to work with. In general, then, static relative gains will be evaluated, with dynamics considered later.

### Properties of the Array

It is convenient to arrange the relative-gain terms in an array

$$
\Lambda =
\begin{array}{c}
c_1 \\
c_2 \\
\\
c_i \\
\\
\end{array}
\begin{bmatrix}
\lambda_{11} & \lambda_{12} & \cdots & \lambda_{1j} & \cdots \\
\lambda_{21} & \lambda_{22} & \cdots & \lambda_{2j} & \cdots \\
\cdots & \cdots & \cdots & \cdots & \cdots \\
\lambda_{i1} & \lambda_{i2} & \cdots & \lambda_{ij} & \cdots \\
\cdots & \cdots & \cdots & \cdots & \cdots
\end{bmatrix}
\quad\quad (8.2)
$$

$$\overset{m_1 \quad\quad m_2 \quad\quad\quad m_j}{}$$

One property of the array $\Lambda$ is that the relative gains in each column and row add up to unity. This reduces the number of terms which must be evaluated to fill an array. In a $2 \times 2$ system, for example, only $\lambda_{11}$ must be found, $\lambda_{22}$ being equal to it and the other terms being its complement. In a $3 \times 3$ array, four relative gains must be evaluated, with the others filled in by difference.

Because of this property, numbers will tend to fall in certain combinations. For example, all the numbers in a given row or column may lie between 0 and 1. If a number greater than 1.0 appears, there must also be a negative number in the same row and in the same column. In a $2 \times 2$ system, if any number is between 0 and 1, all must be, and if any lies outside 0 and 1, so must all the others. There is a substantial difference between processes having relative gains in the 0 to 1 region and those having gains outside that region. For example, the most severe degree of interaction will be exemplified when all relative gains have the same absolute value. For a $2 \times 2$ process, this could be realized either for $\lambda_{11} = 0.5$ or $\infty$. A value of 0.5 is characteristic of an interacting but stable system, whereas $\infty$ indicates that both variables cannot be controlled at the same time.

A relative gain is a dimensionless number since it is, in effect, a parameter divided by itself under a different set of conditions. Therefore it is insensitive to such things as scales and ranges. For the same reason, it is not affected by nonlinearities; presumably the same non-linearity appears in both numerator and denominator of Eq. (8.1) and so cancels. Then if we know that one variable is a singular function of another, for example, $y = f(x)$, we can substitute $x$ for $y$ in both numerator and denominator. This procedure can also be followed if that function includes another variable, for example, $y = f(x,F)$, as long as that other variable, for example, $F$, is not manipulated by the controls.

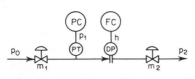

**FIG. 8.1** Both valves affect flow and pressure to a similar but not necessarily equal degree.

## Procedures for Obtaining Gains

The most satisfactory procedure is that of evaluating numerator and denominator of (8.1) by differentiating a mathematical representation of the process. Then relative gains can be left in terms of the controlled or manipulated variables themselves, which enables the interaction to be evaluated across a range of operating conditions. This method is best illustrated by an example, such as the problem of controlling both pressure and flow by two valves in series, as in the process of Fig. 8.1. Because relative gain is insensitive to nonlinearities, we can linearize the process model.

**example 8.1** To see how the matrix is derived, the pressure-flow process of Fig. 8.1 will be examined, using the following simplified equations to express the flowing differential $h$ and the controlled presure $p_1$:

$$h = m_1(p_0 - p_1) = m_2(p_1 - p_2) = \frac{m_1 m_2 (p_0 - p_2)}{m_1 + m_2} \tag{8.3}$$

The gain with both loops open is the partial derivative in terms of $m$

$$\left.\frac{\partial h}{\partial m_2}\right|_{m_2} = \frac{m_2(p_0 - p_2)}{m_1 + m_2} - \frac{m_1 m_2 (p_0 - p_2)}{(m_1 + m_2)^2} = (p_0 - p_2)\left(\frac{m_2}{m_1 + m_2}\right)^2 \tag{8.4}$$

The gain with the pressure loop closed is in terms of $p_1$

$$\left.\frac{\partial h}{\partial m_1}\right|_p = p_0 - p_1 = (p_0 - p_2)\frac{m_2}{m_1 + m_2} \tag{8.5}$$

Then

$$\lambda_{h1} = \frac{m_2}{m_1 + m_2}$$

Gain can also be expressed in terms of pressure, by substituting for $m_1$ and $m_2$ from Eq. (8.3)

$$\lambda_{h1} = \frac{p_0 - p_1}{p_0 - p_2} \tag{8.6}$$

In the same manner, $\lambda_{h2}$ is found to be

$$\lambda_{h2} = \frac{p_1 - p_2}{p_0 - p_2} \tag{8.7}$$

If the following expressions for $p_1$ are used, the other gains can be determined:

$$p_1 = p_0 - \frac{h}{m_1} = p_2 + \frac{h}{m_2} = \frac{m_1 p_0 + m_2 p_2}{m_1 + m_2}$$

The resulting matrix is

$$\Lambda = \begin{array}{c} h \\ p_1 \end{array} \begin{array}{cc} m_1 & m_2 \\ \left[\begin{array}{cc} \dfrac{p_0 - p_1}{p_0 - p_2} & \dfrac{p_1 - p_2}{p_0 - p_2} \\ \dfrac{p_1 - p_2}{p_0 - p_2} & \dfrac{p_0 - p_1}{p_0 - p_2} \end{array}\right] \end{array} \tag{8.8}$$

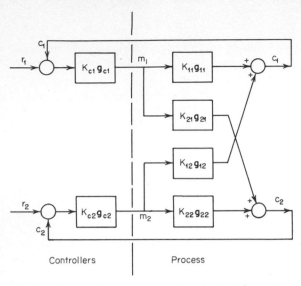

**FIG. 8.2**    A generalized representation of a 2 x 2 interacting process with single-loop controls.

Choice of particular $c$-$m$ pairs for this application depends on the pressure distribution. The matrix indicates that the valve with the greater pressure drop is better for flow control. But if $p_1$ is midway between $p_0$ and $p_2$, all elements in the matrix are 0.5, so it does not matter which pairs are chosen.

Note that in the example all relative gains fall between 0 and 1. This is typical of processes with dissimilar controlled variables (flow and pressure) although their manipulated variables are identical. The same result will be obtained when controlled variables are identical, e.g., compositions, but manipulated variables differ, e.g., mass flow and heat flow.

Relative gains can also be found from a set of open-loop gains with all loops open. Consider the general block diagram of Fig. 8.2, in which steady-state gains are described by coefficients $K_{ij}$ and dynamic vectors by $\mathbf{g}_{ij}$. The steady-state relationships appear as

$$c_1 = K_{11}m_1 + K_{12}m_2 \qquad (8.9)$$
$$c_2 = K_{21}m_1 + K_{22}m_2 \qquad (8.10)$$

The numerator in the relative-gain calculation is simply

$$\frac{\partial c_1}{\partial m_1}\bigg|_{m_2} = K_{11}$$

But the denominator must be found by substituting $c_2$ from (8.10) for $m_2$ in (8.9)

$$c_1 = K_{11}m_1 + K_{12}\frac{c_2 - K_{21}m_1}{K_{22}}$$

$$\left.\frac{\partial c_1}{\partial m_1}\right|_{c_2} = K_{11} - \frac{K_{12}K_{21}}{K_{22}}$$

Then

$$\lambda_{11} = \frac{1}{1 - K_{12}K_{21}/K_{11}K_{22}} \qquad (8.11)$$

As a general solution for the $2 \times 2$ system, Eq. (8.11) allows a prediction of the range of $\lambda_{11}$ based on the signs of the $K$ terms. If the $K$ terms have an odd number of positive signs, $\lambda_{11}$ must lie between 0 and 1; if there are an even number, $\lambda_{11}$ falls outside the 0 to 1 range. This relationship can be demonstrated for the process in Fig. 8.1. Opening valve 1 raises both flow and pressure, whereas opening valve 2 raises flow while lowering pressure. Three of the four open-loop gains are positive, in which case all relative gains fall between 0 and 1.

By way of contrast, consider the process in Fig. 8.3, featuring parallel streams under flow control from a common header. Opening valve 1 causes flow 1 to increase and flow 2 to decrease because of the consequent reduction in header pressure. Therefore $K_{11}$ is positive and $K_{21}$ is negative. Since the flow loops are identical, their gains are also identical, giving

$$\lambda_{11} = \frac{1}{1 - (K_{21}/K_{11})^2} \qquad (8.12)$$

Note that $\lambda_{11}$ cannot fall between 0 and 1. These results are consistent with an earlier observation that similar manipulated *and* controlled variables will give relative gains outside the 0 to 1 range. If $K_{11}$ and $K_{21}$ were equal, $\lambda_{11}$ would approach infinity; if the pump were so self-regulating that header pressure were virtually constant at all flow rates, $\lambda_{11}$ would approach 1.0. For a process like this, it will typically fall between 1 and 2.

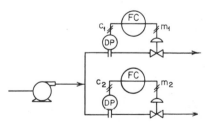

**FIG. 8.3** Processes with parallel elements usually have relative gains outside the 0 to 1 range.

Process equations can also be written to express manipulated variables in terms of controlled variables

$$m_1 = H_{11}c_1 + H_{12}c_2 \qquad (8.13)$$
$$m_2 = H_{21}c_1 + H_{22}c_2 \qquad (8.14)$$

Coefficients $H_{ji}$ are the reciprocals of open-loop gains

$$H_{ji} = \frac{\partial m_j}{\partial c_i}\bigg|_c \qquad (8.15)$$

If both $H_{ji}$ and $K_{ij}$ are known,

$$\lambda_{ij} = K_{ij}H_{ji} \qquad (8.16)$$

If only the set of $H_{ji}$ are known, the same procedure can be followed as with the set of $K_{ij}$ giving

$$m_1 = H_{11}c_1 + H_{12}\frac{m_2 - H_{21}c_1}{H_{22}}$$

$$\frac{\partial m_1}{\partial c_1}\bigg|_{m_2} = H_{11} - \frac{H_{12}H_{21}}{H_{22}}$$

$$\lambda_{11} = \frac{(\partial m_1/\partial c_1)_{c_2}}{(\partial m_1/\partial c_1)_{m_2}} = \frac{1}{1 - H_{12}H_{21}/H_{11}H_{22}} \qquad (8.17)$$

Observe that the form of (8.17) is identical to that of (8.11).

## Matrix Methods

These procedures can be extended to higher-order systems using matrix inversion. The set of equations (8.9) and (8.10) can be expressed for an $n$-dimensional process in matrix-vector notation

$$\mathbf{c} = \mathbf{Km} \qquad (8.18)$$

Equations (8.13) and (8.14) can be expressed in a similar manner

$$\mathbf{m} = \mathbf{Hc} \qquad (8.19)$$

From the above it is apparent that the matrix of gains $\mathbf{H}$ is the inverse of matrix $\mathbf{K}$

$$\mathbf{H} = \mathbf{K}^{-1} \qquad (8.20)$$

Equation (8.16) expresses the relative-gain terms as the product of each element in the $\mathbf{K}$ matrix by the corresponding element in the *transposed* $\mathbf{H}$ matrix (note the reversal of subscripts). Then the matrix $\Lambda$ of relative gains can be expressed as the multiplication (dot product) of each element in the $\mathbf{K}$ matrix by the corresponding element in its transposed inverse

$$\Lambda = \mathbf{K} \cdot [\mathbf{K}^{-1}]^T \qquad (8.21)$$

If only matrix $\mathbf{H}$ is known,

$$\Lambda = \mathbf{H}^T \cdot \mathbf{H}^{-1} \qquad (8.22)$$

The following procedure is suggested for inverting a $3 \times 3$ matrix. Given

$$\mathbf{K} = \begin{bmatrix} K_{11} & K_{12} & K_{13} \\ K_{21} & K_{22} & K_{23} \\ K_{31} & K_{32} & K_{33} \end{bmatrix} \tag{8.23}$$

First calculate the determinant

$$\begin{aligned} D = {}& K_{11}K_{22}K_{33} + K_{13}K_{21}K_{32} + K_{12}K_{23}K_{31} - K_{13}K_{22}K_{31} \\ & - K_{12}K_{21}K_{33} - K_{11}K_{23}K_{32} \end{aligned} \tag{8.24}$$

The inverse consists of the elements

$$\mathbf{H} = \mathbf{K}^{-1} = \begin{bmatrix} H_{11} & H_{12} & H_{13} \\ H_{21} & H_{22} & H_{23} \\ H_{31} & H_{32} & H_{33} \end{bmatrix}$$

It is only necessary to solve for four of them

$$H_{11} = \frac{K_{22}K_{33} - K_{23}K_{32}}{D} \tag{8.25}$$

$$H_{22} = \frac{K_{11}K_{33} - K_{13}K_{31}}{D} \tag{8.26}$$

$$H_{33} = \frac{K_{11}K_{22} - K_{12}K_{21}}{D} \tag{8.27}$$

$$H_{12} = \frac{K_{13}K_{32} - K_{12}K_{33}}{D} \tag{8.28}$$

They are used to calculate the appropriate relative gains

$$\begin{aligned} \lambda_{11} &= K_{11}H_{11} & \text{(8.29)} \\ \lambda_{22} &= K_{22}H_{22} & \text{(8.30)} \\ \lambda_{33} &= K_{33}H_{33} & \text{(8.31)} \\ \lambda_{21} &= K_{21}H_{12} & \text{(8.32)} \end{aligned}$$

All the other relative gains can be found by difference.

**example 8.2** Calculate the relative gain array given the following set of open-loop gains:

$$\mathbf{K} = \begin{bmatrix} 0.10 & 0.79 & 0 \\ 14 & -4.1 & -0.8 \\ 0.0031 & 0.054 & -0.015 \end{bmatrix}$$

$$D = 0.1744 \qquad H_{11} = 6.00 \qquad H_{22} = -0.854$$

$$H_{33} = -65.7 \qquad H_{12} = 0.0684$$

$$\boldsymbol{\Lambda} = \begin{bmatrix} 0.600 & 0.400 & 0 \\ 0.951 & 0.035 & 0.014 \\ -0.551 & 0.435 & 0.986 \end{bmatrix}$$

## Decomposing Large Systems

It is often convenient to reduce high-order processes into $2 \times 2$ subsets for individual examination. Relative gains for the $2 \times 2$ subsets can easily be calculated by the methods shown without matrix inversion. This has the very important advantage of making it possible to express relative gains in terms of the manipulated or controlled variables, as in Example 8.1. The performance can be predicted at various sets of conditions, and general trends developed. Matrix methods do not allow this; the determinant itself contains so many terms that only numerical solutions are realizable.

A case in point is the distillation column. Typically there are five variables to be controlled, compositions of top and bottom product, pressure, and levels in the column base and reflux accumulator. The five manipulated variables are heat input and output, flow rates of both products, and reflux flow. The number of possible control configurations of $n$ single loops is $n!$; for the distillation column, 120 choices are presented.

The variables most difficult to control are the compositions of the products; responses are slow, and the process is quite sensitive to upsets. Therefore, the two composition loops can be removed from the $5 \times 5$ process and evaluated on a $2 \times 2$ basis. For example, relative gains of product compositions can be calculated for manipulation of top product and heat input. Liquid levels and column pressure will be assumed to be controlled without significant interaction by the other flow rates. This is a reasonable assumption, in that they respond at least an order of magnitude faster than the compositions. If the relative gains of the $2 \times 2$ subset are favorable, that pairing would be selected for control. This would be the case where the relative gain of top product composition responding to its own flow is close to unity. If it were close to zero, interaction would be unfavorable, and another $2 \times 2$ subset should be evaluated. This might have bottom-product flow and reflux as the two manipulated variables. Examples of this technique are found in Chap. 11.

## EFFECTS OF INTERACTION

Interaction produces several undesirable effects. The loops upset each other, to be sure, but more than that, there are hidden feedback loops that can destabilize the system. One or both controllers can usually be readjusted to restore stability but not without a certain loss in performance. If the loops are incorrectly configured, however, control can be lost entirely when the last controller is placed in automatic although individually loop performance is adequate. This can be mystifying to the engineer who is unfamiliar with the concepts of interaction.

### Influence of the Other Controller

The evaluation of relative gains considered the extreme cases of no control and complete control of the other variables upon the loop in question. While these limits are necessary to characterize the degree of interaction, they do not completely specify the problem. The influence of controllers that are in automatic over the loop being examined depends on their settings as well as the gains and dynamic elements in the process. To evaluate their effects, turn to the block diagram of Fig. 8.2.

If controller 2 is in manual,

$$\frac{dc_1}{dm_1} = K_{11}\mathbf{g}_{11} \tag{8.33}$$

But if controller 2 is in automatic, $m_1$ upsets $c_2$, causing it to readjust $m_2$, which then affects $c_1$. This second loop adds to the first

$$\frac{dc_1}{dm_1} = K_{11}\mathbf{g}_{11} - \frac{K_{21}\mathbf{g}_{21}K_{12}\mathbf{g}_{12}}{K_{22}\mathbf{g}_{22} + 1/K_{c2}\mathbf{g}_{c2}} \tag{8.34}$$

Equation (8.34) is ideally what the denominator of (8.1) should be.

If the gain of controller 2 is extremely high at the natural period of loop 1, owing to a relatively short integral time, the term $1/K_{c2}\mathbf{g}_{c2}$ approaches zero. This situation could develop if, for example, $c_2$ were flow and $c_1$ were composition, as in Fig. 8.4. Then, typically $\mathbf{g}_{11}$ and $\mathbf{g}_{12}$ would be equal long lags, and $\mathbf{g}_{21}$ and $\mathbf{g}_{12}$ would be short, approaching 1.0 at the period of the composition loop. Then Eq. (8.34) would reduce to

$$\frac{dc_1}{dm_1} \approx \mathbf{g}_{11}\left(K_{11} - \frac{K_{21}K_{12}}{K_{22}}\right) = \frac{K_{11}\mathbf{g}_{11}}{\lambda_{11}} \tag{8.35}$$

If the composition controller were tuned with the flow controller in manual, its proportional band would have to be readjusted by the factor $1/\lambda_{11}$ to restore the

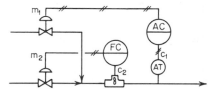

**FIG. 8.4** The composition of the blend is easily upset by the faster flow loop.

same stability with the flow controller in automatic. If $\lambda_{11}$ is near 1.0, little readjustment is necessary, but as it approaches zero, a great increase may be required to restore stability; this is indicative of improper pairing. If $\lambda_{11}$ approaches zero, $\lambda_{12}$ approaches unity and $c_1$ should therefore be controlled with $m_2$.

Next, consider the case where all dynamic elements are equal and $K_{11} = K_{22}$. This could be the case in both Figs. 8.2 and 8.3. Then both controllers should have the same mode settings with the other in manual. What we hope to discover is how the relative gain affects the mode settings required to obtain the same degree of stability when both controllers are in automatic. The procedure used is to multiply $dc_1/dm_1$ in Eq. (8.34) by the gain of controller 1 and set the gain product to $-0.5$ (for quarter-amplitude damping)

$$K_{c1}g_{c1}\left(K_{11}g_{11} - \frac{K_{21}g_{21}\,K_{12}g_{12}}{K_{22}g_{22} + 1/K_{c2}g_{c2}}\right) = -0.5$$

Next, in accordance with the above assumptions, let the controller gains be equal, and all dynamic vectors equal $g_{11}$

$$K_{c1}g_{c1}\left(K_{11}g_{11} - \frac{K_{21}g_{11}\,K_{12}g_{11}}{K_{22}g_{11} + 1/K_{c1}g_{c1}}\right) = -0.5$$

If this expression is solved for the single-loop gain product, a quadratic is formed

$$\frac{(K_{c1}g_{c1}K_{11}g_{11})^2}{\lambda_{11}} + 1.5(K_{c1}g_{c1}K_{11}g_{11}) + 0.5 = 0$$

The solution is

$$K_{c1}g_{c1}K_{11}g_{11} = \frac{-1.5 \pm \sqrt{2.25 - 2/\lambda_{11}}}{2/\lambda_{11}} \qquad (8.36)$$

Where $\lambda_{11}$ lies between 0 and 2/2.25, or 0.889, the roots are imaginary, indicating that loop phase is not $-180°$; in this range, coupling changes the period of the loops. Above this range, all roots are real and negative, indicating that coupling has no effect on loop period. Below this range, i.e., negative relative gains, one root is positive and the other negative. Table 8.1 lists solutions for selective values of relative gain. Each range requires a different interpretation.

**TABLE 8.1   Single-Loop Gain Product as a Function of Relative Gain**
Meaningful roots are shown in boldface

| $\lambda_{11}$ | $K_{c1}g_{c1}K_{11}g_{11}$ |
| --- | --- |
| $-2.0$ | **$-0.302$**,   $+3.30$ |
| $-1.0$ | **$-0.280$**,   $+1.78$ |
| $-0.5$ | **$-0.250$**,   $+1.00$ |
| $-0.2$ | **$-0.200$**,   $+0.500$ |
| $0.2$ | **$0.316\angle-118.3°$**, $-241.7°$ |
| $0.5$ | **$0.500\angle-138.6°$**, $-221.4°$ |
| $0.7$ | **$0.592\angle-152.5°$**, $-207.5°$ |
| $0.8$ | **$0.632\angle-161.7°$**, $-198.3°$ |
| $0.889$ | **$-0.667$**,   $-0.667$ |
| $1.0$ | **$-0.500$**,   $-1.000$ |
| $1.2$ | **$-0.442$**,   $-1.358$ |
| $1.5$ | **$-0.406$**,   $-1.843$ |
| $2.0$ | **$-0.382$**,   $-2.618$ |
| $5.0$ | **$-0.350$**,   $-7.15$ |

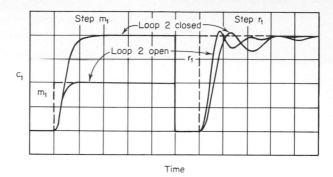

**FIG. 8.5**  Open and closed-loop step responses for a 2 x 2 system with $\lambda_{11} = 0.5$ and equal dynamic elements.

## Relative Gains 0 to 1

To understand the significance of the numbers in Table 8.1, consider two loops having a relative gain of 0.5. Let controller 1 be adjusted so that $K_{c1}g_{c1}K_{11}g_{11}$ is 0.5 at $-180°$ with controller 2 in manual. Table 8.1 says that $K_{c1}g_{c1}K_{11}g_{11}$ must be *readjusted* to 0.5 at $-138.6°$ or $-221.4°$ to retain quarter-amplitude damping when controller 2 is in automatic at the same settings. One of the phase angles must be discarded because both loops will cycle at the same period. If the open-loop phase lag were to exceed $180°$, the period of the coupled loops would have to be less than that of the individual single loops. Simulation results (as shown in Fig. 8.5) indicate that coupling *extends* the loop period; therefore the true phase lag is less than $180°$.

The period $\tau_c$ of the coupled loops will differ from the period $\tau_o$ of the uncoupled loops as a function of this $41.4°$ reduction in phase lag

$$\frac{\tau_c}{\tau_o} = \frac{\phi_d}{\phi_d + 41.4°} \tag{8.37}$$

where $\phi_d$ is the phase shift developed by the dead time in $g_{11}$ at $\tau_o$.

If $g_{11}$ is pure dead time and $g_{c1}$ is set for $30°$ phase lag in a PI controller, $\phi_d$ will be $-150°$. Then $\tau_c/\tau_o = 1.38$; that is, coupling will increase the loop period by 38 percent. To retain the phase angles of both controllers at $-30°$ their integral times will have to increase by the same 38 percent when both are in automatic. No readjustment in proportional band is required in this case because loop gain is still 0.5 and process gain does not change with period.

If $g_{11}$ is dead time plus capacity and $g_{c1}$ is set for $30°$ phase lag, $\phi_d$ will be $-60°$. Then $\tau_c/\tau_o = 3.23$. The integral times of both controllers will have to be increased by this factor, and so will the proportional-band settings, because the gain of the capacity increases with period. If the controllers are set for zero phase lag, $\tau_c/\tau_o = 1.85$. This figure is consistent with reports of engineers adjusting interacting controllers in the field: proportional bands and integral times may have to be doubled to restore stability when the second controller is put into automatic [2].

Figure 8.5 describes the open- and closed-loop step responses for two loops with identical dynamics and $\lambda_{11} = 0.5$. The process was simulated on an analog computer using three equal noninteracting lags for each dynamic element. Coupling apparently increases the period by about 50 percent for these processes.

At other values of relative gain, loop gain is not 0.5, so that a readjustment in proportional band is needed, even with dead-time processes. At $\lambda_{11} = 0.7$, for example, loop gain can increase to 0.592, allowing a reduction in proportional band by the factor 0.5/0.592 if there is no change in process gain. At $\lambda_{11} = 0.889$, loop period is unaffected by coupling, so that process gain remains constant and the proportional band of the controllers can be *reduced* by 0.5/0.667. This slight amount of interaction appears to be favorable.

Figure 8.6 repeats the earlier block diagram but with gains eliminated to allow signs to be followed more readily. The distribution of signs in the process will produce relative gains in the 0 to 1 range. Both controllers provide negative feedback. There is a third loop formed from $m_1$ to $c_2$, through controller 2 to $c_1$, which is also negative in sense. Presumably, when $\lambda_{11}$ falls below 0.889, coupling elements $K_{21}$ and $K_{12}$ begin to contribute significantly to loop gain and their dynamic components $\mathbf{g}_{21}$ and $\mathbf{g}_{12}$ introduce some phase lag.

### Relative Gains Outside 0 to 1

For $\lambda_{11}$ to exceed unity, there must be an even number of positive signs in the process, as shown in Fig. 8.7. In this case, the third feedback loop, from $m_1$ through loop 2 and to $c_1$, is positive in sense. As such, it does not contribute the phase lag of a negative feedback loop, leaving loop period unaffected by coupling. This is confirmed by the simulation results shown in Fig. 8.8 for loops with $\lambda_{11} = 2.0$. Notwithstanding the constancy of period, set-point response is held back by coupling. Integration of the deviation from set point shows that

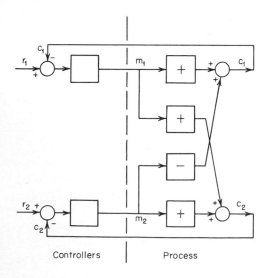

**FIG. 8.6** With an odd number of positive signs in the process, the third feedback loop formed by the cross-coupling elements is negative.

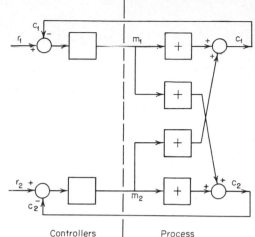

**FIG. 8.7** With an even number of positive signs in the process, the third feedback loop is positive.

better response is achieved when $\lambda_{11} = 0.5$, which is the worst case in the 0 to 1 regime. As $\lambda_{11}$ increases beyond 2.0, response deteriorates further.

Experience in adjusting controllers for processes of this class has shown that proportional band settings must always be increased to restore stability when both loops are closed. Therefore the larger roots in Table 8.1 must be rejected in favor of the smaller ones. For the case where $\lambda_{11} = 2.0$, proportional bands ought to be expanded by 0.5/0.382. However, Toijala and Fagervik [2] found it necessary to more than double the band settings of the controllers on their simulated distillation column. Their relative-gain terms were 5.23.

In a 2 × 2 process, a pair of relative gains below 0 will accompany the pair greater than 1.0. If these loops are connected, i.e., if $\lambda_{11} < 0$, the product of the cross-coupling elements $K_{21}K_{12}$ will exceed that of the selected elements $K_{11}K_{22}$. Then the positive feedback loop formed by the cross-coupling elements will exceed the negative feedback loops in gain, and control will be lost altogether.

However, it can be restored if the action (sign) of *one* of the controllers is reversed. Then the third feedback loop becomes negative although the loop with the reversed controller has become positive. Phase reversal in one controller is indicated by the difference in sign between the real roots for negative values of relative gain in Table 8.1. Apparently the negative-feedback controller must be reduced in gain and the positive-feedback controller increased. This is a precarious situation, however, for if the negative-feedback controller should be placed in manual or otherwise constrained, the remaining loop will have only positive feedback.

While this conditional stability would be enough to keep one from pairing variables with negative relative gains, dynamic performance is also poor. Figure 8.9 shows response curves for a simulated process in which $\lambda_{11} = -1.0$. With loop 2 open, the familiar monotonic step response is obtained. But when loop 2 is closed, its controller responds to the upset in $m_1$ by driving $c_1$ in the *opposite* direction, hence the negative relative gain. The type of open-loop behavior

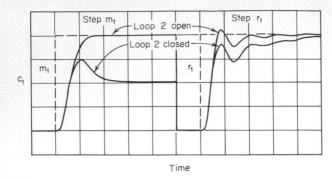

**FIG. 8.8** Open- and closed-loop step responses for a 2 x 2 system with $\lambda_{11} = 2.0$ and equal dynamic elements.

caused by the closure of loop 2 is known as *inverse response*. It introduces a phase shift similar to dead time, such that loop period will be approximately 4 times the *inversion time*, the time required for the controlled variable to cross its original value. More is said about inverse response in reference to boiler drum-level control in Chap. 9. Suffice it to say at this point that inverse response so extends the period of the coupled loops that performance becomes completely unacceptable.

### Dynamic Coupling

The process shown in Fig. 8.10 is the headbox of a paper machine. Its function is to spread the stock uniformly on the moving wire by matching its spouting velocity to that of the wire. Spouting velocity is controlled by maintaining a constant *total head* above the spout, total head being the sum of liquid head and air pressure in the box. Both total head and level are to be controlled. The two manipulated variables are stock feed rate and air-valve position. Air pressure in the headbox is determined by the position of the air valve relative to that of a manual bleed valve exhausting to atmosphere.

An increase in the position of the air valve will raise total head, causing outflow to increase. The flow increase will then cause liquid level to fall. In the absence of control, the falling level will reduce total head until outflow once again equals inflow, as shown in Fig. 8.11. In the steady state, the air valve has had no effect on total head.

The steady-state gain of total head with respect to the air valve appears to be zero, which would make its relative gain also zero. Yet total head *can* be controlled by the air valve because of its dynamic response *if* level is controlled by the feed valve. In fact, total head will respond faster than liquid level to the air valve because it must change in order to change outflow, which only then affects liquid level. The preferred configuration for this system is therefore control of total head with the air valve, even though its steady-state relative gain is zero.

The justification for controlling total head with the air valve requires a dynamic calculation of relative gain. For simplicity, let air pressure $p$ respond to air valve position as

$$p = K_{11}\mathbf{g}_{11}m_1 \qquad (8.38)$$

Total head $c_1$ is the sum of pressure and level $c_2$

$$c_1 = p + c_2 = K_{11}\mathbf{g}_{11}m_1 + c_2 \qquad (8.39)$$

whose derivative is

$$\left.\frac{\partial c_1}{\partial m_1}\right|_c = K_{11}\mathbf{g}_{11} \qquad (8.40)$$

Liquid level is the integral of inflow from $m_2$ less outflow, which is proportional to total head

$$c_2 = \mathbf{g}_{22}(m_2 - K_2 c_1) \qquad (8.41)$$

where $\mathbf{g}_{22}$ represents the integrating function. Substitution of (8.41) into (8.39) gives

$$c_1 = \frac{K_{11}\mathbf{g}_{11}m_1 + \mathbf{g}_{22}m_2}{1 + K_2\mathbf{g}_{22}} \qquad (8.42)$$

whose derivative is

$$\left.\frac{\partial c_1}{\partial m_1}\right|_m = \frac{K_{11}\mathbf{g}_{11}}{1 + K_2\mathbf{g}_{22}} \qquad (8.43)$$

The relative gain of total head $c_1$ with respect to air-valve position $m_1$ can then be calculated from the two derivatives

$$\lambda_{11} = \frac{1}{1 + K_2\mathbf{g}_{22}} \qquad (8.44)$$

Substituting the gain of the integrator having time constant $\tau$ allows solution for $\lambda_{11}$ as a function of period $\tau_o$

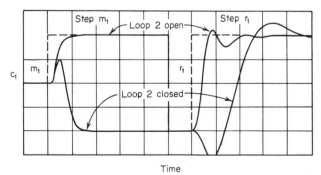

FIG. 8.9  Open- and closed-loop step responses for a 2 x 2 system with $\lambda_{11} = -1.0$ and equal dynamic elements.

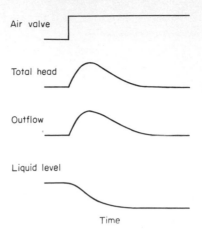

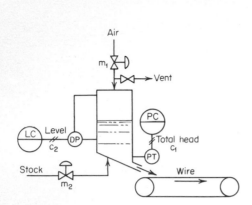

**FIG. 8.10** The headbox discharging paper stock onto a traveling wire is a favorite interaction problem in paper mills.

**FIG. 8.11** The air valve has no steady-state effect on total head.

$$\lambda_{11}(\tau_o) = \frac{1}{\sqrt{1 + (K_2\tau_o/2\pi\tau)^2}}, \angle \tan^{-1}\frac{K_2\tau_o}{2\pi\tau} \tag{8.45}$$

Observe that as $\tau_o/\tau$ approaches zero, $\lambda_{11}$ approaches 1.0. Then if $\mathbf{g}_{11}$, which determines $\tau_o$, has a much smaller time constant than $\mathbf{g}_{22}$, $\lambda_{11}$ will approach 1.0 and the control of total head with the air valve is justified.

At extremely long periods, $\lambda_{11}$ approaches zero, as indicated by the tail of the step response, but this condition is inconsequential as long as liquid level is being controlled.

## DECOUPLING

In some instances interaction can become sufficiently severe to preclude satisfactory control, even with the best loop pairing. Identical loops are particularly troublesome because of their sympathetic dynamic response. Detuning of one or more controllers may be acceptable if the loops are fast, as in flow, but not for slow composition loops. And in some systems having interacting liquid-level loops, stability may be unattainable *without* decoupling.

The essence of decoupling is the imposition of a computing network which will cancel the interaction naturally existing in the process, permitting independent single-loop control. In theory, it would seem that the decoupling system need only be an inverse of the process to cancel its interaction. In practice, the inverse is not realizable because it would require leads and time advances to compensate lags and dead times in the process. Even without dynamic compensation, a steady-state decoupler poses operational and stability problems. For these reasons the various possible means of applying decoupling are carefully explored, along with an evaluation of their effectiveness under real conditions.

## Decoupling Methods

Decoupling may be applied either to the input of a process or its output. If applied to the input, each controller will be capable of manipulating *all* valves, in order to control its own variable without upsetting any others. If a decoupler is applied to the process output, it generates a set of new controlled variables, each of which responds only to *one* manipulated variable. Whether input or output decoupling is likely to be more effective in a given situation depends primarily on the significance of the new set of variables.

This can best be illustrated through an application of decoupling to an existing fractionator having parallel reboilers, as shown in Fig. 8.12. Two liquid levels and one temperature need to be controlled by manipulating the common bottom-product valve and the two hot-oil valves. Opening one hot-oil valve would increase the rate of boiling in one reboiler, raising the pressure drop across its vapor line. This action depresses the liquid level in that reboiler while raising the level in the other. Eventually both levels fall due to a reduction in liquid inventory.

Temperature does not respond directly to bottom-product flow, and so in the plant it was controlled with one of the oil valves. Bottom-product flow was used to control one level. But the other level could not be controlled by the other oil valve; closing that loop threw all three into oscillation.

Decoupling was applied to the level measurements, as shown in Fig. 8.13. Since bottom-product flow affected both levels, it was used to control their sum, representing total liquid inventory. To balance the two levels, one is used as the set point and the other as the measurement to a controller which biases the hot-oil valves in opposite directions. The temperature controller moves both valves in the same direction. If the system is properly calibrated, it effectively decouples the three controlled variables. Moving the bottom-product valve affects the two levels the same and therefore should have no influence over their difference. And the level-difference controller, in moving the hot-oil valves in opposite directions, should not change the total heat input, which would then not affect either temperature or total liquid inventory. Although the tempera-

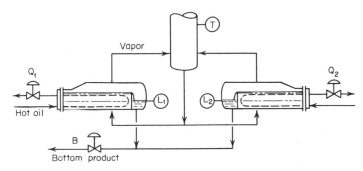

**FIG. 8.12** Liquid levels in the parallel reboilers interact severely.

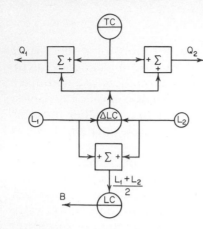

**FIG. 8.13** Decoupling is applied to the inputs of the level controllers and to the output of the temperature controller.

ture controller can still upset liquid inventory, its influence should be slight because of the faster response of the level-control loop.

Observe that decoupling is applied to the inputs of the level controllers and the output of the temperature controller. In this application, the particular combination shown seemed to be the simplest and least costly.

Decoupling can be applied directly to controller output signals, as shown in Fig. 8.14. The function required of the decouplers can be derived from the process equations

$$c_1 = K_{11}\mathbf{g}_{11}m_1 + K_{12}\mathbf{g}_{12}m_2 \tag{8.46}$$
$$c_2 = K_{21}\mathbf{g}_{21}m_1 + K_{22}\mathbf{g}_{22}m_2 \tag{8.47}$$

It is desired to make $dc_1/dm_2$ and $dc_2/dm_1$ equal zero. Solving Eq. (8.46) for $m_1$ gives

$$m_1 = \frac{c_1 - K_{12}\mathbf{g}_{12}m_2}{K_{11}\mathbf{g}_{11}} \tag{8.48}$$

For $dc_1/dm_2$ to be zero we must have

$$\frac{dm_1}{dm_2} = -\frac{K_{12}\mathbf{g}_{12}}{K_{11}\mathbf{g}_{11}} \tag{8.49}$$

Similarly, for $dc_2/dm_1$ to be zero we must have

$$\frac{dm_2}{dm_1} = -\frac{K_{21}\mathbf{g}_{21}}{K_{22}\mathbf{g}_{22}} \tag{8.50}$$

There are two problems associated with the arrangement of Fig. 8.14, initialization and constrained operation. Initialization involves finding the correct values of the two controller outputs $m_{c1}$ and $m_{c2}$ to begin automatic control bumplessly. Calculation of $m_{c1}$ depends not only on the known value of $m_1$ but also on the unknown value of $m_{c2}$. If $m_{c1}$ is initialized on the basis of an arbitrary value of $m_{c2}$, $m_1$ will be bumped when $m_{c2}$ is initialized. Proper initialization involves solving the decoupling equations in reverse to generate $m_{c1}$ and $m_{c2}$ simultaneously from known values of $m_1$ and $m_2$.

While this can be overcome by programming, the problem of constrained operation is not as easy to resolve. Should a constraint be applied to either $m_1$ or $m_2$, variables $c_1$ and $c_2$ cannot *both* be controlled. But in the system of Fig. 8.14, both controllers will *attempt* to control by manipulating the remaining unconstrained variable. Neither will be satisfied, and the unconstrained variable will be driven to a limit in the attempt.

These problems are avoided with the arrangement shown in Fig. 8.15. Here, $m_{c1}$ and $m_{c2}$ are calculated from known values of $m_1$ and $m_2$, simplifying initialization. Furthermore, the imposition of a constraint completely opens one control loop; the constrained manipulated variable continues to be sent to the other loop as feedforward compensation for disturbances.

Although the structure of Fig. 8.15 differs substantially from that of 8.14, the decouplers themselves are identical. Yet Fig. 8.15 contains a feedback loop which could be unstable; it is formed by the two decouplers. The gain of this loop is the product of the two decoupler gains. If $\lambda_{11}$ is between 0 and 1, the decouplers will have opposite signs and their feedback loop will be negative. If $\lambda_{11} > 0.5$, the steady-state gain of the feedback loop is less than 1.0, so it will be stable.

If $\lambda_{11} > 1$, the feedback loop through the decouplers is positive but its gain is less than 1.0. Stability then depends on $\mathbf{g}_{12}\mathbf{g}_{21}$ being less than $\mathbf{g}_{11}\mathbf{g}_{22}$ at all periods. If $\lambda_{11} < 0$, the decoupling feedback loop will be unstable under all conditions.

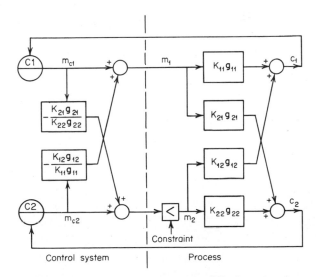

**FIG. 8.14** This decoupling system is difficult to put into automatic operation and is subverted by constraints.

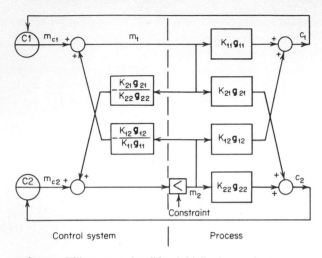

**FIG. 8.15** This system simplifies initialization and accommodates constraints.

### Stability of Decoupled Systems

While it is easy to specify what the decoupler functions ideally ought to be, obtaining and maintaining their ideal values is quite another matter. Processes tend to be nonlinear and time-variant. As a result, decoupler gains for most applications should not be constant. The decouplers must be nonlinear or even adaptive if optimality is to be achieved. If they are linear and constant then imperfect decoupling must be expected. In some instances, decoupler errors can lead to unstable operation [2, 3]. To predict the likelihood of this happening, the relative gain of the *decoupled* process is derived. (In the following derivations, dynamic vectors are omitted for simplicity; they can be inserted in the result where their corresponding steady-state gains appear.)

Let

$$m_1 = m_{c1} + d_{12}m_2 \tag{8.51}$$

$$m_2 = m_{c2} + d_{21}m_1 \tag{8.52}$$

where $d_{12}$ and $d_{21}$ are the two decoupler gains, which may or may not be set at the ideal values. When Eqs. (8.51) and (8.52) are combined with the process steady-state equations, the relative gain of $c_1$ with respect to $m_{c1}$ can be found. This is identified as $\lambda_{11d}$ the relative gain of the decoupled process

$$\lambda_{11d} = \cfrac{1}{1 - \cfrac{(K_{21} + d_{21}K_{22})(K_{12} + d_{12}K_{11})}{(K_{22} + d_{12}K_{21})(K_{11} + d_{21}K_{12})}} \tag{8.53}$$

The loops will be effectively decoupled when $\lambda_{11d} = 1.0$, which will be accomplished when *either* $d_{21} = -K_{21}/K_{22}$ or $d_{12} = -K_{12}/K_{11}$. Other limits of performance exist, however. For example, $\lambda_{11d} = 0$ when either $d_{12} = -K_{22}/K_{21}$ or $d_{21} = -K_{11}/K_{12}$. Another possibility is $\lambda_{11d} \rightarrow \infty$ should $d_{12} = 1/d_{21}$.

To evaluate the tolerance of the process for decoupler errors, let both decoupling functions depart from their ideal values by a common factor $\delta$

$$d_{12} = (1 + \delta)\left(\frac{-K_{21}}{K_{11}}\right) \tag{8.54}$$

$$d_{21} = (1 + \delta)\left(\frac{-K_{21}}{K_{22}}\right) \tag{8.55}$$

Then $\lambda_{11d}$ can be calculated as a function of $\lambda_{11}$ and $\delta$

$$\lambda_{11d} = \frac{[1 - (\lambda_{11} - 1)\delta]^2}{1 - (\lambda_{11} - 1)\delta(\delta + 2)} \tag{8.56}$$

For the limits of zero and infinity, the numerator and denominator of (8.56) must respectively equal zero

$$\lambda_{11d} = 0 \qquad \text{when } \delta = \frac{1}{\lambda_{11} - 1} \tag{8.57}$$

$$\lambda_{11d} \rightarrow \infty \qquad \text{when } \delta = \sqrt{\frac{\lambda_{11}}{\lambda_{11} - 1}} - 1 \tag{8.58}$$

Figure 8.16 plots the decoupled relative gain against decoupler error for several values of $\lambda_{11}$. Negative errors, i.e., insufficient decoupling, produce the expected result of forcing the relative gain partway toward unity for all loops

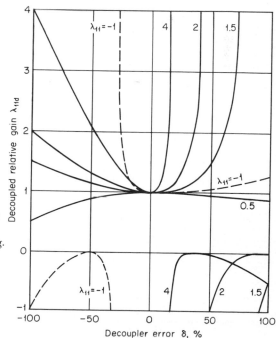

**FIG. 8.16** Loops having relative gains greater than unity are destabilized by excessive decoupling.

with positive gains. However, positive errors can precipitate instability in those loops having relative gains exceeding unity. Negative errors can do the same for loops having negative relative gains. No amount of decoupler error can degrade the performance of loops having relative gains between 0 and 1. This point is important to remember in multiloop processes where a choice exists between loops with gains greater or less than unity. An example is given in Chap. 11.

The upper limit of stability is reached when $\lambda_{11d}$ approaches infinity, which always occurs at an error smaller than $\lambda_{11d}$ of zero. As $\lambda_{11}$ increases, the margin of error tolerance decreases, according to Eq. (8.58). This error limit, identified as $\delta_\infty$, appears with $\lambda_{11}$ in Table 8.2. Note that $\delta_\infty$ for loops in the 0 to 1 range is imaginary.

Error tolerance is an important consideration in designing decouplers because few processes have constant gains. If process gain varies with the magnitude of the controlled variable, decoupling with fixed-gain devices will cause the decoupler error to vary, too. Then a large excursion could precipitate instability, even though decoupling is perfect at the set point. The process described in Example 8.1 exhibits this characteristic, as does the blending system in Fig. 8.4. Fortunately both have relative gains between 0 and 1.

## Partial Decoupling

The adverse effects of decoupling errors can be minimized by using nonlinear decouplers, designed after the process in the manner of feedforward controls. Destabilization can be eliminated altogether by applying only partial decoupling; in the $2 \times 2$ system, only one of the decouplers would be implemented. This effectively breaks the feedback loop formed by the two controllers, as indicated by Eq. (8.53) and the discussion following. Furthermore, upsets from one loop are prevented from entering the second, although those originating in the second loop may still propagate to the first but cannot be returned.

Partial decoupling has the following advantages:

1.  It breaks the third feedback loop.
2.  It keeps upsets out of the decoupled loop.
3.  It avoids instability in the feedback loop through the decouplers.
4.  It avoids instability caused by decoupler errors.
5.  It is easier to design and adjust than a complete decoupler.

**TABLE 8.2   Decoupler Error Causing Infinite Relative Gain**

| Relative gain $\lambda_{11}$ | Error $\delta_\infty$, % | Relative gain $\lambda_{11}$ | Error $\delta_\infty$, % |
|---|---|---|---|
| $-5$ | $-8.7$ | 2 | 41.4 |
| $-2$ | $-18.4$ | 4 | 15.5 |
| $-1$ | $-29.3$ | 8 | 6.9 |
| 1.5 | 73.2 | 16 | 3.3 |

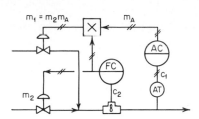

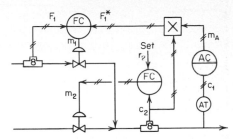

**FIG. 8.17** Partial decoupling is adequate for the composition-control loop.

**FIG. 8.18** This system decouples using $c_2$ rather than $m_2$ because of the increased precision it affords.

In fact, the design, implementation, and adjustment of a partial decoupler is identical to that described for feedforward controls.

The principal decision to be made concerns which of the decouplers to implement. There may be a priority basis for this decision; $c_1$ may be more important to control than $c_2$, in which case $d_{12}$ would be implemented and $d_{21}$ omitted. Another consideration is the imposition of constraints. Figure 8.15 shows constraints applied to $m_2$ but not to $m_1$. In this case, the constraint could result in loss of control of $c_2$ and an upset of $c_1$; however, decoupler $d_{12}$ can prevent the constraint from affecting $c_1$ on a feedforward basis and is therefore useful in both constrained and unconstrained modes of operation.

Another consideration is the relative speed of response of the two variables. If $c_2$ responds more rapidly than $c_1$ to both $m_1$ and $m_2$, as in the blending process of Fig. 8.4, it cannot be upset by the other controller. Then there is no need to decouple $c_2$ from $m_1$, although decoupling the slower $c_1$ from flow effects is very important. Figure 8.17 shows how this might be done.

Note that the decoupler is nonlinear, being a multiplier rather than a summer. In effect, the position of valve 1 is set in ratio to that of valve 2, so that the two flows will be held in constant ratio, regardless of their individual rates. This is because composition is a function of their ratio rather than their sum or difference. The relative gains of this process bear this out; $\lambda$ varies with composition. If liquid level instead of composition were controlled, which responds to total flow, summation would be appropriate. Again, these guidelines were presented for feedforward control in Chap. 7.

Note that if the compositions of the two streams should change or the set point of the composition controller, it will adjust the ratio of $m_1$ to $m_2$ to return its deviation to zero. Subsequent changes in flow rate should not affect composition at the new ratio. Should a higher degree of precision than provided by the valve characteristics be necessary, flowmeters are used, as shown in Fig. 8.18.

This is a typical blending system where ingredients are proportioned to total flow. A positive-feedback loop is formed, in that an increase in flow $F_1$ will cause total flow $c_2$ to increase, which raises set point $F^*_1$ in direct ratio. The gain of the positive-feedback loop is $F_1/c_2$, which also happens to be $1 - \lambda_{11}$. If $1 - \lambda_{11}$ is

0.5 or less, the positive-feedback loop will not cause a stability problem. As it approaches 1.0, positive feedback will extend the settling time of the loop, but the low value of $\lambda_{11}$ also indicates that the opposite pairing should be used. The positive feedback can be eliminated by using set point $r_2$ instead of measured $c_2$ for decoupling.

Partial decoupling has also been applied successfully between composition loops on distillation columns. This application is described in Chap. 11.

### Dynamic Compensation

Now that we have reduced the problem of decoupling to one of feedforward control, there is little need to elaborate further on implementation. Steady-state accuracy, as with feedforward systems, is of paramount importance. Dynamic compensation, while helpful, is ordinarily added after steady-state calibration has been achieved. It typically takes the form of a lead-lag function, with dead time where required. Tuning is the same as for a feedforward system; step changes can be introduced by a manually operated controller or by applying a constraint, where available.

Consider partial decoupling applied to the paper-machine headbox described by Eq. (8.39). Let total head be decoupled from level because it directly influences paper quality. As in a feedforward system, the manipulated variable is calculated in terms of the controlled set point and the disturbance variable. Rearranging Eq. (8.39) and replacing $c_1$ with $r_1$, we have

$$m_1 = \frac{r_1 - c_2}{K_{11}\mathbf{g}_{11}} \tag{8.59}$$

where $K_{11}$ is an adjustable steady-state gain representing $(\partial c_1/\partial m_1)_c$ and $\mathbf{g}_{11}$ is the lag in the response of $c_1$ to $m_1$. The inverse of a lag is a lead, which is an unrealizable function. A lead-lag unit would then be applied, with lead time set to match $\mathbf{g}_{11}$ and lag time long enough to reduce noise response to an acceptable level.

Feedback must also be added. Again, as with most feedforward systems, it is conveniently inserted where the set point appears in the process model. Although Eq. (8.59) shows dynamic compensation applied to both set point and disturbing variable, the set point needs it only if it is subject to frequent

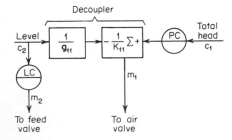

**FIG. 8.19** As with feedforward systems, dynamic compensation is kept out of the feedback loop.

variation. Because $g_{11}$ is already a small lag with respect to the response of the paper machine itself, this feature is probably unnecessary.

In the system shown in Fig. 8.19, note that the dynamic compensator is kept out of the feedback loop, as was done with feedforward systems. The feedback controller has the necessary adjustments to achieve loop stability.

## REFERENCES

1. Bristol, E. H.: On a New Measure of Interaction for Multivariable Process Control, *IEEE Trans. Autom. Control*, January 1966.
2. Toijala, K., and K. Fagervik: A Digital Simulation Study of Two-Point Feedback Control of Distillation Columns, *Kem. Teollisuus*, January 1972.
3. Shinskey, F. G.: The Stability of Interacting Control Loops with and without Decoupling, *IFAC Conf. Multivariable Technol. Syst. Univ. New Brunswick, Can., July 5–8, 1977.*

## PROBLEMS

**8.1**  Derive the relative gain terms for the blending system described in Fig. 8.4. Start with a material balance on the process, assuming that $m_1$ and $m_2$ are mass flow rates of streams having compositions $x_1$ and $x_2$ of the component analyzed. Leave your results in terms of controlled composition $x$.

**8.2**  Prepare a relative-gain array for the process described in Fig. 8.12. Use it to explain why single-loop control of all variables could not be achieved.

**8.3**  A certain process is described by the following matrix of open-loop gains with all loops open:

$$\mathbf{K} = \begin{bmatrix} 0.58 & -0.36 & -0.36 \\ 0.73 & -0.61 & 0 \\ 1 & 1 & 1 \end{bmatrix}$$

Derive the relative-gain array and select the best control loops. Should decoupling be considered?

**8.4**  It is desired to control both the temperature $T$ and the pressure $p$ in a chemical reactor by manipulating coolant temperature $T_c$ and reagent flow $F$. Gain $\partial T/\partial T_c$ at constant flow is 1, and $\partial p/\partial T_c$ is 0.4 (lb/in$^2$)/°F; $\partial T/\partial F$ at constant $T_c$ is 12°F/(gal/min), and $\partial P/\partial F$ is 4.8 (lb/in$^2$)/(gal/min). Calculate the relative gains and explain your result.

**8.5**  Evaluate the decoupled relative gain of the system shown in Fig. 8.18.

**8.6**  Devise a partial decoupler for the process of Fig. 8.1, assuming flow to be the more critical variable. Is dynamic compensation required?

**Part Four**

# Applications

# Chapter Nine

# Energy Transfer and Conversion

The principles governing energy transfer and conversion apply to a broad spectrum of processes from combustion to heat transfer to pumping and compression. In some applications, energy appears in the form of heat, whereas in others it exists as work. In any process, the sum of work and heat in and out must balance, according to the *first law of thermodynamics*. The *second law of thermodynamics* is concerned with the relationship between heat and work. Work, defined as that form of energy which can be used to lift a weight, can be quantitatively converted into heat by means of friction and impact. However, only a portion of the thermal energy in a hot fluid or body is convertible into work. Work, therefore, is the superior form of energy, and care should be taken to avoid degrading it into heat insofar as possible. Electricity is essentially pure work, in that it can be converted by means of a motor into mechanical energy with very little loss. On the other hand, the conversion of fuel energy into electricity in a fossil-fueled power plant is only about 34 percent; the balance leaves as waste heat.

This chapter describes control over processes in which work is converted into heat and vice versa as well as simply transferred from one medium to another.

## HEAT TRANSFER

In most heat-transfer processes, a conductive barrier separates the fluids which are exchanging heat, but there are also direct-contact exchangers, where the fluids mix to form a single discharge. Because this is the simplest to explain, it is covered first.

### Direct Contact

This process is typified by the blending of hot and cold streams to form a single stream having an intermediate temperature. This temperature is ordinarily the controlled variable, although the flow of the blend may also be controlled. Mass and energy balances are

$$W_H + W_C = W \tag{9.1}$$
$$W_H H_H + W_C H_C = WH \tag{9.2}$$

where $W$ is the mass flow and $H$ the enthalpy of the blend and the subscripts identify the hot $H$ and cold $C$ inlet streams. Enthalpy of the blend is found by combining the two balances

$$H = H_C + \frac{W_H}{W_H + W_C}(H_H - H_C) \tag{9.3}$$

If the two fluids have the same heat capacity and there is no change of phase, temperatures can be substituted for enthalpies

$$T = T_C + \frac{W_H}{W_H + W_C}(T_H - T_C) \tag{9.4}$$

The solution to Eq. (9.4) is plotted in Fig. 9.1.

Observe the nonlinearity of the relationship between the controlled and manipulated variables. Process gain will change both with temperature set point and flow. Dead time from the point of mixing to the temperature-measuring element will also vary with flow. These properties are typical of most temperature-control loops.

In addition, there is the prospect of interaction between total-flow and blend-temperature loops. The relative gain of blend temperature $T$ in response to hot-fluid $W_H$, as derived from Eq. (9.4), is

$$\lambda_{TH} = \frac{T_H - T}{T_H - T_C} = \frac{W_C}{W_H + W_C} \tag{9.5}$$

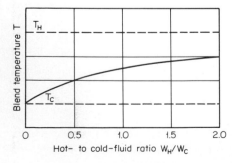

**FIG. 9.1** The sensitivity of temperature to either manipulated flow varies both with the control point and with the other flow.

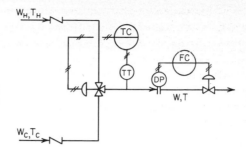

**FIG. 9.2** A three-way mixing valve can be used to decouple temperature from flow.

Like the component-blending interaction described in the previous chapter, relative gain varies with the fraction of flow delivered by the other valve.

Decoupling can be achieved by using a three-way mixing valve for temperature control, along with a two-way valve in the common line for flow control. Check valves should be added in the supply lines as shown in Fig. 9.2, however, to ensure that one fluid does not enter the other line at low demand rates.

### Fluid-Fluid Heat Exchangers

Heat transfer from one fluid to another through a barrier surface is determined by driving force and resistance

$$Q = UA\,\Delta T_m \tag{9.6}$$

Control of heat flow $Q$ can thus be effected by manipulating the heat-transfer coefficient $U$, surface area $A$, or the mean temperature difference $\Delta T_m$ between the fluids.

Even if $U$ and $A$ could be maintained constant, Eq. (9.6) still contains two variables. The objective of most heat exchangers is the control of temperature, which varies with heat-transfer rate but which also affects the rate of heat transfer, as Eq. (9.6) indicates. Consequently most heat-transfer processes are highly self-regulating.

Further equations are necessary to close the loop, by relating fluid temperatures to heat flow. But a heat exchanger involves two fluids whose temperature distributions from inlet to outlet are both subject to change, both affecting $\Delta T_m$. For the general case, consider heat transfer between two fluids with no change in phase, as shown in Fig. 9.3.

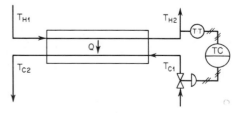

**FIG. 9.3** The general case is heat transfer between hot and cold fluids in counterflow.

The temperature difference affecting heat transfer between the two fluids in Fig. 9.3 is actually a logarithmic mean

$$\Delta T_{lm} = \frac{(T_{H1} - T_{C2}) - (T_{H2} - T_{C1})}{\ln\left[(T_{H1} - T_{C2})/(T_{H2} - T_{C1})\right]} \tag{9.7}$$

In most cases, fortunately, the arithmetic mean is sufficiently accurate for indicating the relationships between the variables, if not for use in equipment design:

$$\Delta T_{am} = \frac{(T_{H1} - T_{C2}) + (T_{H2} - T_{C1})}{2} \tag{9.8}$$

The error approaches zero as the temperature differences at the ends of the exchanger approach each other and is less than 10 percent with a 4:1 ratio of temperature difference.

Each of the two fluids will be assigned a mass flow $W$ and a specific heat $C$. One of the flows ordinarily is wild and represents the load on the exchanger; the other is often manipulated in some way to control the exit temperature of the first. Temperature changes in both streams are interrelated

$$Q = W_H C_H (T_{H1} - T_{H2}) = W_C C_C (T_{C2} - T_{C1}) \tag{9.9}$$

Equations (9.6), (9.8), and (9.9) contain four expressions with four unknowns, $Q$, $\Delta T_{am}$, $T_{H2}$, and $T_{C2}$. They can be solved simultaneously for any of the four unknowns. The solution for heat-transfer rate is the least complicated

$$Q = \frac{T_{H1} - T_{C1}}{1/UA + \tfrac{1}{2}(1/W_H C_H + 1/W_C C_C)} \tag{9.10}$$

Heat-transfer rate can be normalized by dividing by its maximum possible value, which would occur with both streams at infinite flow such that $T_{H2} = T_{H1}$ and $T_{C2} = T_{C1}$

$$Q_{max} = UA(T_{H1} - T_{C1}) \tag{9.11}$$

$$\frac{Q}{UA(T_{H1} - T_{C1})} = \frac{1}{1 + (UA/2)(1/W_H C_H + 1/W_C C_C)} \tag{9.12}$$

Figure 9.4 is a plot of normalized heat-transfer rate vs. normalized flow of

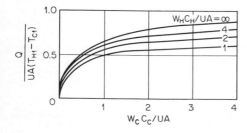

**FIG. 9.4** Manipulation of flow has little effect on heat transfer at high flow rates.

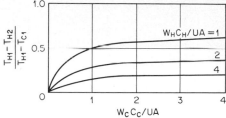

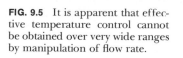

**FIG. 9.5** It is apparent that effective temperature control cannot be obtained over very wide ranges by manipulation of flow rate.

cold fluid with the flow of hot fluid as a parameter. Observe the extreme nonlinearity of the curves and how ineffective the manipulation of flow is over a wide operating range.

Substitution of Eq. (9.9) into (9.11) yields the following formulas describing dimensionless temperatures as a function of flow rates:

$$\frac{T_{H1} - T_{H2}}{T_{H1} - T_{C1}} = \frac{1}{W_H C_H / UA + \frac{1}{2}(1 + W_H C_H / W_C C_C)} \tag{9.13}$$

$$\frac{T_{C2} - T_{C1}}{T_{H1} - T_{C1}} = \frac{1}{W_C C_C / UA + \frac{1}{2}(1 + W_C C_C / W_H C_H)} \tag{9.14}$$

To envision what effect flow rates have upon exit temperature, Eq. (9.13) is plotted in Fig. 9.5 with the same abscissa and parameter that were used in Fig. 9.4.

Not only does the slope of the curves change with temperature, but it also changes with load $W_H$. Any horizontal line drawn across Fig. 9.5 will present the conditions required for temperature control. Doubling the load at any given temperature requires the manipulated variable $W_C$ to be much more than doubled.

In practice, the overall heat-transfer coefficient also varies with the flow rates, which improves the controllability somewhat. Although the film coefficient on each side of the heat-transfer surface varies at about the 0.8 power of the fluid velocity, for simplification it will be assumed that the relationship is linear. Furthermore the reciprocal of the overall heat-transfer coefficient will be assumed to be the sum of the reciprocals of the individual film coefficients

$$\frac{1}{U} = \frac{1}{W_H k_H} + \frac{1}{W_C k_C}$$

The terms $k_H$ and $k_C$ are the flow indexes of their respective heat-transfer coefficients. Combining with Eq. (9.13) yields

$$\frac{T_{H1} - T_{H2}}{T_{H1} - T_{C1}} = \frac{1}{C_H / A k_H + \frac{1}{2} + (W_H C_H / W_C C_C)(C_C / A k_C + \frac{1}{2})} \tag{9.15}$$

A plot of Eq. (9.15) for conditions of $C_H / A k_H = C_C / A k_C = 1$ is given in Fig. 9.6. Compare it with the curves of Fig. 9.5.

The point of the foregoing analysis has been to demonstrate the nonlinear

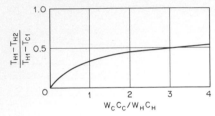

**FIG. 9.6** Variation of the heat-transfer coefficient with flow somewhat eases the nonlinearity of the process.

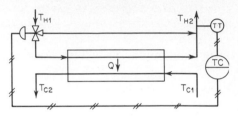

**FIG. 9.7** Bypassing the exchanger will not improve linearity but does reduce response time.

properties associated with heat transfer. Even under the most favorable conditions, manipulation of flow is far from satisfactory for temperature control. There are practical considerations, too. Throttling streams which may contain impurities (river water, for example) can cause deposits to accumulate, fouling the heat-transfer surfaces. Furthermore, manipulation of flow causes variable loop gain through the variation of dead time. In the event that there is no alternative to the manipulation of flow, an equal-percentage valve characteristic should be chosen.

Part of the stream whose temperature is to be controlled may be allowed to bypass the exchanger, as shown in Fig. 9.7. But Fig. 9.4 indicates that the rate of heat transfer is scarcely affected by the flow of either stream for reasonable rates of flow. If the heat-transfer rate is nearly constant, the final temperature of the process stream after reunion with the bypassed flow will also be nearly constant; consequently the linearity of response is not noticeably improved.

Bypassing can help the dynamic response, however, in that the flow of coolant is maintained at a high rate, rather than being throttled, as it would be if it were the manipulated variable. Furthermore, the bypass stream shortens the time delay between a change in valve position and the response of final temperature.

In fact, *without* the bypass shown in Fig. 9.7, the temperature-control loop tends to limit-cycle, as described in Ref. 1. The dead time in the response of exit temperature to manipulated flow varies with the flow, developing a nonsinusoidal cycle. A falling temperature will cause the controller to increase flow; this shortens the dead time and causes a rapid correction. However, when the temperature is too high, flow will be reduced, retarding the response to control action. Furthermore, a changing load will cause a subsequent change in manipulated flow, bringing about a proportional variation in period and process gain. Then the controller cannot be adjusted to any reasonable degree of satisfaction.

### Boiling Liquids and Condensing Vapors

The control situation is much more favorable when there is a change in phase. Because the latent heat of vaporization $H_v$ predominates, a measurement of the mass flow $W$ of the boiling or condensing medium is also a measure of the rate of heat transfer

$$Q = WH_v \tag{9.16}$$

Furthermore, the temperature of the boiling or condensing medium scarcely changes from inlet to outlet of the exchanger.

Whenever steam is used as a heating medium, manipulation of its flow to bring about temperature control of the process fluid is effective. If the process fluid is boiling, steam flow directly infers its rate of vaporization. The pressure of the steam in the exchanger is only an indication of steam temperature and is not a particularly useful measure of heat transfer; it can be used to estimate the heat-transfer coefficient, however.

Exchangers supplied with steam as a heating medium exhibit a strong tendency toward self-regulation. Since the film-transfer coefficient for condensing steam is much greater than for a flowing gas or liquid, the rate of heat transfer is principally governed by the film coefficient of the process fluid. Since this coefficient varies almost linearly with fluid velocity, heat transfer will vary almost linearly with flow if steam temperature is maintained. The latter is achieved simply by regulating the pressure of the steam in the exchanger. Thus without being directly controlled, the exit temperature of the process fluid will nonetheless be well regulated. Process-fluid temperature can be controlled very effectively by setting steam pressure in cascade.

The flow of steam to a process heater or reboiler can also be manipulated by a valve in the condensate line. The rate of heat transfer is actually changed by partially flooding the exchanger with condensate. Because a change in condensate level is necessary to influence steam flow, this system may respond more slowly than direct manipulation of steam flow, but it has the distinct advantage of requiring a much smaller valve.

Whether sufficient heat has been removed to condense a vapor totally can be determined by the temperature of its condensate if constant pressure prevails or, more accurately, by vapor pressure if the vessel is closed. Control of condensate temperature or vapor pressure is complicated by the flow of the condensing vapor being the load and not the manipulated variable. The relationship between heat transfer and coolant flow $W_C$ can be found simply by solving the equations developed earlier using constant temperature $T_v$ for the condensing vapor

$$Q = \frac{T_v - T_{C1}}{1/UA + 1/2W_C C_C} \tag{9.17}$$

Notice the similarity between Eqs. (9.17) and (9.10). This indicates that the response of heat transfer to coolant flow will be identical to the curve $W_H C_H / UA = \infty$ of Fig. 9.4. For the manipulation of coolant flow, then, the nonlinearity problem is just as severe as it is when there is no phase change.

Under conditions of constant condensate temperature, the heat-transfer rate is entirely dependent upon coolant flow. If coolant flow is maintained constant, bypassing part of the vapor around the condenser will not affect the rate of heat transfer unless the condensate becomes appreciably subcooled. Under these conditions, the condenser begins to act more like the liquid-liquid heat exchanger, described in Fig. 9.7.

The most effective way to control a condenser is to vary its heat-transfer area. This is done by manipulating the flow of condensate so as to flood the condenser partially, thereby reducing the surface available for condensation. The level of condensate within the condenser is an indication of the heat load on the process. The system is described in Fig. 9.8.

In a flooded condenser a certain amount of subcooling always takes place in whatever area is not used for condensing. The amount of subcooling varies with the flow of vapor, so condensate temperature cannot be used for control. If the heat-transfer coefficients for condensing and subcooling were equal, this system would have no control over vapor pressure at all because heat-transfer rate would not depend on liquid level. Fortunately, heat-transfer coefficients of condensing vapors are generally much greater than those of condensate, particularly if the velocity of the condensate is low, as it would be in the shell of the condenser.

On the other hand, manipulation of liquid level is a slow process, with 90° phase lag between valve position and heat-transfer area. Since vapor pressure is a fast measurement, however, the loop generally performs well dynamically, except perhaps for severe load changes requiring the condenser to be filled or emptied. Linearity and rangeability are important factors in its favor.

Flooded heat exchangers are generally non-self-regulating in that they depend on liquid level for control. Consequently the controller, whether it is a pressure, temperature, or steam-flow controller, must be adjusted like a liquid-level controller. However, the heat capacity in the system may filter noise well enough to permit the use of derivative. A valve positioner will be quite helpful in stabilizing the loop and improving response, even for steam-flow control.

Air-cooled condensers are in common use for distillation columns and refrigeration systems. Heat transfer is quite sensitive to ambient conditions, particularly rainfall. Coolant-side manipulations include louvers, variable-pitch fans, and variable-speed drives, listed in order of both increasing cost and capability for saving fan power. Two-speed motors can be used in conjunction with other controls to extend their operating range while saving energy.

Process-side manipulations will be found more satisfactory from a control standpoint, with flooding and bypassing both commonly used. Chapter 11 describes a method for taking advantage of the influence of weather over air-cooled condensers to save energy in distillation.

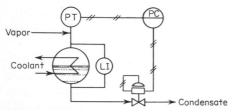

**FIG. 9.8**  The heat-transfer area available for condensation can be changed by manipulating the flow of condensate.

## COMBUSTION CONTROL

When a fuel burns, the products of combustion, along with whatever other vapors may be present, are raised to a flame temperature determined by the energy content of the fuel. Since heat of combustion is rated in Btu per pound or Btu per cubic foot, the actual quantity of fuel involved does not affect its flame temperature. To estimate the flame temperature, the sensible heat of either the combustion products or the fuel and air may be used since the energy balance can be satisfied in either case. The rate of heat generated by the combustion of a given mass flow of fuel $W_F$, whose heat of combustion is $H_c$, is

$$Q = W_F H_c \tag{9.18}$$

This flow of heat must equal what is necessary to raise the flows of fuel and air $W_A$ to the flame temperature $T$

$$Q = W_F C_F (T - T_F) + W_A C_A (T - T_A) \tag{9.19}$$

The terms $C_F$, $T_F$, $C_A$, and $T_A$ represent the average specific heat and the inlet temperature of fuel and air, respectively.

To ensure complete combustion, a specified ratio of air to fuel $K_A$ must be selected, based upon the chemical constituents in the fuel. Substitution of $K_A$ for $W_A/W_F$ will allow the solution of Eqs. (9.18) and (9.19) for flame temperature

$$T = \frac{H_c + C_F T_F + K_A C_A T_A}{C_F + K_A C_A} \tag{9.20}$$

Equation (9.20) must be recognized as being valid only for conditions where there is no excess fuel. Because fuel is more expensive than air, and because incomplete combustion can cause soot and carbon monoxide, furnaces are invariably operated with excess air. But it should be apparent that the maximum flame temperature will be reached only with no excess of either. Equation (9.20) also gives an indication of the effect air temperature can have on the flame. The nitrogen, of course, does not participate in combustion and acts as a diluent, reducing the flame temperature. If oxygen is used instead of air, $K_A$ can then be reduced fivefold, producing a sizable effect on flame temperature.

The flame temperature estimated in Eq. (9.20) will be higher than what would actually be measured, because some of the energy contained in the combustion products partially ionizes them. This ionization increases with temperature, but the energy is recovered when the ions cool sufficiently to recombine into molecules.

### Control of Fuel and Air

Since the temperature of the flame falls with either an excess or a deficiency of air, it is not a particularly good controlled variable. The most universally used indication of combustion efficiency is a measurement of oxygen content in the combustion products. The amount of excess air required to ensure complete combustion depends on the nature of the fuel. Natural gas, for example, can be burned efficiently with 5 percent excess air (0.9 percent excess oxygen), while oil requires 6 percent excess air (1.1 percent excess oxygen) and coal, 10 percent excess air (1.9 percent excess oxygen). The reason for the differences is the relative state of the fuel.

Since the amount of heat transferred by radiation varies with the fourth power of the absolute flame temperature, the greatest furnace efficiency will always be realized with maximum flame temperature. But the distribution of the heat is also important. Increasing the amount of excess air will reduce the flame temperature, thereby reducing the heat-transfer rate in the vicinity of the burner. Since the net flow of thermal power into the system has not changed, the rate of heat transfer farther away from the burner tends to increase.

Safety dictates certain operating precautions for fuel-air controls. A deficiency of air can allow fuel to accumulate in the furnace, which upon ignition, may explode. Care must therefore be taken to ensure that the fuel rate never exceeds what is permissible for given conditions of airflow. Fuel and airflow both can be set from a master firing-rate control, but automatic selection is necessary to achieve this safety feature [2]. A system for control of fuel and air is shown in Fig. 9.9 [2].

Notice that the fuel-air ratio is adjusted through manipulation of the span of the air measurement by the oxygen controller. Normally the set point would be adjusted, but in order for the selection system to operate, the set points of both controllers must have the same values. If airflow is lost, its measurement is preferentially selected to lower the fuel-flow set point. If fuel flow is higher than called for, on the other hand, its measurement is automatically selected to raise the airflow. The furnace is thereby protected not only from blower or controller failure but also from lags in the set-point response of either loop.

Gilbert [3] recommends control over carbon monoxide in the flue gas rather than its oxygen content. His argument is that air infiltration or poor mixing at a burner could cause a high oxygen reading even with incomplete combustion. By contrast, carbon monoxide concentration is virtually unaffected by infiltration and is a true measure of the condition of the flue gas.

**Fired Heaters**

Heaters fired directly by the combustion of gas or oil are common in refineries, particularly where high temperatures are needed. The control problem is one

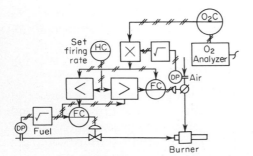

**FIG. 9.9** This system automatically protects against a deficiency of air.

of manipulating fuel rate to achieve the desired exit temperature of the heated fluid. Air is usually inspirated into the burner in proportion to the fuel; therefore regulation of its flow is inherent.

Because of the many hundreds of feet of tubing enclosed within a heater, dead time is in the order of minutes, varying with flow. Where sudden load changes are encountered and close control of outlet temperature is necessary, feedforward systems have proved effective. The heat-balance equation is similar to that solved for the heat exchanger in Fig. 7.3. The only difference is that fuel flow is manipulated instead of steam, and heat of combustion takes the place of latent heat of vaporization. Although the loss of heat out the stack may be significant, it varies directly with load and can be readily accommodated by the action of the feedback temperature controller, as in Fig. 7.24.

Should the fuel be gas at a variable temperature or pressure, computation of mass flow may be warranted, particularly if these variations are frequent or rapid.

## STEAM-PLANT CONTROL SYSTEMS

In order to apply controls to the generation of steam successfully, a thorough familiarity with its thermodynamic properties is essential. The most important point to remember is that steam is valued principally for its work content, which is the ability to drive turbines, pump heat, move fluids, etc. Its available work increases with the logarithm of pressure and with temperature [4a]. Pressure is controlled by applying heat to the evaporative heat-transfer surfaces. Superheat is added by applying more heat to the steam after it is removed from equilibrium with the boiling water.

Steam is also superheated by passing through a control valve or orifice, but this results in a loss in available work. Therefore throttling steam to reduce its pressure should be avoided wherever possible.

The mass flow of steam can be measured with an ordinary orifice meter, but the reading must be corrected if pressure or temperature deviate from the conditions under which calibration was specified. In the case of saturated steam, pressure and temperature are not independent of each other, so either one is capable of indicating density. It happens, however, that pressure is a linear function of density, with an intercept of 0 lb/in² gage

$$W = k \sqrt{hp} \qquad (9.21)$$

where $W$ = mass flow
$k$ = orifice scaling factor
$h$ = differential pressure across flowmeter
$p$ = static gage pressure

The density of superheated steam varies inversely with temperature and directly with pressure to make the mass-flow calculation [5] more complicated and less accurate. But if a steam flowmeter is used to indicate the actual delivery of thermal power, an interesting phenomenon appears: temperature causes the enthalpy of superheated steam to vary in a way which offsets its effect upon density. Thus thermal power $Q$ only varies with differential and pressure

$$Q = WH = kH_0 \sqrt{hp} \tag{9.22}$$

Coefficients $H$ and $H_0$ represent steam enthalpy at flowing and calibration conditions, respectively.

### Drum-Level Control

In a drum boiler, water is circulated at a rapid rate upward through the furnace tubes, in which it partially vaporizes. Upon reaching the drum, the liquid disengages from the vapor and returns through relatively cool downcomers to the bottom of the furnace to begin another pass upward. The most characteristic feature of drum boilers is the difficulty of controlling the level of liquid in the steam drum. A feedforward-feedback system for its control was described briefly in Chap. 7.

Introduction of feedwater at a temperature below that of the boiling water in the drum causes some internal condensation. A sudden increase in flow can then momentarily reduce the rate of boiling. The liquid in the drum is supported by a rising current of bubbles from the evaporating tubes. When these bubbles collapse as a result of a reduction in boiling rate, the liquid level in the drum falls. Consequently, an increase in feedwater flow can actually cause liquid level to fall momentarily before the increasing liquid inventory begins to raise it, as shown in Fig. 9.10.

Dynamically, inverse response is characterized by a lag accompanied by a negative lead. The negative lead produces the familiar gain characteristics of a positive lead but develops a phase lag. The gain and phase of the combination of negative lead $\tau_1$ and lag $\tau_2$ are

$$G_i = \sqrt{\frac{1 + (2\pi\tau_1/\tau_0)^2}{1 + (2\pi\tau_2/\tau_0)^2}} \tag{9.23}$$

$$\phi_i = -\tan^{-1}\frac{2\pi\tau_1}{\tau_0} - \tan^{-1}\frac{2\pi\tau_2}{\tau_0} \tag{9.24}$$

For the particular case where $\tau_1 = \tau_2 = \tau_i$, $G = 1.0$ for all periods, and

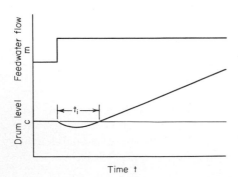

FIG. **9.10** Inverse response is characteristic of liquid level in a drum boiler.

$$\phi_i = -2 \tan^{-1} \frac{2\pi\tau_i}{\tau_o} \tag{9.25}$$

If this function is combined with an integrating element, as it would be in a level-control application, the natural period of the loop will be that at which $\phi_i = -90°$. Under these conditions,

$$\tau_n = 2\pi\tau_i \tag{9.26}$$

The relative magnitude of the time constant $\tau_i$ can be inferred from a step response by noting the time required for the controlled variable to cross its original position. This time is designated as the *inversion time* $t_i$ in Fig. 9.10. The step response for this process is

$$c(t) = m \frac{t - 2\tau_i(1 - e^{-t/\tau_i})}{\tau_I} \tag{9.27}$$

where $t$ is time and $\tau_I$ is the time constant of the integrator. Time $t$ is equal to the inversion time $t_i$ when $c(t) = 0$. A trial-and-error solution of (9.27) at $c(t) = 0$ yields $t_i = 1.594\tau_i$. If $t_i$ is then substituted for $\tau_i$ in Eq. (9.26), the natural period is expressed in terms of inversion time

$$\tau_n = \frac{2\pi}{1.594} t_i = 3.94 t_i \tag{9.28}$$

Observe that in this application, inverse response is virtually equivalent to dead time; its dynamic gain is unity, and it develops a natural period about 4 times its observed value. This explains why the period of the drum-level loop is much longer than what could be caused by hydraulic resonance and measurement and valve lags.

If the boiler must operate under varying steam pressure, the calibration of the liquid-level transmitter will vary with steam density [6]. But pressure has a transient effect too. If a load increase (withdrawal of steam) is sufficient to cause drum pressure to fall, some of the water in the tubes will flash into steam, temporarily increasing the flow of both liquid and vapor into the drum. This effect is called *swell* because it causes a transient rise in liquid level, even though the rate of steam withdrawal may momentarily exceed that of feedwater flow. Conversely, upon a pressure increase, the liquid level tends to *shrink*. This effect is more prominent in low-pressure boilers because of the greater difference between the densities of steam and water. The most favored method of coping with shrink and swell is to ignore them, by letting the forward loop carry the load, while maintaining loose settings on the level controller. Drum-level controllers customarily require a proportional band near 100 percent and several minutes' integral time.

## Drum-Pressure Control

Pressure in a saturated or even a superheated boiler is a measure of the amount of energy stored in it. The flow of steam from the plant is usually at the demand of the user. Pressure can be maintained, then, only if the flow of energy into the boiler equals the rate of withdrawal. Since the drum-level control system admits feedwater at a rate equal to the flow of steam, the pressure-control system is left to manipulate the input of thermal power. To achieve high-performance control, a feedforward loop should be used to set firing rate proportional to steam flow.

Steam flow is a measure of thermal power and is affected by firing rate. If it alone is used to set firing rate, a positive feedback loop will be formed, a pitfall that was mentioned under decoupling systems in Chap. 8. What is really needed is a steam-flow demand signal, such as a user valve position. Actually such a signal can be generated by appropriately combining steam flow and pressure.

Consider user demand to be represented by a single valve whose thermal power $Q$ varies with opening $a$ and pressure $p$

$$Q = ap \qquad (9.29)$$

Then $a$ is the demand for thermal power at constant pressure. When Eq. (9.22) for the steam flowmeter is combined with (9.29), a solution can be obtained for power demand

$$a = kH_0 \sqrt{\frac{h}{p}} \qquad (9.30)$$

A feedforward system using Eq. (9.30) as an index of load is shown in Fig. 9.11.

A sudden increase in demand will raise steam flow immediately. The feedforward system will respond by increasing firing rate at the same time. But before the effect of the increased heat input can be felt, steam flow and pressure will begin to fall; however, they will tend to fall in a constant ratio, so that firing will be maintained. If pressure were not incorporated in the calculation, or if it were used in a multiplier rather than a divider, falling steam flow and pressure would cause firing rate to fall again. The resulting response would be sluggish and marginally stable.

## Steam-Temperature Control

The temperature of superheated steam is controlled by varying the intensity of heat directed toward the superheaters as opposed to the evaporating tubes.

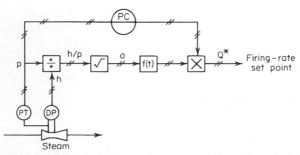

**FIG. 9.11**  Demand for thermal power is calculated from measured flow and steam pressure.

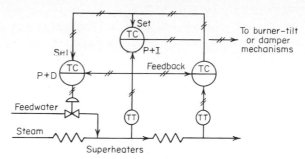

**FIG. 9.12**   Control action is split, with proportional and derivative being applied to the attemperator valve.

This can be accomplished either by tilting burners or by recirculating or bypassing flue gas. Trim control is often applied by attemperating the superheated steam with feedwater. This practice results in lost work, however, in that the evaporating heat-transfer surface is bypassed by the attemperating water, thereby raising internal temperatures higher than they need to be [4b].

The control system shown in Fig. 9.12 minimizes the flow of attemperating water while maintaining responsive temperature control. The superheater-outlet temperature controller sets inlet temperature in cascade, with secondary measurement as feedback for the primary's integral mode. This avoids windup at low loads when high temperatures cannot be maintained.

Inlet temperature control is split, with proportional and derivative action applied to attemperation. At zero deviation, the attemperator valve will always return to the same position, equal to the controller's output bias. The PI controller modulates burner or damper positions; their effect on superheat is slower than the attemperating valve, but they are necessary for long-term control. Split control functions are also shown for evaporators in Chap. 12.

## PUMPS AND COMPRESSORS

There are many ways to control the flow and pressure of streams discharging from pumps and compressors, but all are not equally efficient. Throttling a valve in the discharge line of a centrifugal pump may be convenient, but it may also be wasteful if long periods of low flow are encountered, and it cannot be done at all with a positive-displacement pump because pressure ratings would be exceeded.

The point is that the type and size of a prime mover dictate the means which are to be used to control it. This is particularly true of compressors, which are capable of exhibiting unstable characteristics.

### Positive-Displacement Pumps

There are two principal classes of pumps, positive-displacement and centrifugal. In positive-displacement pumps, a given volume of fluid is mechanically forced from the suction port to the discharge with every rotation of the shaft. In reciprocating pumps, this is done in a periodic fashion, such that outflow pulsates; if multiple cylinders are used, they are phased so as to diminish the amplitude and period of the pulsations, smoothing the flow. Often an air chamber is attached to the discharge line to help absorb these pulsations.

If either the stroke or the speed of a reciprocating pump is adjustable, it can be used for metering an accurate amount of fluid. The accuracy of these

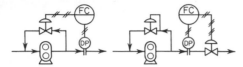

**FIG. 9.13**  Two methods for controlling the flow from a gear or vane pump.

metering pumps requires that they be free from leakage, particularly backflow from discharge to suction. This means that their valves must be tight-sealing, and the fluid free from particles which might interfere with their action. The fluid must also be incompressible; these pumps often vapor-lock when the fluid contains dissolved or entrained gas. When used with clean liquids, metering pumps are valuable for flow control, particularly where high discharge pressures are encountered. They are available with a pneumatic operator to adjust the stroke automatically from a set station or a primary controller.

Other positive-displacement pumps include those which move liquid with rotating gears or vanes. Their output is continuous, although noisy. But as the name implies, positive-displacement pumps must be allowed to discharge their rated flow. They must always be protected by a relief valve connected from discharge to suction; otherwise, if the discharge line is restricted, high enough pressure can be developed to rupture the line or overload the motor.

Control of flow from a gear or vane pump can be achieved by manipulating a bypass or by regulating the discharge pressure with a bypass, both of which are shown in Fig. 9.13.

Any of the positive-displacement pumps can be driven by a variable-speed dc motor. Variable-speed metering pumps are commonly used to deliver reagents in water and wastewater treatment plants. Their rangeability varies from 10 to 20. Typically the motor will stall when its speed setting is reduced below 5 percent, but it will not resume driving until the setting is raised above 10 percent. Under automatic control, the pump will develop a sawtooth limit cycle in the controlled variable when the plant load falls below 10 percent.

### Centrifugal Pumps

Centrifugal pumps exhibit slippage. They impart momentum to the fluid, which is converted to velocity head. At no-flow conditions, rotation of an impeller of a given diameter at constant speed produces the maximum head which the pump is capable of delivering. As flow increases, the head falls by an amount equivalent to frictional losses within the pump itself. It should be noted that the pressure which a centrifugal pump is capable of delivering varies with the density of the fluid, since pressure equals head times density.

The characteristics of a centrifugal pump are usually plotted as head vs. flow on linear coordinates. Contours of speed, horsepower, and efficiency are often included. The choice of linear coordinates is unfortunate in the sense that only the horsepower contours are straight lines. If speed is an important variable, plotting head vs. flow squared will yield straight contours. The two plots are compared in Fig. 9.14. Discharge pressure $p$ varies with speed $N$, flow $F$, and density $\rho$ as follows:

$$p = \rho(k_1 N^2 - k_2 F^2) \tag{9.31}$$

The coefficients $k_1$ and $k_2$ are functions of the mechanical parameters of the pump, i.e., impeller diameter, clearance, etc.

Small pumps are usually driven by constant-speed electric motors. Flow can be controlled by throttling a valve in the discharge line. The suction should never be throttled because a centrifugal pump requires a positive suction head to operate. Low suction pressure causes cavitation and loss of flow.

The hydraulic horsepower (hhp) imparted to a fluid is defined as the product of the pressure (not head) developed and the flow delivered

$$\text{hhp} = \frac{Fp}{1714} \tag{9.32}$$

The units of flow are gallons per minute, and the units of pressure are pounds per square inch. (Replacing 1714 with 98 yields power in kilowatts for flow expressed in cubic meters per second and pressure in kilograms per square centimeter.) If flow is shut off, or if pressure is lost, hhp falls to zero. But the pump nevertheless absorbs power from the drive, which indicates that its efficiency has gone to zero. Like most machinery, centrifugal pumps operate most efficiently in the middle of their pressure-flow range. The point of maximum hhp can be found by multiplying the right side of Eq. (9.31) by flow to give hhp, then differentiating with respect to flow. Solving the equation for zero derivative gives the optimum flow $F_o$

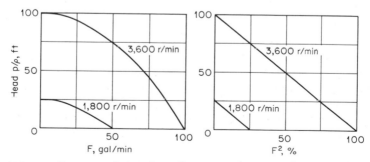

FIG. 9.14  By proper choice of coordinates, speed contours for a centrifugal pump are made into straight lines.

$$F_o = N \sqrt{\frac{k_1}{3k_2}}$$

For the characteristics given in Fig. 9.14, maximum hhp would occur at 58 gal/min for 3600 r/min, and 29 gal/min for 1800 r/min. This confirms that speed should be varied with flow if the most efficient conditions are to be maintained. Manipulation of speed is therefore economically justified with large pumps.

Many pumps are driven by steam turbines, which are equipped with governors capable of being set pneumatically or electrically. Others are driven by constant-speed electric motors through hydraulic or magnetic couplings. The speed of the pump shaft can usually be varied to 100 percent of motor speed by adjusting the degree of coupling.

To fully assess the characteristics of speed manipulation, the flow process must be combined with the pump parameters. Flow through a process whose flow coefficient is $k_3$ is given by

$$F^2 = k_3 \frac{p}{\rho} \tag{9.33}$$

Substituting for $F^2$ in Eq. (9.31) yields the response of discharge pressure to speed

$$p = \frac{k_1 N^2 \rho}{1 + k_2 k_3} \tag{9.34}$$

Solving for flow gives

$$F = N \left( \frac{k_1}{k_2 + 1/k_3} \right)^{1/2} \tag{9.35}$$

Manipulating speed is very much like positioning a linear valve.

Note that hhp varies with speed cubed. This is favorable for variable-speed couplings, in that the load placed on them is quite low when speed is gradually increased from zero.

### Compressor Control

Reciprocating compressors are considered in the same light as pumps except that their outflow can be measured easily. Their higher speed and the compressibility of the fluid help to reduce pulsations. Control methods previously described for gear and vane pumps (Fig. 9.13) can also be used for reciprocating compressors.

Multicylinder compressors can have their flow reduced by *unloading* some of the cylinders sequentially. This consists of holding the suction valve open during the entire stroke, effectively disabling the cylinder. Most multicylinder compressors are equipped with solenoid or pneumatic unloaders which can be operated from the output of a controller. Some reciprocating compressors are also fitted with *clearance pockets,* which reduce the volumetric efficiency of cylinders by withholding some of the compressed gas. They are operated in an on-off manner in the same way as unloaders. To minimize limit cycling in the control of a reciprocating compressor, pockets and unloaders can be sequenced in coordination with a bypass valve, using the system shown in Fig. 6.9. Variable-speed operation is more efficient, however, and should be used where practicable.

Centrifugal compressors are similar to centrifugal pumps in the manner in which energy is imparted from the impeller to the fluid. A plot of compression ratio vs. volumetric suction flow, as in Fig. 9.15, gives the familiar parabolic curve. In the low-flow region, however, the compressibility of the fluid gives rise to an unstable condition known as *surge.* The characteristic curve exhibits negative resistance, causing an uncontrollable limit cycle. When flow is reduced below the surge line, the compression ratio falls, further reducing flow, causing the compressed gas in the discharge line to flow back into the compressor. In a few milliseconds, the flow reverses again, but unless the condition is corrected, another cycle will begin in less than a second.

White [7] has described the surge line as a parabola relating adiabatic head to the square of volumetric suction flow. Adiabatic head is a complex function including gas molecular weight, heat-capacity ratio, temperature, and super-compressibility. For low compression ratios, it is relatively linear with compression ratio. If linearity can be assumed, then

$$\left(\frac{p_d}{p_s} - 1\right)\frac{T_s}{w} = k_s F^2 \tag{9.36}$$

where $T_s$ is absolute suction temperature and $w$ the molecular weight of the gas. Volumetric suction flow $F$ cannot be measured directly but is inferred from the differential pressure across an orifice

$$F = k\sqrt{\frac{hT_s}{p_s w}} \tag{9.37}$$

Substituting (9.37) into (9.36) yields the equation for the surge line in terms of measurable pressures. Note that $T_s$ and $w$ cancel

$$h = K(p_d - p_s) \tag{9.38}$$

Here $K$ is the inverse of the slope of the surge line when plotted as compressor differential pressure $p_d - p_s$ against orifice differential pressure $h$.

In actual practice, the relationship is not linear, and individual compressors deviate substantially from theory. Nevertheless, the simple model of Eq. (9.38) serves as a useful starting point for an antisurge control system. Control-system designers should convert data points from actual compressor curves to values of $p_d - p_s$ and $h$ and plot them as in Fig. 9.16. Then they should draw the best

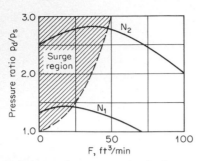

**FIG. 9.15** The surge region is characterized by negative resistance.

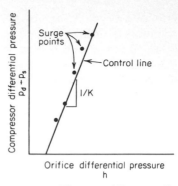

**FIG. 9.16** The control line must lie to the right of all surge points.

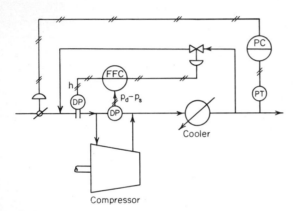

**FIG. 9.17** The bypass valve opens only when the flow drawn into the compressor approaches the minimum calculated for surge protection.

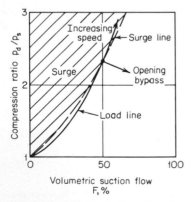

**FIG. 9.18** The load line for the bypass valve rotates normal to the surge line; the bypass valve therefore has more effect on surge than any other manipulation.

straight line to the right of all points. Their control system then must be calibrated to follow low that line.

Figure 9.17 shows the typical arrangement of controls for a centrifugal compressor. The ratio flow controller (FFC) is adjusted to maintain suction flow on the line described in Fig. 9.16. Should flow approach the line, the bypass valve must open immediately to avoid surge. This requires that the flow controller be protected against windup by the circuit shown in Fig. 4.4. Furthermore, the bypass valve should have a linear characteristic, and fail open to protect the compressor against loss of air supply.

The discharge pressure controller is shown manipulating a valve on the suction line; this is preferred to discharge manipulation, in that it affords greater rangeability and lower power consumption at reduced loads. Inlet guide vanes are still more efficient in controlling flow, in that they impart a prerotation to the gas entering the compressor. They also offer a wider range than the suction valve. Speed manipulation is preferred where practical, since it can conserve the most power at low loads.

As may be expected, there is a degree of interaction between the two control loops of Fig. 9.17. The effect of opening the bypass valve is to change the load on the compressor; the valve is represented by a parabolic load line passing through the origin of the coordinates shown in Fig. 9.18. It has very nearly the same shape and origin as the surge curve, rotating around the origin as the valve is stroked. Opening the valve then moves the load line normal to the surge curve, so that most of its effect is in the direction which corrects for surge. By contrast, changing speed at a constant load causes motion up and down the load line, which has little effect over the departure from surge conditions. Therefore the bypass valve should always be used for surge control.

## REFERENCES

1. Shinskey, F. G.: Controlling Unstable Processes, pt. II: A Heat Exchanger, *Instrum. Control Syst.*, January 1975.
2. Manter, D., and R. Tressler: A Coal-Air Ratio Control System for a Cyclone Fired Steam Generator, *ISA Pap.* 1-CI-61.
3. Gilbert, L. F.: Precise Combustion Control Saves Fuel and Power, *Chem. Eng.*, June 21, 1976.
4. Shinskey, F. G.: "Energy Conservation through Control," Academic, New York, 1978, (*a*) p. 24; (*b*) pp. 15–17.
5. Shinskey, F. G.: Analog Computing Control for On-Line Applications, *Control Eng.*, November 1962.
6. Roos, N. H.: Level Measurement in Pressurized Vessels, *ISA J.*, May 1963.
7. White, M. H.: Surge Control for Centrifugal Compressors, *Chem. Eng.*, Dec. 25, 1972.

## PROBLEMS

**9.1** Design a simplified decoupling system for the control of temperature and flow of a mixture of hot and cold water, whose flow rates are the manipulated variables.

**9.2** Feed to a reactor is being preheated countercurrently by oil at a temperature of 500°F. The feed is a liquid, entering the exchanger at 200°F and leaving at a controlled temperature of 400°F. Under normal conditions, the feed rate is 100 lb/min and oil flow is 400 lb/min; the heat capacity of each is 0.8 Btu/(lb)(°F). Estimate the change in oil flow required to maintain control as feed flow varies ±20 percent from normal. Repeat for variations of ±40°F in feed inlet temperature.

**9.3** Calculate the process gain $dT_{c2}/dW_H$ for the normal and the $-20$ percent flow conditions encountered in Prob. 9.2. Then calculate the gain product of process and valve $dT_{c2}/dm$ using an equal-percentage characteristic (let $dW_H/dm = kW_H$).

**9.4** The temperature of condensate leaving a condenser is being controlled by manipulating the flow of cooling water. Suppose $U \approx kW_C$; derive the variation of $Q$ with $W_C$. What are the limitations of this approximation?

**9.5** The level in a drum boiler exhibits an inversion time of 12 s. Estimate the optimum settings of proportional band and integral time if the time constant of the drum is 50 s.

**9.6** Find the values of coefficients $k_1$ and $k_2$ for the pump characteristics given in Fig. 9.14. If 50 gal/min is being drawn as load, what speed is required to control the discharge head at 50 ft? Calculate the hhp required at that speed and also the hhp required to deliver the same flow at 3600 r/min. The fluid is water.

# Chapter Ten

# Controlling Chemical Reactions

The heart of a chemical process is a reaction in which several feedstocks are combined to make one or more products of greater value. Before the plant is designed, process engineers have to determine whether the reaction will proceed at a favorable rate, with sufficient conversion into the principal product to make the operation profitable. Once the optimum conditions for the conduct of the reaction have been established, the control system must be designed to maintain them. But there are many factors affecting the rate of the principal and side reactions, which must be appreciated before an intelligent design can be made.

To a great extent, the success of a reactor control system depends upon how well the reactor is designed. That is, it is possible to design a reactor so that it is unstable no matter what kind of control system is employed. For this reason, a few words will be directed toward reactor design, both to serve as a guide to the process engineer involved in design and to facilitate recognition of an unstable reactor.

In most plants, the reactor is the starting point for the process, and its production rate sets the load. This is fortunate in that the reactor is then free from the rapid or unpredictable load upsets which other units experience. But the reaction system ought to be well regulated; if it is not, it will generate disturbances that will be propagated through the rest of the plant.

## PRINCIPLES GOVERNING THE CONDUCT OF REACTIONS

Unless this section is thoroughly understood, much of the significance of the design procedures that follow will be lost. Many processes can tolerate poorly designed control systems, particularly if they are protected from a rapidly varying load. But most chemical reactors will not. Some need no load upset to break into oscillations sufficiently violent to ruin product, destroy catalyst, damage equipment, and endanger life.

It is therefore not surprising to find many reactors operated in manual control, simply because their automatic systems are not capable of doing their job. Control is usually unsatisfactory in manual, too, but operators tend to have more confidence in themselves than in poorly designed control systems.

### Chemical Equilibrium

A great many chemical reactions are reversible. That is, under certain conditions it is possible to start with the products and make a measurable amount of the reactants. In these cases an equilibrium state can exist in which the reaction comes to a standstill because the forward and reverse rates are equal. This equilibrium point determines how much of the reactants can be converted into products and also what conditions will favor conversion.

As an example, consider the vapor-phase oxidation of sulfur dioxide into sulfur trioxide, one of the steps in the manufacture of sulfuric acid

$$SO_2 + \tfrac{1}{2}O_2 \overset{K}{\rightleftharpoons} SO_3 + \text{heat} \tag{10.1}$$

A certain ratio of product to reactant concentration can be reached which will bring about equilibrium. This ratio is identified by the equilibrium constant $K$, such that

$$K = \frac{[SO_3]}{[SO_2][O_2]^{1/2}} \tag{10.2}$$

Every reversible reaction has an equilibrium constant which is a function of temperature and catalyst.

The existence of an equilibrium state discloses that a certain fraction of the reactants must be withdrawn along with the product. This places a limit on the conversion which can be achieved within a reactor. But several tactics can be used to improve conversion, which can be deduced from Eqs. (10.1) and (10.2):

1. Reduction in product concentration will permit reduction in reactant concentration. Thus, if product can be removed from the reaction zone by condensation, for example, more conversion is possible. Conversion can approach 100 percent if products are easily separated from the reactants as gases or solids from a liquid-phase reaction or as liquids or solids from a vapor-phase reaction.

2. An increase in reactant concentration will also increase conversion. Notice in Eq. (10.2) that $SO_3$ concentration is a function of both the $SO_2$ and $O_2$

concentrations. An increase in either $SO_2$ or $O_2$ will promote conversion; therefore either reactant can be used in excess to augment conversion of the other.

3. For the particular example being used, Eq. (10.1) indicates that the total moles (hence volume) of the reactants (1.5) is greater than that of the product (1). Therefore an increase in pressure will tend to increase the denominator of Eq. (10.2) more than the numerator; this will enhance conversion. This particular reaction should therefore be conducted under pressure.

4. The evolution of heat indicated by Eq. (10.1) also affects equilibrium. Just as separation of product from the reactants will promote conversion, removal of heat will do likewise. In fact, high temperature favors the reverse reaction. Consequently the reaction should be conducted at low temperature with continuous removal of heat.

A catalyst is a substance that has the property of changing the equilibrium constant without actually taking part in the reaction. It may be nothing more than a porous surface onto which the reactants are adsorbed. Or it may serve to establish a token concentration of an intermediate product without which the final product might not be formed. Or it may serve to provide the correct environment, e.g., acidity. Even light catalyzes some reactions.

Catalyst may be packed in a fixed bed within the reactor. Uniformly small particles may also be supported by the upward velocity of the reactant stream (gas or liquid), in which case it is called a *fluidized bed*. Solid catalyst may also be dissolved or suspended in a liquid reaction media, then separated from the products and recycled. Metal catalysts may be made into screens or other shapes across which the reactants flow. It should be remembered, however, that the reaction takes place on the surface of the catalyst; if heat is evolved, cooling should be applied there, or the catalyst could be destroyed or deactivated. Most catalysts also become deactivated due to fouling of the surface with by-products and contamination by impurities in the feedstock, called *poisons*. The conversion within a reactor depends on the active surface area of the catalyst, which can be time-variant. A classic optimization problem involves determining the most efficient schedule for replacing or reactivating catalyst.

### Reaction Rate

Equilibrium occurs when the rates of forward and reverse reactions are equal. These rates are proportional to concentrations of reactants and products respectively. Let $k_f$ and $k_r$ be designated as forward- and reverse-rate coefficients. Then at equilibrium,

$$k_f[SO_2][O_2]^{1/2} = k_r[SO_3] \tag{10.3}$$

From Eqs. (10.2) and (10.3) the equilibrium constant is the ratio of the forward- to reverse-rate coefficients

$$K = \frac{k_f}{k_r} \tag{10.4}$$

The rate of reaction can be identified as the rate of change of concentration of one of the reactants or products in approaching equilibrium

$$-\frac{d[SO_2]}{dt} = \frac{d[SO_3]}{dt} = k_f[SO_2][O_2]^{1/2} - k_r[SO_3]$$

This forward reaction is 1.5-order, indicated by the sum of the exponents, while the reverse reaction is first-order. If one of the reactants, for example, $O_2$, is in considerable excess, the rate of reaction will depend principally on the concentration of the other, and therefore will approach first-order. This is, in fact, a very common occurrence, so the majority of reactions can be treated as first-order. Furthermore, if any of the four steps previously given to promote conversion is employed, the rate of the reverse reaction is usually negligible. So a general equation can be applied to describe the rate of most reactions, relative to the concentration $x$ of the controlling reactant

$$-\frac{dx}{dt} = kx \tag{10.5}$$

A *batch* of reactant will change its concentration exponentially with time from an initial value $x_0$ to a current value $x$, according to the integration of Eq. (10.5)

$$\int_{x_0}^{x} \frac{dx}{x} = \int_{0}^{t} - k \, dt$$

$$\ln x - \ln x_0 = kt$$

Converting the natural logarithms to exponents yields

$$x = x_0 e^{-kt}$$

Note that the time constant for a first-order reaction is $1/k$; thus the units of the reaction-rate coefficient are in inverse time.

Fractional conversion of reactant into product will be identified as $y$, varying with time:

$y \equiv$ fractional conversion of reactant $x$ into product

$$y = \frac{x_0 - x}{x_0} = 1 - e^{-kt} \tag{10.6}$$

In a *continuous plug-flow reactor* the reaction mixture flows through a pipe without back mixing. This type of reactor is dominated by dead time. The residence time of the reactants traveling through a volume $V$ at a flow $F$ is $V/F$. Thus the concentration at the exit of the vessel is

$$x = x_0 e^{-kV/F}$$

Conversion also varies exponentially with flow

$$y = 1 - e^{-kV/F} \tag{10.7}$$

A *continuous back-mixed reactor* is one throughout which the reactant is uniformly distributed by means of agitation. It approaches a single-capacity sys-

tem. Reaction rate is constant throughout, given by Eq. (10.5). [The rate of consumption of the reactant is the volume of the vessel times the reaction rate, which equals the flow times the loss in concentration between inlet and outlet]
[1]

$$-\frac{V\,dx}{dt} = kVx = F(x_0 - x)$$

*where,* $V =$ *vessel volume*
$F =$ *flow*
$x =$ *concentration of controlling reactant (instantaneous)*
$x_0 =$ *initial concentration of $x$.*
$k =$ *reaction rate coeff*

Solving for exit concentration gives

$$x = \frac{x_0}{1 + kV/F}$$

Conversion in a back-mixed reactor varies inversely with residence time

$$y = 1 - \frac{1}{1 + kV/F} = \frac{kV/F}{1 + kV/F} \tag{10.8}$$

A plug-flow reactor is dominated by dead time equal to the residence time. A back-mixed reactor, however, has a time constant which is a function of both $k$ and $V/F$. To illustrate this, a differential equation will be written to describe the dynamic material balance

$$F(x_0 - x) - Vkx = V\frac{dx}{dt}$$

Solving for $x$ gives

$$x + \frac{V}{F + Vk}\left(\frac{dx}{dt}\right) = \frac{Fx_0}{F + Vk}$$

The concentration time constant is the coefficient of the second term

$$\tau_x = \frac{V}{F + Vk} \tag{10.9}$$

The reaction-rate coefficient $k$ increases sharply with temperature, perhaps its most outstanding characteristic

$$k = ae^{-E/RT} \tag{10.10}$$

where $a, E =$ constants peculiar to reaction
$\quad\quad R =$ universal gas constant
$\quad\quad T =$ absolute temperature

To illustrate this strong dependency, $k$ is plotted vs. $T$ in Fig. 10.1 for a typical reaction whose parameters are $a = e^{29}$ min$^{-1}$, $E/R = 20,000°R$.

The conversion in a plug-flow reactor varies with temperature in a double exponential, combining Eqs. (10.7) and (10.10). Conversion vs. temperature for a back-mixed reactor is found by combining Eqs. (10.8) and (10.10). In Fig. 10.2 conversion is plotted against temperature for three values of $V/F$ in both types of continuous reactors using values taken from Fig. 10.1. The plug-flow reactor delivers higher conversion than a back-mixed reactor operating under the same conditions.

Differentiation of the conversion-vs.-temperature relationships for each reactor yields expressions for their slopes. For the plug-flow reactor

$$\frac{dy}{dT} = \frac{E}{RT^2}(1 - y) \ln(1 - y)$$

$y = $ fractional conversion of controlling reactant $x$ into product    (10.11)

$y = \dfrac{x_0 - x}{x_0} = 1 - e^{-kt}$

And for the back-mixed reactor

since $x = x_0 e^{-kT}$

$$\frac{dy}{dT} = \frac{E}{RT^2} y(1 - y)$$

(10.12)

*This page is discussing Continuous reactors both Plug Flow and Back Mixed Reactors*

These expressions will be useful in determining temperature stability later in the chapter. The maximum slope of a conversion-vs.-temperature curve always occurs at $kV/F = 1$, which corresponds to 63 percent conversion in the plug-flow reactor and 50 percent in the back-mixed. For the $V/F = 1$ curves of Fig. 10.2, the maximum slopes are 1/64.7°F and 1/95°F for the two reactors at 230°F.

The curves of Fig. 10.2 and Eqs. (10.11) and (10.12) describe steady-state conditions only, however. In a departure from steady state because of a heat-transfer upset, temperature will change the reaction-rate coefficient in advance of a change in reactant concentration. Thus the reaction rate will increase with temperature above the new steady-state level until reactant concentration is accordingly reduced. A partial derivative of conversion with respect to temperature at constant concentration describes the instantaneous conversion that exceeds the steady-state conversion relative to the amount unconverted [2]

$$\left.\frac{\partial y}{\partial T}\right|_x = \frac{dy}{dT}\frac{1}{1 - y}$$

(10.13)

This means that the maximum dynamic slope of the plug-flow reactor is

$$\left.\frac{\partial y}{\partial T}\right|_x = \frac{E}{RT^2} \ln(1 - y)$$

(10.14)

And for the back-mixed reactor

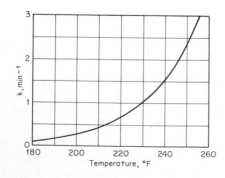

FIG. 10.1 Reaction rate is profoundly influenced by temperature.

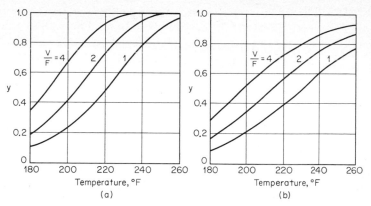

FIG. 10.2 Comparison of conversion-vs.-temperature characteristics for (*a*) plug-flow and (*b*) back-mixed reactors.

$$\frac{\partial y}{\partial T}\bigg|_x = \frac{E}{RT^2} y \qquad\qquad (10.15)$$

### The Stability of Exothermic Reactors

An exothermic reaction is one in which heat is evolved. The evolution of heat increases temperature, which increases the rate of reaction. This series of events forms a positive feedback loop which can result in a runaway if other conditions permit. The conditions are (1) that heat cannot be removed to the surroundings as fast as it is evolved and (2) that conversion is sufficiently below 100 percent to ensure that heat evolution is not thereby limited. The rate of heat evolution $Q_r$ is simply the rate of reaction times the heat of reaction $H_r$

$$Q_r = H_r F x_0 y \qquad\qquad (10.16)$$

Because $Q_r$ varies directly with $y$, the curves of Fig. 10.2 can also be plotted as heat evolution against temperature. Figure 10.3 shows the heat evolution of the back-mixed reactor at $V/F = 1$; the temperature is to be controlled at 230°F to maintain 50 percent conversion.

FIG. 10.3  The slope of the heat-removal line determines whether the reactor will be stable.

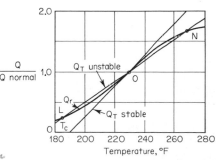

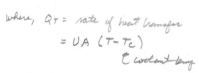

where, $Q_T$ = rate of heat transfer

$= UA (T - T_c)$

$T$ coolant temp

This plot seems to simply say that the reactor cooling system must be able to dissipate more heat than is generated if $T > 230°F$ and can be controlled so that $Q_T < Q_R$ if $T < 230°F$

For the moment, neglect the sensible heat of the reactants; i.e., all the evolved heat is to be transferred to a cooling system. The rate of heat transfer $Q_T$ will approach

$$Q_T = UA(T - T_c)$$

where $T_c$ is the coolant temperature. This describes a straight line of slope $UA$ and intercept $T_c$. Lines representing two possible cooling systems designed for the same heat flow are also shown in Fig. 10.3.

The normal condition for the reactor is described by point $O$ in Fig. 10.3. No matter which heat-removal line is followed, $Q_T = Q_r$ at that point, so that a state of thermal equilibrium can exist. But should the temperature rise, the rate of heat evolution will increase more than the rate of heat transferred by the system designated as unstable [3]. This will cause the temperature to rise farther until the $Q_r$ curve crosses the line again, at point $N$. Points $L$ and $N$ are stable intersections, while point $O$ is an unstable intersection.

*Says the same thing as mentioned by plot on prior page*

The other heat-transfer line demonstrates a capability for removing more heat than is evolved upon a temperature increase, thereby restoring equilibrium. This indicates that an exothermic reactor can be made inherently stable by providing sufficient heat-transfer area. To state it another way, sufficient heat-transfer area will provide negative feedback in excess of the positive feedback of the reaction.

To more explicitly define the relationships involved, an unsteady-state heat balance must be written

*Heat Balance Eqn.*

*heat absorbed by incoming reactant* $\quad$ *heat up rate of heating* $\quad$ where, $C =$ specific heat of stream; $P =$ density of stream

$$\underbrace{H_r F x_0 y}_{Q_r} - \underbrace{UA(T - T_c)}_{Q_T} - \underbrace{F\rho C(T - T_F)}_{Stream} = \underbrace{V\rho C \frac{dT}{dt}}_{System\ Thermal\ Capacity} \tag{10.17}$$

$T_F =$ inlet temp of stream

The first two terms of Eq. (10.17) have already been described. The third term represents the sensible heat absorbed by the reaction stream of density $\rho$ and specific heat $C$ as it rises from inlet temperature $T_F$. The term on the right of the equation represents the thermal capacity of the system.

Since heat evolution is a nonlinear function of temperature, it is necessary to linearize Eq. (10.17) in order to find the thermal time constant of the reactor. Let Eq. (10.18) describe variations about a designated reference temperature $T_r$

*was $T_c$ before* $\quad$ *$T_F$ before*

$$H_r F x_0 \left(\frac{\partial y}{\partial T}\right)_x (T - T_r) - UA(T - T_r) - F\rho C(T - T_r) = V\rho C \frac{dT}{dt} \tag{10.18}$$

*term for instantaneous y about a reference temp $T_r$ [look like an approximation true only near $T_r$*

Arranging in the classical form of a first-order equation allows identification of the time constant

*Here, text begins to talk about the control dynamics*

$$T + \frac{V\rho C}{UA + F\rho C - H_r F x_0 (\partial y/\partial T)_x} \frac{dT}{dt} = T_r$$

The thermal time constant is

$$\tau_T = \frac{V\rho C}{UA + F\rho C - H_r F x_0 (\partial y/\partial T)_x} \tag{10.19}$$

$\Rightarrow ? \quad \tau_T = \frac{T_r}{T \ dT/dt}$

*Followed concepts and derivation to this Point thoroughly. Only brushed over P.255 briefly. Need more information on prior control dynamics theory to understand further. Totie 9/27/84*

This page devoted to concepts of instability and stability and possible ways of obtaining at least pseudo-stability (my word)

**Controlling Chemical Reactions**   255

If $T_c$ is the manipulated variable, the steady-state process gain turns out to be

$$K_T = \frac{T}{T_c} = \frac{UA}{UA + F\rho C - H_r F x_0 (\partial y / \partial T)_x} \qquad (10.20)$$

If the reactor is unstable, both the gain and the time constant will be negative. The denominator in both expressions is the difference between the slopes of the heat-removal and heat-evolution curves, as in Fig. 10.3. If both denominators are positive, the reactor behaves as a simple first-order lag. If both are negative, positive feedback dominates; the dynamic gain is the same as a simple lag, but the phase angle goes from $-90°$ at zero period to $-180°$ at an infinite period

*(margin note: Control dynamics — Jargon for instability)*

$$\phi_T = -180° + \tan^{-1} 2\pi \frac{\tau_T}{\tau_0} \qquad (10.21)$$

The $-180°$ indicates a negative steady-state gain, while the plus sign in front of the $\tan^{-1}$ indicates a negative time constant. Both the time constant and the steady-state gain can also approach infinity, in which case the reactor acts as an integrator whose dynamic gain at period $\tau_0$ is

$$G_T = \frac{K_T \tau_0}{2\pi \tau_T} = \frac{\tau_0}{2\pi V \rho C / UA} \qquad (10.22)$$

Equation (10.22) defines the dynamic asymptote of process gain for all conditions of stability, exhibiting an effective time constant of $V\rho C/UA$.

A stable reactor can operate without temperature control; regulation of $T_c$ alone is ordinarily sufficient. But an unstable reactor will drift away from the control point in either direction at an ever-increasing rate unless feedback control is enforced.

*(margin note: Impt concept! = result of stability; $T_c$ alone must be controlled)*

Unfortunately, it may not always be possible or economical to design for stability. Enough heat-transfer area must be provided for only about 50 to 60°F (28 to 33°C) differential to be required across it to remove the rated flow of heat. (This is an estimate [3] of the $T - T_c$ ordinarily required to exceed $dT/dy$, such as that given in Fig. 10.2.)

*(margin note: Stability not always designed for (at least control) always stable)*

Stability will be assured if heat is removed by boiling one or more of the ingredients in the reaction, since this makes the system almost isothermal. On the other hand, if heat is removed by a mechanism like evaporation of liquid into a dry gas stream, its flow may change very little with temperature. In this case the slope of the heat-removal curve would be slight, and the reactor could be expected to be unstable.

All the foregoing statements on stability were used to describe open-loop situations. Some unstable reactors can be given steady-state stability by applying enough negative feedback from the control system to overcome the positive feed-

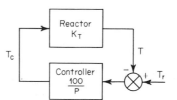

FIG. 10.4 Negative feedback of the controller must overcome positive feedback in the reactor in order to attain steady-state stability.

back of the reaction. To visualize how this is possible, consider the proportional control loop of Fig. 10.4 for steady-state conditions only.

Figure 10.4 can be represented mathematically by

$$T_c = \frac{100}{P}(T_r - T) = \frac{T}{K_T}$$

The closed-loop steady-state gain is found by solving for $T/T_r$

$$\frac{T}{T_r} = \frac{1}{1 + P/100K_T} \tag{10.23}$$

Steady-state stability is identified by positive gain. In order for $T/T_r$ to be positive,

$$\frac{P}{100K_T} > -1$$

If $K_T$ were positive, $P$ could have any value because the reactor would be stable with the loop open. But with $K_T$ negative, $P$ cannot be greater than $-100\,K_T$: if $K_T = -2$, $P < 200$ percent. This sets an upper limit on $P$.

The dynamic properties of the rest of the loop set a lower limit on $P$. The natural period of the temperature-control loop is found by equating the sum of the phase lags of all dynamic elements to $180°$. Since the phase of a negative lag is between $-90$ and $-180°$, there is little room for other elements. This clearly rules out integral control.

If all the other dynamic elements in the loop can be lumped together as dead time $\tau_d$, the period of oscillation can be found by equating the sum of the phase lags to $-180°$

$$-180° = -360° \frac{\tau_d}{\tau_0} - 180° + \tan^{-1} 2\pi \frac{\tau_T}{\tau_0}$$

Having found $\tau_0$, we can determine the dynamic gain of the unstable reactor

$$G_T = \left[ 1 + \left( 2\pi \frac{\tau_T}{\tau_0} \right)^2 \right]^{-1/2}$$

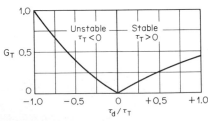

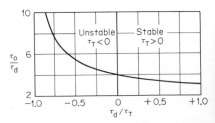

**FIG. 10.5** The dynamic gain varies almost linearly with the amount of dead time in the loop.

**FIG. 10.6** The response of an unstable reactor is both slower and more variable than a stable reactor with the same parameters.

A plot of dynamic gain vs. the ratio of $\tau_d/\tau_T$ is given in Fig. 10.5. The dynamic gain of a stable reactor is included for comparison.

The upper limit on $P$ has been established at $-100K_T$. But if $P$ were set exactly at that limit, the reactor would have no net feedback and so could scarcely be considered stabilized. A realistic value for $P$ would apply twice the necessary negative feedback

$$P < -50K_T$$

But in order to provide quarter-amplitude damping at $\tau_o$

$$P = 200K_T G_T$$

The combination of these two conditions can only be realized if

$$G_T < 0.25$$

This corresponds to a $\tau_d/\tau_T$ of 0.35 or less in Fig. 10.5. If $\tau_d > 0.35\tau_T$, the reactor will be either poorly damped at $\tau_o$ or it will be prone to float in the long term, depending on which limit the proportional band is set to favor. An unstable reactor whose dead time approaches its thermal time constant cannot be controlled with any confidence. The temperature tends to limit-cycle in a sawtooth manner, rising slowly to the set point as the cooling is reduced, then descending rapidly when cooling is applied by the controller. Often the only remedy for this situation is a reduction in throughput until stability is achieved.

Instability also affects the natural period of a reactor in a closed loop. Based on the same equations as Fig. 10.5, the periods of unstable and stable reactors are compared in Fig. 10.6. All the foregoing should provide ample incentive for designing a reactor which will be stable in the open loop. Because an unstable reactor is such a difficult control problem, engineers should use every advantage at their command. For example:

1. Use cascade control from reactor temperature to coolant temperature for fast response.
2. Maintain the maximum flow of coolant to minimize dead time.
3. Use derivative action in both controllers.

## CONTINUOUS REACTORS

Continuous reactors are designed to operate under conditions of constant feed rate, withdrawal of product, and removal or supply of heat. If properly controlled, they are ordinarily invariant; i.e., the distribution of composition and temperature is constant with respect to time and space. (Gradual degradation of catalyst, fouling of heat-transfer surfaces, etc., often are encountered, but their time scale is beyond the control spectrum.) The goal of the control system is to ensure that the operating conditions do remain constant at the design specifications while minimizing the losses of both product and reactants.

There are so many types of reactions that it is not possible to discuss them all, yet a general classification can be quite helpful. The distinction between plug-flow and back-mixed reactors has already been made, the former being capable of greater conversion and the latter being easier to control. Beyond this, some reactions are carried virtually to completion in a single pass; others are forced to be conducted at low conversion due to low reaction rates, reversibility of the reaction, or occurrence of side reactions.

When conversion is incomplete, the excess reactant(s) must be recycled; therein lies a major distinction—single-pass vs. recycle operation. Some reactions are conducted in an essentially inert media such as a solvent, which also may be recycled. Finally, reactions are occasionally moderated by dilution with product, which then is recycled. To summarize, reactors may be classified as

1. Single-pass
2. Recycle
    a. Reactant(s)
    b. Inert vehicle
    c. Product

Each group has its own distinctive arrangement of flow- and inventory-control loops.

### Apportioning Reactant Flows

Whenever one of the reactants differs in phase from products and other reactants, it may be automatically added at the same rate as it is consumed, by controlling its inventory within the reactor. Figure 10.7 shows single-pass reactors with either a liquid or a gas as the manipulated flow.

Note that flow recorders in Fig. 10.7 will tell just what the average reaction rate is if allowance is made for evaporation or absorption and purge flow. The purge need not be continuous but is an absolute necessity to rid the reactor of inert contaminants. A trace of nitrogen, for example, in a hydrogen feed stream can soon accumulate sufficiently to impede the reaction unless it is periodically or continuously discharged. By the same token, lubricating oils from pumps and compressors can accumulate in a liquid dead end unless purged.

In single-phase reactions carried out in one pass, as a neutralization would be,

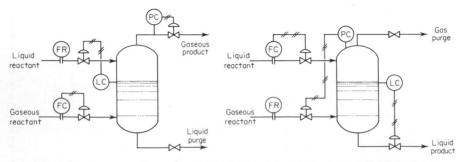

**FIG. 10.7**  One reactant can be automatically added as it is consumed if it differs in phase from the other materials.

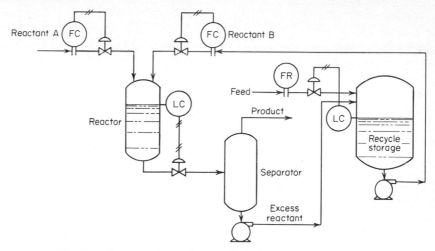

**FIG. 10.8** The flow of reactant B may be set to limit the concentration of reactant A or to fix the residence time.

accurate control of the ratio of the reactants is of paramount importance. Excess of any reactant is not only wasted but may cause undesirable side reactions, including corrosion. If an endpoint analyzer, such as pH, is available, it must be used for feedback trim of the reactant ratio. Manually set ratio control of the feed streams is ordinarily not accurate enough, particularly if the composition of one stream is variable.

If one of the reactants is recycled, control of its flow is not critical at all, because it is always in excess. And since it is ordinarily separated entirely from the products, addition of fresh reactant can be manipulated by inventory control. Figure 10.8 shows how a liquid reactant is added to the recycle stream to make up what is consumed in the reaction.

Product is recycled for the purpose of moderating a reaction. If no reactant is recycled along with it, the requirements for control of the ratio of the feed streams is as stringent as in the single-pass reactor.

Many reactions occur in the presence of some vehicle favorable to both reactants, such as a solvent. An inert diluent acts also to moderate a reaction, facilitating control of temperature and product distribution. If the reactants are soluble in the vehicle, whatever amount goes unreacted will ordinarily be recycled with it. An example of a process with solvent recycle is shown in Fig. 10.9.

In a single-pass reactor, an excess of any reactant is lost; but with solvent recycle, excess accumulates if there is no feedback loop to control its concentration. In the system of Fig. 10.9, reactants X and Y are to be added in equal quantities, the reaction going to completion. But solvent enters at a rate $F$, carrying with it recycled reactant X at a concentration $x_1$. The concentration of X in the solvent leaving the reactor is designated $x_2$ and is found from a mass balance, neglecting holdup within the reactor

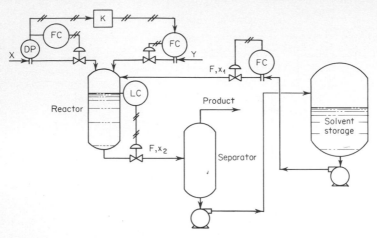

**FIG. 10.9**   Excess reactant is ordinarily recycled with the solvent.

$$Fx_2 = Fx_1 + X - Y$$

The content of reactant in volume $V$ of stored solvent is found by an unsteady-state mass balance on the storage tank

$$Fx_2 = Fx_1 + V \frac{dx_1}{dt}$$

Combining the two equations yields the variation of $x_1$ with respect to the imbalance in reactant flows

$$x_1 = \frac{F}{V} \int \frac{X - Y}{F} \, dt \tag{10.24}$$

Equation (10.24) shows that because of the recycle loop, the process is non-self-regulating [4]. Consequently an integral controller cannot be used to regulate composition. This rules out any kind of feedback-optimizing control system, but because of the lack of self-regulation, endpoint control is essential.

### Temperature Control

Endothermic reactors present no problem regarding temperature control, since they exhibit a marked degree of self-regulation. The exothermic reactors, which have already been introduced, pose the real problem. Their negative self-regulation has been demonstrated.

One facet of an exothermic reaction that has not yet been discussed is its initiation. Because reaction rate increases with temperature, heat must be applied before any conversion is obtained; then heat must be removed. So the heat-transfer system must be able to operate in either direction. This creates something of a problem; steam, for example, is most often used for heating, but is worthless for cooling. There are two general approaches to this problem: (1)

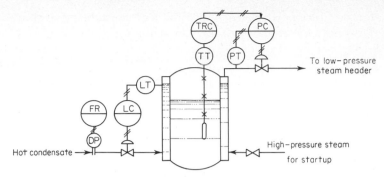

**FIG. 10.10**  This system will allow easy startup as well as efficient cooling.

employ a two-way cooling system, i.e., one capable of heating too, or (2) split the duties by preheating the reactants and cooling the reactor.

A very effective two-way cooling system uses a boiling liquid to which heat can be applied externally for startup. An example is pictured in Fig. 10.10. Because the rate of heat transfer in the system shown in Fig. 10.10 is directly proportional to coolant temperature, that variable should be manipulated to control reactor temperature. Since the boiling point of the condensate is a function of its pressure, the manipulated variable is the set point of the jacket back-pressure controller.

Another system commonly used features a liquid coolant rapidly circulated past the heat-transfer surfaces. Figure 10.11 shows the arrangement when water is used as the coolant. Coolant temperature is chosen as the manipulated variable since it is linear with both heat-transfer rate and reactor temperature. A high rate of circulation allows maximum heat transfer and speed of response.

If recirculation of coolant is not used, the dead time in the secondary temperature loop will vary with the coolant flow. Combined with the nonlinear variation of temperature with flow (Eq. 9.14), it results in a limit cycle, even with an equal-percentage valve. The cycle has a distinctive appearance, the high-

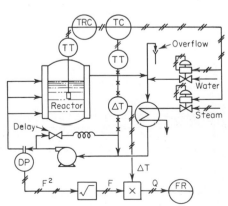

**FIG. 10.11**  Manipulation of exit temperature is effective when coolant is circulated at a rapid rate.

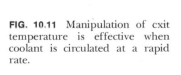

temperature portion, when the flow is greatest, being of short duration and the low-temperature part of the cycle being longer.

Heat-removal systems can be used for monitoring conversion. The record of condensate flow to the reactor in Fig. 10.10 should provide a reliable indication of heat evolution. With a circulating liquid coolant, however, flow must be multiplied by temperature difference from inlet to outlet in order to determine the rate of heat evolution. There are some pitfalls in this system: (1) When the primary controller calls for more cooling, the temperature at the jacket inlet will fall before that at the outlet. The difference between inlet and outlet is ordinarily only about 5°F, which could be less than a transient in inlet temperature. This makes the record of heat transfer appear very erratic, unless dynamic compensation is applied to the inlet measurement. (2) There are many sources of error, the principal one being a difference in temperature between the reactants and products.

Dynamic compensation requires that the inlet temperature measurement be delayed behind its actual value an amount equal to the delay through the jacket. It can be applied most effectively by simulating the jacket by a length of tubing whose dead time can be adjusted by changing the flow. The system in Fig. 10.11 uses this compensation.

If the reactants are to be preheated, it should be done before mixing, unless a catalyst is necessary for the reaction to take place. Otherwise there is no assurance that the reaction would not begin inside the preheater, where it cannot be controlled. Adding heat in the preheater(s) and removing it in the reactor is hardly economical. But once a reaction is initiated, it is often possible to bypass the preheaters without adverse effect.

Some reactors have a regenerative preheating system, in which heat is transferred from the product stream to the reactants through an exchanger. Although this is economically advantageous, unless preheat temperature is controlled, a positive feedback loop is formed which can destroy whatever self-regulation the reactor might have had. Temperature control of regenerative preheat can be accomplished as shown in Fig. 10.12.

Whenever a liquid-phase reaction is conducted at a temperature near the boiling point of one of the reactants or products, heat of vaporization may be used for control. If one of the reactants vaporizes, it can be refluxed back to the

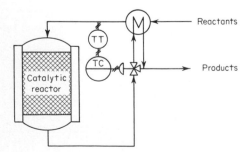

**FIG. 10.12** Control of preheat temperature is necessary for this reactor to be stable.

reactor after condensing. If a product vaporizes, it can be removed as a vapor. This type of heat-removal system is highly self-regulating, but it is also pressure-sensitive. In fact, pressure control should be applied rather than temperature control, since it is a more responsive measurement. Throttling the reflux from the condenser, or the vapor leaving the reactor if there is no reflux, is an effective means of controlling pressure.

## pH CONTROL

While a discussion on endpoint control in general might be in order, pH is used far more widely than any other measurement to sense the state of a reaction. So while pH has some peculiarities of its own, principally its logarithmic character, much of the following commentary applies to other endpoint measurements.

In several instances in earlier chapters, pH has been cited as a difficult control problem. It has, in addition to the usual properties of a composition loop, a severely nonlinear measurement. This very characteristic imposes exceptional demands in flow rangeability of the valves and control system.

### Defining the pH Curve

The outstanding property of each acid-base system is its pH curve; the one shown in Fig. 2.17 is typical of a base neutralized by an acid. The shape of the curve is related to the equilibrium constants for ionization of the acid and base and the concentrations of each of the ions. But the basis of the coordinate system is logarithmic, in that pH is defined as the negative logarithm of the hydrogen-ion concentration, in gram ions per liter (normality):

$$pH \equiv - \log [H^+] \quad \text{or} \quad [H^+] = 10^{-pH} \qquad (10.25)$$

Pure water ionizes into hydrogen and hydroxyl ions of equal concentration

$$H_2O \overset{K_w}{\rightleftharpoons} H^+ + OH^-$$

The equilibrium constant for the ionization of water is $10^{-14}$ at 25°C

$$K_w = [H^+][OH^-] = 10^{-14}$$

This is a useful relationship, because it defines the hydroxyl-ion concentration of any aqueous solution whose pH is known

$$[OH^-] = 10^{pH - 14} \qquad (10.26)$$

The neutral point for water is where hydrogen and hydroxyl ions are at equal strength, i.e., at pH 7.

Strong acids and bases ionize completely; i.e., all their hydrogen or hydroxyl groups appear in the ionized form in solution and therefore are measurable by a pH electrode. For example, consider the addition of hydrochloric acid (HCl) and caustic soda (NaOH) to water in concentrations $x_A$ and $x_B$, respectively,

$$x_A HCl + x_B NaOH + H_2O \rightarrow [H^+] + [Cl^-] + [Na^+] + [OH^-] \qquad (10.27)$$

**TABLE 10.1    Acid and Base Concentrations vs. pH**

| pH | $x_A - x_B$, $N$ | pH | $x_A - x_B$, $N$ |
|---|---|---|---|
| 0 | 1.0 | 8 | $-0.99 \times 10^{-6}$ |
| 2 | $1.0 \times 10^{-2}$ | 10 | $-1.0 \times 10^{-4}$ |
| 4 | $1.0 \times 10^{-4}$ | 12 | $-1.0 \times 10^{-2}$ |
| 6 | $0.99 \times 10^{-6}$ | 14 | $-1.0$ |
| 7 | 0 | | |

The resulting solution must be balanced in charge

$$[H^+] + [Na^+] = [Cl^-] + [OH^-] \tag{10.28}$$

and material

$$[H^+] + x_B = x_A + [OH^-] \tag{10.29}$$

Substituting Eqs. (10.25) and (10.26) into the last expression yields the relationship between acid and base concentrations and solution pH

$$x_A - x_B = 10^{-pH} - 10^{pH-14} \tag{10.30}$$

When $x_A = x_B$, the solution is neutral at pH 7; each unit change in pH is brought about by approximately a decade change in concentration difference. Table 10.1 lists solutions of Eq. (10.30) for several values of pH.

Control in the neutral range of pH 6 to 8 is particularly difficult when only strong acids and bases are present because a very slight difference in their concentration will cause a large change in pH. Consider, for example, a solution of a strong acid at pH 2 which must be neutralized with caustic. To bring the solution within pH 6 to 8, caustic must be added to match the acid concentration of $10^{-2}$ $N$ within $\pm$ $10^{-6}$ $N$, or to an accuracy of 1 part in 10,000. Control valves, metering pumps, or flowmeters cannot be made to this accuracy. Therefore feedback control in a well-mixed vessel with a responsive measuring system must be relied on to achieve the required precision. While feedforward may be helpful in reacting to flow upsets, it is not nearly sufficient by itself.

Nitric acid is also strong, although sulfuric and hydrofluoric are not, in that they are incompletely ionized, notwithstanding their respective reputations of dehydrating carbohydrates and etching glass. Potassium hydroxide is another strong base. Calcium hydroxide (lime) is moderately strong, despite its limited solubility, which gives a saturated solution of pH 12.3.

Weak acids and bases are incompletely ionized

$$HA \overset{K_A}{\rightleftharpoons} H^+ + A^- \qquad BOH \overset{K_B}{\rightleftharpoons} B^+ + OH^- \tag{10.31}$$

where HA and BOH are a monoprotic acid and base. The concentration of the unionized acid or base varies with pH in relationship to the value of its ionization constant

$$[HA] = \frac{[H^+][A^-]}{K_A} \qquad [BOH] = \frac{[B^+][OH^-]}{K_B} \qquad (10.32)$$

Equation (10.32) can be restated using pH in place of $[H^+]$ and $[OH^-]$ and expressing $K_A$ and $K_B$ as their negative base 10 logarithms $pK_A$ and $pK_B$. This places the ionization constants in the same terms as pH

$$[HA] = [A^-]\, 10^{pK_A - pH} \qquad [BOH] = [B^+]\, 10^{pK_B + pH - 14} \qquad (10.33)$$

At the point where $pH = pK_A$, the acid is half ionized; i.e., the concentration of $A^-$ ions is equal to that of the unionized acid HA; similarly at $pH = 14 - pK_B$, the base is half ionized. Table 10.2 lists ionization constants for common weak acids and bases.

The ionization constants of bases are often expressed as $pK_A$, which equals $14 - pK_B$, instead of as $pK_B$. Note, for example, that the ionization of $FeOH^{2+}$ to ferric ion $Fe^{3+}$ has a $pK_B$ of 11.5. The reverse of this reaction is the hydration of $Fe^{3+}$ to form $FeOH^{2+}$ and $H^+$, whose $pK_A$ is $14 - 11.5 = 2.5$. The $pK_B$ of 11.5 indicates that ferric hydroxide is such a weak base that its product, the ferric ion, is a moderately strong acid.

A weak acid may replace HCl in the reaction described by Eq. (10.27)

**TABLE 10.2   Ionization Constants of Common Acids and Bases**

| Acid | Equilibrium | $pK_A$ |
|---|---|---|
| Acetic acid | $CH_3COOH \rightleftharpoons CH_3COO^- + H^+$ | 4.75 |
| Carbon dioxide | $CO_2 + H_2O \rightleftharpoons HCO_3^- + H^+$ | 6.35 |
| | $HCO_3^- \rightleftharpoons CO_3^{2-} + H^+$ | 10.25 |
| Hydrogen fluoride | $HF \rightleftharpoons F^- + H^+$ | 3.17 |
| Hydrogen sulfide | $H_2S \rightleftharpoons HS^- + H^+$ | 7.0 |
| | $HS^- \rightleftharpoons S^{2-} + H^+$ | 12.9 |
| Hypochlorous acid | $HClO \rightleftharpoons ClO^- + H^+$ | 7.5 |
| Sulfuric acid | $H_2SO_4 \rightleftharpoons HSO_4^- + H^+$ | −3 |
| | $HSO_4^- \rightleftharpoons SO_4^{2-} + H^+$ | 1.99 |
| Sulfur dioxide | $SO_2 + H_2O \rightleftharpoons HSO_3^- + H^+$ | 1.8 |
| | $HSO_3^- \rightleftharpoons SO_3^{2-} + H^+$ | 6.8 |

| Base | Equilibrium | $pK_B$ |
|---|---|---|
| Aluminate ion | $AlO_2^- + 2H_2O \rightleftharpoons Al(OH)_3 + OH^-$ | 1.6 |
| Ammonia | $NH_3 + H_2O \rightleftharpoons NH_4^+ + OH^-$ | 4.75 |
| Calcium hydroxide | $Ca(OH)_2 \rightleftharpoons CaOH^+ + OH^-$ | 1.40 |
| | $CaOH^+ \rightleftharpoons Ca^{2+} + OH^-$ | 2.43 |
| Ethylamine | $C_2H_5NH_2 + H_2O \rightleftharpoons C_2H_5NH_3^+ + OH^-$ | 3.3 |
| Ferric hydroxide | $Fe(OH)_2^+ \rightleftharpoons FeOH^{2+} + OH^-$ | 9.3 |
| | $FeOH^{2+} \rightleftharpoons Fe^{3+} + OH^-$ | 11.5 |
| Hydrazine | $N_2H_4 + H_2O \rightleftharpoons N_2H_5^+ + OH^-$ | 5.5 |
| Hydroxylamine | $NH_2OH + H_2O \rightleftharpoons NH_3OH^+ + OH^-$ | 7.97 |
| Magnesium hydroxide | $Mg(OH)_2 \rightleftharpoons MgOH^+ + OH^-$ | 2.6 |

$$x_A HA + x_B NaOH + H_2O \rightarrow$$
$$[H^+] + [A^-] + [HA] + [Na^+] + [OH^-] \quad (10.34)$$

To satisfy the charge balance we must have

$$[H^+] + [Na^+] = [A^-] + [OH^-] \quad (10.35)$$

and the mass balance we must have

$$x_A = [HA] + [A^-] \qquad x_B = [Na^+] \quad (10.36)$$

Combining these expressions with (10.33) and placing the result in terms of pH yields

$$x_B = 10^{pH-14} - 10^{-pH} + \frac{x_A}{1 + 10^{pK_A - pH}} \quad (10.37)$$

This establishes the relationship between solution pH and the amount of strong base $x_B$ added to an initial concentration $x_A$ of weak acid. Figure 10.13 compares this relationship for acetic acid against a similar titration of hydrochloric acid. Observe that the slope of the curve is lowest where the weak acid is half-neutralized, i.e., where pH = $pK_A$. For acetic acid, this occurs at pH 4.75. This is the *buffering point,* where variations in acidity affect pH least, an inherent regulating property of the process.

The gain of the process at the pH control point can be determined by differentiating Eq. (10.37)

$$\frac{d\,pH}{dx_B} = \frac{0.434}{10^{-pH} - 10^{pH-14} + x_A 10^{pK_A - pH}/(1 + 10^{pK_A - pH})^2} \quad (10.38)$$

Where pH > $pK_A$, the squared term reduces to unity and the slope of the titration curve approaches

$$\frac{d\,pH}{dx_B} \simeq \frac{0.434}{x_A 10^{pK_A - pH}} \quad (10.39)$$

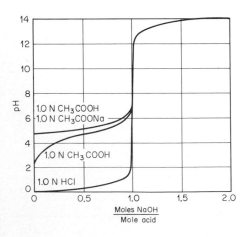

FIG. 10.13  A weak acid or base will require more reagent for neutralization from a given pH, but control is easier; buffering provides regulation.

That is to say, the concentration of the weak acid and its $pK_A$ principally determine process sensitivity. The sensitivity of the strong-acid–strong-base process described in Table 10.1 was approximately $10^6$ pH/$N$ in the region of pH 6 to 8. By comparison, an acetic acid solution of 0.01 $N$ has a sensitivity of only $0.78 \times 10^4$ pH/$N$ to the addition of strong base, making its pH over 100 times easier to control.

The same method can be used to determine the pH curve for a weak base such as ammonia. Table 10.2 also lists some diprotic acids and bases, which have two steps of ionization. Carbonates, for example, buffer at both pH 6.35 and 10.25, giving a titration curve with two inflection points. The low gain afforded by carbonates in the region of pH 6 to 7 greatly facilitates control over their neutralization. Ground and surface waters are often rich in carbonates, tending to regulate their pH. Many solutions are mixtures of weak acids and bases and are characterized by pH curves without a region of high gain.

A nonlinear controller of the form shown in Fig. 5.28 has been found quite successful in controlling the pH of a wide variety of solutions. Although some solutions are so highly buffered that a linear controller produces satisfactory results, in others the buffering is variable, requiring feedback adaptation of the nonlinear characteristic. In these applications the adaptive controller described in Fig. 6.17 has been used to adjust the width of the low-gain region in the nonlinear controller [5]. A further discussion of the nonlinear controller in pH applications and a complete development of the equations for diprotic titration curves are given in Ref. 6.

### Rangeability Requirements for pH-Control Systems

When pH control is exercised on a chemical reaction in which a product is being made, conditions can be expected to be well defined. For example, the required ratio of reagent acid (or base) flow to that of the product or other reactants ordinarily would change but little. Furthermore, the pH curve ought to be known and invariant.

Because of the tremendous sensitivity of the curve in the region of neutrality, it is always necessary to trim the ratio with a feedback loop. In addition, the nonlinearity of the measurement should be compensated by using the afore-mentioned nonlinear controller. A diagram of the recommended system appears in Fig. 10.14. The flow signals are linearized to maintain loop gain constant over the full range of flow.

Most pH applications involve neutralization of plant waste from a combination of drains, sumps, vent scrubbers, etc. The demands of these waste-treating systems complicate the control problem in several dimensions:

1. The flow of the effluent stream may vary as much as four- or fivefold.
2. The stream may alternate between acidic and basic, requiring two reagents.
3. Its acid or base content may vary over several decades.

4. The type of acid or base in solution may vary from weak to strong; thus the pH curve is variable.

The flow range required of the reagents is moderated by whatever tolerance is set on final pH. The scheme shown in Fig. 10.14 is wholly unsuited to such an application because the differential meters are accurate to 1 percent of span only from 25 to 100 percent, a 4:1 range. Metering pumps are limited to the vicinity of 20:1 rangeability, while control valves start at 35:1 and reach 100:1; even higher rangeability is claimed for some equal-percentage styles. But often a rangeability in excess of 1000:1 is required, in which case two or more final elements must be sequenced. Failure to provide sufficient rangeability will result in insufficient reagent delivery to meet high loads and/or limit cycling when the load falls below the throttling range. Momentary peaks can sometimes be shaved by blocking outflow and recirculating effluent, allowing levels to rise. This is often a better solution than incorporating oversized valves that are rarely used.

Sequencing control valves to achieve wide range can be difficult. If linear valves or metering pumps are used, rangeability can be expanded only by a factor of 2 or so. The selection of two pumps whose capacities differed by 10:1 would require that the small pump operate only over the first 10 percent of the control range; this usually is not satisfactory.

However, two equal-percentage valves having a rangeability of 50:1 can be sequenced to provide a turndown approaching 2500:1. Figure 10.15 shows how this is done. The large valve is selected to deliver the maximum required flow, and the small valve is selected so that its flow is slightly greater than the minimum capability of the large valve. To calibrate the valve positioners properly, the points corresponding to maximum flow for the large valve and minimum flow for the small valve are plotted at 100 and 0 percent controller output, as shown. A straight line is then drawn connecting them. The value of controller output corresponding to full flow from the smaller valve then specifies the calibration of its positioner. Similarly, where the minimum flow

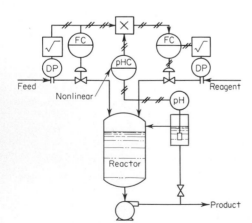

**FIG. 10.14** When controlling a reactor, flow ratio should be trimmed by a nonlinear controller.

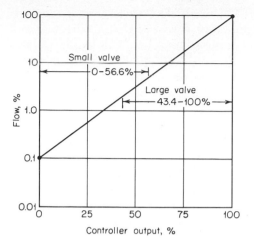

**FIG. 10.15** Two 50:1 equal-percentage valves are sequenced to provide a combined rangeability of 1000:1.

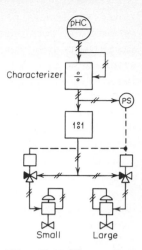

**FIG. 10.16** The valves must not be opened at the same time; the 1:1 repeater avoids cycling of the pressure switch due to loss of air during transfer.

from the large valve falls on the line, determines the calibration of its positioner. The actual rangeability achieved depends on the ratio of the valve sizes, which are available only in a limited set of $C_v$'s.

It is imperative that only one valve be opened at any time. If the small valve remains open while the large valve is throttling, their overall characteristic will be distorted. Moreover, too much flow will be delivered at the point where the large valve opens, causing limit cycling in that load range. To avoid these difficulties, the logic shown in Fig. 10.16 should be provided. A pressure switch is adjusted to close the small valve and open the large one when the controller output *increases* to the point corresponding to full flow from the small valve. Upon a *decrease* in output to the point corresponding to minimum flow from the large valve, the solenoid valves are deenergized, returning the system to its low-flow state. The transfer requires only about a second for completion, so that solution pH is not noticeably upset by it.

In most cases, a linear overall characteristic is desired for the manipulation of reagent flow. The exception is the neutralization of a single weak acid or base whose titration curve changes in slope with concentration, and hence reagent demand, according to Eq. (10.39). Usually, however, there are also strong acids and bases present or buffers such as carbonates whose concentration is constant, in which case the slope of the curve cannot be correlated with reagent demand. Where it is necessary to use equal-percentage valves to achieve wide range, a characterizer must be added to linearize the system, as described in Figs. 2.13 and 2.14 and Eqs. (2.25) to (2.28).

Should rangeability requirements exceed what can be obtained with two valves, it is usually practical to manipulate the smallest valve from a separate proportional controller. The gain of the valve is so low that a fairly narrow proportional band can be used, thereby minimizing offset. This loop is usually sufficient to eliminate the limit cycle that would otherwise develop at loads below the throttling range of the next larger valve.

When both acid and basic reagents are required, their valves are usually sequenced so that there is no flow of either at 50 percent controller output. Acid reagent is added as output increases above 50 percent, and base is added on a decreasing signal below 50 percent. This assumes, however, that the two reagent systems have identical capability for neutralization. Their capability is determined by the product of reagent normality and maximum flow. (The most common reagents are 98% sulfuric acid at 36 $N$, and 10% caustic and lime, both at approximately 2.8 $N$.) If the product of flow and normality for the two reagents differ, their respective positioners should be calibrated so as to keep loop gain constant. If a lime valve, for example, can deliver twice the equivalent reagent concentration as the acid valve with which it is sequenced, its positioner should have twice the operating range. In this case the point where both valves would close would be 66.7 percent.

This arrangement of sequencing allows the base reagent valve to open upon loss of control signal. This is usually preferable to opening the acid valve, which could cause extensive corrosion. Furthermore, lime is the most common basic reagent, and its limited solubility keeps solution pH from exceeding 12.3. However, the system can be protected against loss of air supply by selecting *both* reagent valves to fail closed. Then the positioner for the base valve must be reverse-acting for proper sequencing. This allows all reagent flow to be stopped on command, by shutting off the air supply to the positioners with a solenoid valve. Then reagent addition can be coordinated with feed-pump controls, although the pH controller must be transferred to manual whenever the system is shut down. Reference 5 describes this system in detail, along with protective features to prevent discharge of off-specification effluent.

If the reagent valves are equal-percentage and a linear characteristic is desired, a characterizer calibrated with the appropriate S-shaped curve must be inserted at the controller output.

### Feedforward Control of pH

Figure 10.14 showed a feedforward-feedback system based on feed flow. In most effluent-treatment plants, flow is reasonably constant, yet pH can vary rapidly over a wide range. Every effort should be made to see that the pH feedback system is as responsive as possible by providing adequate back mixing and carefully locating the electrodes to minimize dead time. A reservoir upstream of the neutralization vessel helps, by smoothing potential load upsets. Reference 6 lists many guidelines for vessel sizing and layout, agitation, etc., to obtain satisfactory performance with minimum controls. But when, for reasons of inadequate holdup or mixing, feedback is incapable of coping with the large and rapid variations in solution pH encountered, feedforward may be the only recourse.

The feedforward system attempts to supply the reagent needed to balance the acid or base content of the waste stream. If the waste contains only strong agents, the relationship between reagent demand and influent pH is logarithmic, as indicated in Table 10.1. If the waste contains only a single weak agent, its pH changes 0.5 unit for each decade in concentration, which is the solution of Eq. (10.32) for the condition of $[H^+] = [A^-]$ and $[B^+] = [OH^-]$. Unfortunately, most wastes contain a mixture of strong and weak agents, causing the relationship between pH and reagent demand to be quite variable.

The system shown in Fig. 10.17 accommodates the logarithmic nature of the pH function by means of an equal-percentage valve. The feedback controller biases the relationship as necessary to reach neutrality, leaving feedforward gain $K$ and intercept $r$ as manual adjustments

$$m_A = 2m - K_A(r_A - pH) \qquad (10.40)$$
$$m_B = 2(1 - m) - K_B(pH - r_B) \qquad (10.41)$$

Here $m_A$ and $m_B$ are the positions of the acid and base valves, respectively, and $m$ is the output of the feedback controller. Intercepts $r_A$ and $r_B$ are the values of influent pH at which full reagent delivery is achieved when the feedback controller's output is 50 percent. As the pH moves toward neutrality, the valves are closed. Equation (10.41) describes how basic reagent is added in a two-reagent system. In a single-reagent system, it is not necessary to reverse the feedback signal $m$ to $1 - m$; instead, the controller's action would be reversed.

Although equal-percentage valves provide the proper feedforward characterization, they cause a variable feedback gain. To overcome this limitation, Fig. 10.17 shows the width of the low-gain zone in the nonlinear controller being adaptively adjusted as a function of the feedforward signals. An extreme value of influent pH will expand the zone width either due to a high reagent demand or absence of buffering, both of which result in a high loop gain. A situation of

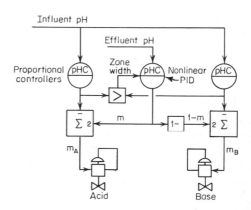

**FIG. 10.17**  This system combines feedforward with adaptive nonlinear feedback.

high reagent demand with a moderate pH would be caused by heavy buffering, in which case a narrow zone would be desirable.

The adaptive signal must not form a feedback loop around the pH controller, another reason for its being derived from the feedforward signals. The feedforward controllers shown have only the proportional mode. Adjustments of $K$ and $r$ are through their proportional band and set point. Adjustments to $K$ are made to minimize the effect of rapid changes in influent pH. Set points $r_A$ and $r_B$ should be coordinated in such a way that transfer between acid and base addition proceeds without overlap. Any dead zone, however, should be small enough to ensure that $m$ does not have to change appreciably during transfer.

*Picked up here.*
*9/27/84*

## BATCH REACTORS

*MW of polymer usually easier to control in batch reactor*

Although the progress of the chemical industry has been toward continuous processes, some reactions will inevitably be conducted batchwise. [The bulk of commercial batch reactions are polymerizations involved in the production of rubber and many types of plastics.] Distribution of molecular weight is an important parameter in polymer manufacture, and it seems to be the most easily controlled batchwise. Another consideration is the great change in viscosity frequently encountered between the reactants and products.

The process consists of the several steps listed below, although considerable variation exists from one product to another:

1. Charge the reactor with reactants and catalyst.
2. Heat to operating temperature.
3. Allow the reaction to proceed to completion, normally several hours.
4. Heat or cool to cure temperatures.
5. Cool and empty the reactor.

Production reactors are stirred, jacketed vessels of several thousand gallons capacity. If the reaction is first-order, conversion varies with time according to Eq. (10.6)

*same equation 10.6 on page 250*
*y = fractional conversion of reactant x into product*

$$y = 1 - e^{-kt} \tag{10.6}$$

The rate of conversion is the derivative of Eq. (10.6)

$$\frac{dy}{dt} = ke^{-kt} \tag{10.42}$$

The rate is greatest when the conversion is least, i.e., at time zero.

*Impt chemical principle of Polymerization reaction = 2nd order or higher*

[Polymerization reactions are second-order or higher, because they depend on the simultaneous combination of two or more monomer molecules to form a polymer. In a second-order reaction, the rate depends on the square of reactant concentration

*2nd order reaction*

$$-\frac{dx}{dt} = kx^2 \tag{10.43}$$

Dividing both sides by $-x^2$ and integrating gives

$$\int_{x_0}^{x} \frac{dx}{x^2} = \int_{0}^{t} - k \, dt$$

$$\frac{1}{x} - \frac{1}{x_0} = kt$$

Conversion and its rate can be found by substituting for $x$

$$y = \frac{1}{1 + 1/ktx_0} = \frac{ktx_0}{1 + ktx_0} \qquad \qquad (10.44)$$

*These equations are valid for 2nd order reactions only*

$$\frac{dy}{dt} = \frac{kx_0}{(1 + ktx_0)^2} \qquad \qquad (10.45)$$

The rate of conversion is also the rate of production in a batch reactor and is proportional to heat evolution if the reaction is exothermic. The rate of conversion of first- and second-order reactions is plotted against time in Fig. 10.18.

## Temperature Control

In the early stages of a batch reaction, temperature control is most important because the rate of conversion is usually at its highest. Exothermic reactions pose a real control problem because heat must be applied to raise the batch to reaction temperature and then be removed. The cooling system most frequently used is that shown in Fig. 10.11. A proportional controller is used for coolant exit temperature because the proportional band is ordinarily only about 10 percent and offset is not harmful. But the primary controller is three-mode, with special features to permit:

*See p.261 for T control strategies*

1. Maintenance of optimum settings for operation at reaction temperature
2. Delivery of the batch to reaction temperature without overshoot
3. Conduction of the reaction in a minimum of time

Thanks to its large volume and rapid agitation, a batch reactor is usually not difficult to control. The reactor will respond rapidly, with a period of perhaps 20 min, and 10 percent proportional band may be sufficient for effective damping. All three control modes should be adjusted while at the operating temperature.

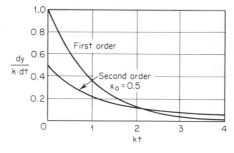

**FIG. 10.18** The rate of conversion of higher-order reactions varies less with time, particularly at low concentration levels.

*Impt:*
*about overshoot*    In order to avoid overshoot, the primary controller must be equipped with
    antiwindup protection with preload applied to the integral circuit. It is the
*See page 93*    preload which determines the magnitude of overshoot (see Fig. 4.5). The    *P 93*
*for preload*    correct value for preload is not difficult to estimate for a known reaction. The
*information*    initial rate of conversion will release a predictable flow of heat, all of which must
    be removed through the heat-transfer surface. Both the batch and the cooling
    fluid are circulated at very high rates to ensure good heat transfer; thus each
    has little temperature gradient and constant flow. The rate of heat transfer is
    therefore directly proportional to the temperature difference between primary
    and secondary measurements. The output of the primary controller, which is
    the set point of the secondary, is predictable, and its predicted value can be
    introduced as preload. As pointed out in Chap. 4, however, it is necessary to set
    the preload a few percent below the predicted value to allow for the integral
    action of the controller from the time the proportional band is entered until the
    set point is reached.

    A batch reactor can be unstable, in which case its natural period will be
perhaps twice as long and its proportional band requirement twice as great as a
physically similar stable one. The control system described above loses its
effectiveness when a wide proportional band is required. In order to avoid
overshoot, the heat input must be throttled early, which can add considerable
time to the length of the operation.

*See page 131*    For a problem such as this, the dual-mode control system described in Fig.
    5.23 is extremely effective. The preload is estimated as before, but no correc-
*(P 131)* tion is required for integration, because PID action is not initiated until the
error is nearly zero. Full heating can be applied to within 1 or 2 percent of the
set point, far beyond the capabilities of a 25 percent proportional band. Yet full
cooling need only be applied for a time delay of perhaps a minute to dissipate
the energy stored in the jacket. As pointed out in Fig. 5.24, the switching
parameters are easy to adjust and tolerant of maladjustment.

    Figure 10.19 shows the relationships between primary and jacket tempera-
tures and the dual-mode output for a typical reactor. If the settings are correct,
jacket temperature will fall to meet its set point at the preload value when the
time delay is over. Notice how the rate of heat transfer diminishes to zero as
conversion is completed. There is no heat evolution during the cure phase.

    Equation (10.19) gave the thermal time constant of a continuous reactor.
Among other things, it depended on reactant concentration and conversion.

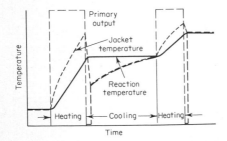

**FIG. 10.19**  The dual-mode system
requires different values of preload for reaction and cure.

Since these are both variable in a batch reactor, it is entirely possible to proceed from an unstable to a stable condition with the passage of time. Control settings necessarily must favor the more difficult situation.

In some batch reactions, one of the ingredients is introduced continuously until the other reactants are entirely consumed. The rate at which this ingredient is added is usually limited by the rate of heat removal. In order to carry out the reaction in minimum time, the batch must be heated to reaction temperature as rapidly as possible, where full cooling is applied. Reagent is added by a valve-position controller acting on the coolant valve. If the valve is less than fully open, the valve-position controller increases the flow of reagent, thereby increasing the rate of heat evolution. Should coolant supply be incapable of maintaining temperature control, the signal to the valve will exceed 100 percent, causing the controller to decrease feed rate. A PID controller on valve position has been found to give excellent results. An interlock should be provided to inhibit reagent addition until the batch has reached reaction temperature.

This same technique can be applied to provide supplementary cooling by refrigeration. As long as the coolant valve is throttling, no refrigeration is needed; but when it reaches the full-open position, the valve-position controller starts to add refrigeration to restore temperature control.

## Endpoint Control

At the completion of the reaction, heat evolution will subside, causing the temperature controller to increase reactant flow. Since this is obviously incorrect action, some logic must be arranged to override temperature control upon the termination of the reaction. If there is no way of measuring the completion of the reaction, reactant flow can be stopped by a flow totalizer once a predetermined quantity has been introduced. The termination of some reactions is indicated by a fall in pressure; others are indicated by a rise in pressure. In either case, a pressure switch may be used to shut off flow.

Endpoint control of a batch reaction has one outstanding characteristic: integral action must not be used. The bulk of the reaction is conducted away from the set point; integral action would try to overcome this offset and ultimately result in overshoot. The controller should be proportional-plus-derivative, with zero bias, so that the valve is shut when the set point is approached. If the measurement is pH, reagent should be added through an equal-percentage valve (without a cascade flow loop), to match the process characteristic.

## REFERENCES

1. Corrigan, T. E., and E. F. Young: General Considerations in Reactor Design, *Chem. Eng.*, October 1955.
2. Harriott, P.: "Process Control," p. 311, McGraw-Hill, New York, 1964.
3. Shinskey, F. G.: Temperature Control for Gas-Phase Reactors, *Chem. Eng.*, Oct. 5, 1959.

4. Shinskey, F. G.: Designing Stability into Plant Operations, *Petro/Chem. Eng.*, October-November 1962.
5. Shinskey, F. G.: Adaptive pH Controller Monitors Nonlinear Process, *Control Eng.*, February 1974.
6. Shinskey, F. G.: "pH and pIon Control in Process and Waste Streams," Wiley-Interscience, New York, 1973.
7. Shinskey, F. G., and J. L. Weinstein: A Dual-Mode Control System for a Batch Exothermic Reactor, *20th Annu. ISA Conf., October 1965.*

## PROBLEMS

**10.1**  The reaction whose characteristics appear in Fig. 10.2 is to be conducted at 50 percent conversion with $V/F = 1$ h in a back-mixed reactor. Temperature is to be controlled at 230°F by manipulation of coolant temperature. Feed is preheated to reaction temperature. What is the lowest coolant temperature that will ensure stability in manual control?

**10.2**  For the same reactor, the heat of reaction is 5000 Btu/lb, $F$ is 4000 lb/h, and $x_0$ is 0.2. Coolant temperature required for control is 190°F. The reactor contains 4000 lb of material with a specific heat of 0.4 Btu/lb. What is its thermal time constant? If the dead time in the loop is 1 min, what is the natural period and the proportional band of the primary loop? Is it stable in the steady state?

**10.3**  Reduce the flow to the reactor above to 2000 lb/h, with all other conditions except coolant temperature constant. Calculate its thermal time constant, period, and proportional band. Reduce $x_0$ to 0.1 with $F$ at 4000 lb/h and repeat the calculation.

**10.4**  In the process shown in Fig. 10.9, reactants X and Y are each soluble in the solvent, but the product they form is gaseous. The endpoint of the reaction is to be controlled, as measured by the electrolytic conductivity of the solvent. The conductivity increases with the amount of whichever reactant is in excess and is therefore to be controlled at zero. Devise a way to accomplish this; modify the process if necessary.

**10.5**  Calculate the pH at neutrality for a 1 $N$ solution of acetic acid neutralized by caustic, as shown in Fig. 10.13. Estimate the process gain $d\text{pH}/dx_B$ at that point.

**10.6**  The concentration of reactant in a batch reactor decreases with time; it may be desirable to gradually increase the reaction temperature to hasten completion. Will either of these factors cause the dynamic gain of the process to change? Explain.

# Chapter Eleven

# Distillation

Distillation is an integral part of the operations in most chemical plants and all oil refineries. Considerable attention is devoted to controlling distillation columns more effectively because:

1. Consisting of many stages, they are slow to respond to control action.
2. They are sensitive to upsets in ambient conditions.
3. Stringent specifications must be met on final products.
4. As much as 40 percent of the energy consumed in a plant is allocated to distillation.

The usual operating policy is to maintain product quality within specifications at all costs. Because the process is difficult to control and sensitive to upset, operators tend to overpurify as a matter of course. As will be demonstrated, the sensitivity of product quality to disturbances decreases as impurities are reduced, so that the operation seems more stable under these conditions. Unfortunately, overpurification is very costly in terms of product losses, energy consumption, and production capacity. There is much to be gained by holding product compositions closer to specification limits. The first step in achieving this goal is understanding the factors that determine product quality.

## FACTORS AFFECTING PRODUCT QUALITY

Most texts on distillation start with a design procedure which determines the number of trays needed for a given separation. In control work, however, the column already exists, and speculation over theoretical trays and equilibrium diagrams is of no consequence. A technique especially devised for control application is necessary. This technique begins with a simple block diagram of the tower, which has already been designed to perform a given separation (Fig. 11.1). Although the figure indicates only a binary separation, the concept will be advanced later to multicomponent towers.

The block diagram reveals two extremely important facts: (1) Energy is necessary for separation. In fact, it can be assumed that no separation will take place if no energy is introduced. (2) The relative composition of the two product streams is intimately bound up with their relative flow rates. More of a given component cannot be withdrawn than is being fed to the tower: the material balance must be satisfied. To be sure, tray efficiency, loading, etc., also color the picture, but the two factors above are so outstanding in their effects that they must be the prime consideration in any control system.

### The Material Balance

In the steady state, as much material must be withdrawn as enters a tower

$$F = D + B \tag{11.1}$$

where $F$ = feed rate
$D$ = distillate flow
$B$ = bottoms flow

A material balance on each component must also be closed, using $z_i$, $y_i$, and $x_i$ to represent the fraction of any component $i$ in $F$, $D$, and $B$:

$$Fz_i = Dy_i + Bx_i \tag{11.2}$$

From the overall material balance it is evident that the flow of only one of the product streams can be set independently. The flow of the other is determined by the feed rate and is therefore a dependent variable. But one flow must be set by some criterion, since they cannot both be allowed to drift. For the moment, distillate flow will be chosen to be manipulated by the control system, either directly or indirectly. (The reason for this choice will be discussed later.) Bottoms flow must then be manipulated by a controller which senses liquid level at the bottom of the tower in order to close the material balance by maintaining constant liquid inventory. Bottoms flow is thus dependent on current values of feed and distillate

$$B = F - D$$

Substituting for $B$ in the material balance of component $i$ permits expressing the relationship between the quality of both products in terms of distillate flow

$$Fz_i = Dy_i + (F - D)x_i$$

The ratio $D/F$ determines the relative composition of each product

$$\frac{z_i - x_i}{y_i - x_i} = \frac{D}{F} \tag{11.3}$$

Figure 11.2 shows one way in which this relationship might be pictured for the light key $l$. Remember that $z_l$ is an uncontrolled variable, like $F$. Therefore if $z_l$ should change, $D/F$ must be adjusted to maintain constant values of $x_l$ and $y_l$.

Unfortunately, the set of material-balance equations alone is insoluble; there are more unknowns than equations. A second relationship, based on liquid-vapor equilibrium, is the Fenske equation [1]

$$\frac{y_i/y_j}{x_i/x_j} = \alpha_{ij}^n \qquad (11.4)$$

where $i,j$ = any two components being separated
$\qquad \alpha_{ij}$ = relative volatility of components
$\qquad n$ = number of theoretical equilibrium stages
The Fenske equation is used to determine the minimum number of trays required to effect a given separation at total reflux. For control purposes, this relationship is extended to conditions other than total reflux

$$\frac{y_i/y_j}{x_i/x_j} = S \qquad (11.5)$$

where $S$, the separation factor, is a function of $\alpha$, $n$, and the energy input to the column, as detailed later.

For the typical multicomponent column, the desired split is between the light $l$ and heavy $h$ keys, so that the separation factor is

$$S = \frac{y_l/y_h}{x_l/x_h} \qquad (11.6)$$

We can generally assume that all components $ll$ lighter than the light key leave with the distillate and all components $hh$ heavier than the heavy key leave with

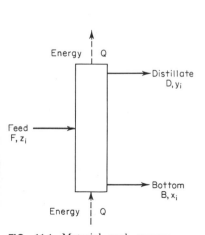

FIG. 11.1 Material and energy balances play key roles in distillation control.

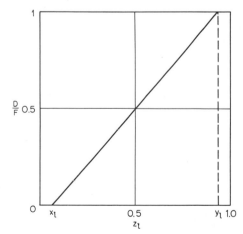

FIG. 11.2 If $x_l$ and $y_l$ are to be controlled, $D/F$ must vary proportionately to $z_l$.

the bottom product. Then combining Eq. (11.3) for the key components with (11.6) produces a quadratic, whose solution is

$$y_h = \frac{-b + \sqrt{b^2 - 4ac}}{2a} \tag{11.7}$$

where $a = \dfrac{D}{F}(S - 1)$ $\qquad b = Sz_l - (S - 1)\left(\dfrac{D}{F} - z_u\right) + z_h$

$$c = -z_h\left(1 - \frac{z_u}{D/F}\right)$$

Once $y_h$ has been found, $x_h$ can be determined from $D/F$ and $z_h$ by rearranging (11.3)

$$x_h = \frac{z_h - y_h D/F}{1 - D/F} \tag{11.8}$$

Next, $y_u$ can be calculated from (11.3) under the stated assumption that $x_u$ is zero

$$y_u = \frac{z_u}{D/F} \tag{11.9}$$

Then $y_l$ is the balance in the distillate

$$y_l = 1 - y_h - y_u \tag{11.10}$$

Having found $y_h$, $x_h$, and $y_l$, we can use Eq. (11.6) to find $x_l$

$$x_l = \frac{x_h y_l}{S y_h} \tag{11.11}$$

Thus the complete set of compositions can be found for any value of $D/F$ and $S$. For a binary system, the solution is simpler

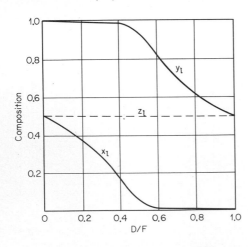

**FIG. 11.3** Increasing $D/F$ reduces both $x$ and $y$, but to different degrees.

$$S = \frac{y(1 - x)}{x(1 - y)} \qquad (11.12)$$

where $y$ and $x$ are given in terms of the more volatile product. Equation (11.12) can be solved for either $y$ or $x$

$$y = \frac{Sx}{1 + x(S - 1)} \qquad x = \frac{y}{y + S(1 - y)} \qquad (11.13)$$

For a given separation, $y$ can be solved for a stipulated value of $x$, or vice versa, and both entered into the material balance to determine $D/F$. This procedure was used to generate the curves shown in Fig. 11.3, based on initial conditions of $x = 0.05$, $y = 0.95$, and $z = 0.50$.

**example 11.1**   An ethylene fractionator has the following feed and product compositions:

|  | Feed z, % | Distillate y, % | Bottoms x, % |
|---|---|---|---|
| Methane (*ll*) | 0.05 | 0.1 | — |
| Ethylene (*l*) | 47.0 | 99.7 | 2.0 |
| Ethane (*h*) | 51.1 | 0.2 | 94.6 |
| Heavies (*hh*) | 1.8 | — | 3.4 |

Calculate $D/F$ and $S$. Then increase $D/F$ by 0.01 and calculate the resulting product compositions using Eqs. (11.7) to (11.11).

$$\frac{D}{F} = \frac{0.47 - 0.02}{0.997 - 0.02} = 0.461 \qquad S = \frac{0.997/0.002}{0.02/0.946} = 23,579$$

New $D/F = 0.471$

$$a = 0.471(23,578) = 11,105$$
$$b = 23,579(0.47) - (23,578)(0.471 - 0.0005) + 0.511$$
$$= -10.81$$
$$c = -0.511\left(1 - \frac{0.0005}{0.471}\right) = -0.510$$

$$y_h = \frac{10.81 + \sqrt{(10.81)^2 + 4(11,105)0.510}}{2(11,105)} = 0.00728$$

$$x_h = \frac{0.511 - 0.00728(0.471)}{1 - 0.471} = 0.9590$$

$$y_u = \frac{0.0005}{0.471} = 0.00106$$

$$y_l = 1 - 0.00728 - 0.00106 = 0.9917$$

$$x_l = \frac{0.959(0.9917)}{23,579(0.00728)} = 0.00554$$

$$x_{hh} = 1 - 0.00554 - 0.959 = 0.03546$$

Summary of new product compositions:

|  | Distillate y, % | Bottoms x, % |
|---|---|---|
| Methane (*ll*) | 0.10 | — |
| Ethylene (*l*) | 99.17 | 0.55 |
| Ethane (*h*) | 0.73 | 95.90 |
| Heavies (*hh*) | — | 3.55 |

Three conclusions can be drawn from the foregoing discussion:

1. Compositions of *both* product streams are profoundly affected by distillate-to-feed ratio.

2. Changes in feed composition can be offset by appropriate adjustment of *D/F*.

3. If separation is constant, control of composition of either product will also result in control of composition of the other product. (The relationship between *x* and *y* is fixed for a given separation.)

## Energy vs. Feed

A certain amount of energy is required to separate a given feed stream into its components. It is reasonable to assume that energy requirements will be roughly in proportion to feed rate. In a distillation tower, energy is introduced as heat $Q$ to the reboiler, which generates a proportional flow of vapor $V$

$$V = \frac{Q}{H_v} \tag{11.14}$$

The term $H_v$ represents the latent heat of vaporization. Expressing heat input in terms of vapor flow makes it possible to evaluate separation in terms of dimensionless ratio of vapor to feed rate $V/F$.

Douglas, Jafarey, and McAvoy [2] have developed a separation model in which the Fenske equation is adjusted as a function of reflux ratio and the sum of the light components in the feed. The author has modified their relationship to place separation in terms of $V/D$

$$S = \left\{ \frac{\alpha}{\sqrt{1 + 1/[(V/D - 1)z]}} \right\}^{nE} \tag{11.15}$$

where $E$ is tray efficiency and $z = z_l + z_{ll}$. Solving (11.15) for $V/D$ gives

$$\frac{V}{D} = 1 + \frac{1}{[(\alpha/S^{1/nE})^2 - 1]z} \tag{11.16}$$

For a given column with unknown tray efficiency, $nE$ can be estimated from observed conditions of $S$, $z$, $D/F$, and $V/F$. Then the effect of a variation in $V/F$ on separation can be calculated; or conversely, the reduction in $V/F$ corresponding to operating closer to product specifications can be estimated.

The above model is also useful in calculating the savings in energy possible by increasing relative volatility through operation at minimum pressure. This concept is explored in more detail later in the chapter, under the heading Floating Pressure Control.

## Operating Modes

There are two basic operating modes for distillation columns, constant separation and maximum separation. If separation is held constant, the compositions of both products are controlled, either with two composition feedback loops or one in conjunction with a constant $V/F$ ratio. In this mode, no more energy is used than needed to make the specified separation, a factor becoming increasingly important as energy resources dwindle.

The alternate operating mode maximizes separation for all conditions by maximizing boilup. This mode is falling out of favor as energy costs rise but is still used to recover an extremely valuable product or where waste heat is available for boilup. Let *recovery R* be defined as the ratio of the flow of valuable product to the flow of that component present in the feed. For the distillate

$$R_D = \frac{D}{Fz_l} = \frac{z_l - x_l}{z_l(y_l - x_l)}$$

and for the bottom product

$$R_B = \frac{B}{Fz_h} = \frac{z_h - y_h}{z_h(x_h - y_h)}$$

Recovery is maximized when the key impurity in the other product is minimized. Again, this is achieved by holding boilup (or reflux) constant at its maximum value, which maximizes separation.

Table 11.1 lists several sets of conditions for a column operated at constant boilup with distillate composition controlled. As feed rate is reduced, $V/F$ increases, causing a corresponding reduction in $x$ and hence improved recovery. Recognize, however, that the 5 percent increase in recovery at 60 percent feed rate is gained at the expense of a 67 percent increase in energy consumed per unit feed. In most cases, this is no longer justifiable.

## ARRANGING THE CONTROL LOOPS

There are typically five valves available for controlling a distillation column, heat input, heat removal, distillate, bottom product, and reflux. There may also be an equal number of controlled variables, distillate and bottom composition,

**TABLE 11.1   Recovery vs. Feed Rate for Constant Boilup and $y$ Controlled at 0.95**

| Feed rate, % | $V/F$ | $x$ | $D/F$ | $R_D$, % |
|---|---|---|---|---|
| 100 | 3.00 | 0.05 | 0.500 | 100.0 |
| 80 | 3.75 | 0.006 | 0.523 | 104.7 |
| 60 | 5.00 | 0.001 | 0.526 | 105.2 |

pressure, and liquid levels in the reflux accumulator and column base. Figure 11.4 shows their locations. Feed rate is usually considered to be an independent variable in that it sets production; furthermore, in a train of columns, the feed to one is the product from another and so is not available for manipulation.

Figure 11.4 shows temperature measurements located near the top and bottom of the column because temperature is the most frequently used indication of composition. Where a more exact measurement is needed, an analyzer would be located on the product stream itself. If the maximum-recovery mode of operation were selected, only one temperature or composition would be controlled. Then heat input or reflux would be set at the maximum allowable value.

There are $5!=120$ ways of connecting five pairs of manipulated and controlled variables in single loops. Many of these combinations are obviously unworkable, e.g., any including base level controlled by the distillate valve. Still, there is one combination that will produce the best results for any particular column. Because of extensive variations from one column to another in operating characteristics such as $D/F$ and $V/F$ ratios, heating and cooling media, condenser, reboiler, and column constraints, as well as the locations of valves, the optimum loop arrangement varies considerably. It is important to establish a set of guidelines that will help one arrive at the optimum configuration for each column one encounters.

### Interaction between Energy and Material Balances

The influence of the external material balance on product quality has been demonstrated in Figs. 11.2 and 11.3. Logically, then, the external material balance ought to be manipulated through either $D$ or $B$ to achieve quality control. However, this action is insufficient in itself. Observe in Fig. 11.4 that

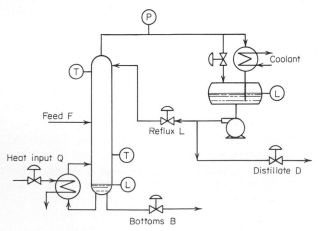

**FIG. 11.4** There are typically five pairs of variables used in controlling a distillation column.

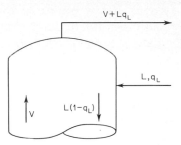

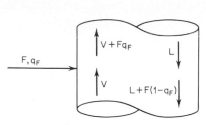

**FIG. 11.5** The material balance at the feed tray is affected by feed enthalpy $q_F$.

**FIG. 11.6** The material balance at the top of the tower is a function of the enthalpy $q_L$ of the reflux stream.

unless other loops are closed, manipulating $D$ and $B$ will only affect liquid levels in the vessels they leave. To influence product compositions requires that liquid and vapor rates within the column itself be adjusted. These rates are sensitive to many upsets, however. They are, of course, primarily determined by the heat input to the reboiler and the flow of reflux. Yet Fig. 11.5 shows that flow rates above and below the feed tray depend also on the feed rate and its enthalpy $q_F$. If the feed is completely vaporized at its dew point, $q_F$ will be 1.0* and the upward flow of vapor will be augmented by all the feed. If the feed is superheated, it will vaporize some of the liquid on the tray, increasing vapor flow above that point still further. If the feed is a liquid at its bubble point, $q_F = 0$ and all the feed adds to the downward flow of liquid. A subcooled feed has a negative $q_F$ and condenses some of the upflowing vapor.

A similar situation occurs at the top tray, where reflux is introduced, as shown in Fig. 11.6. Reflux is commonly cooled below its bubble point, and in such a condition condenses some vapor at the top tray. The degree of subcooling is quite variable, depending both on the heat load placed on the condenser and the temperature of the cooling medium. Air-cooled condensers are subject to the widest variation in temperature and therefore present the most problems in this regard.

The difference between the vapor flow leaving the top tray and the reflux returning to it is the distillate product. If distillate flow is manipulated for quality control, either reflux or boilup must be used to control liquid level in the accumulator. Then the difference between vapor and liquid flow rates within the top of the column will be forced to follow the external material balance.

Alternately, either reflux or boilup could be manipulated for quality control, with distillate under accumulator-level control. In this configuration, the difference between internal vapor and liquid rates, and hence distillate flow, is subject to variations in feed and reflux enthalpies. A further consideration is the relative magnitude of the three flow rates in question. For a reflux ratio $L/D$ of 1.0, since reflux and distillate have the same effect on the external material

*Note that $q_F$ as defined here is the reverse of that commonly used in distillation texts.

balance and accumulator level, it would seem that either could be used for quality control or level control. However, the external material balance is subject to variations in $V$, $q_F$, and $q_L$ when distillate is under level control but not when reflux is under level control. These disturbances are amplified by the reflux ratio. For $L/D = 3$, for example, variations of 1 percent in $L$ and $q_L$ will cause variations of 3 percent in $D$; 1 percent changes in $V$ will affect $D$ fourfold; and similar variations in $q_F$ will cause $D$ to change by the ratio $F/D$. But if $D$ is manipulated for quality control with $L$ or $V$ under level control, these upsets are all returned to the column, virtually eliminating their effect on product quality. Columns having only one product on level (or pressure) control are considered to be under material-balance control.

### Material-Balance Control

Either product may be manipulated for quality control with the other under level control to close the material balance. To maximize the accuracy of the system, the smaller product flow should be chosen to control product quality. This may or may not be the same product whose quality is to be controlled. There are four possible configurations: $D$ can be manipulated to control $x$ or $y$, or $B$ can be manipulated. Controlling $x$ with $B$ leaves base level to manipulate heat input; this is ordinarily a responsive loop, so that satisfactory control can be expected. Furthermore, the entire column responds quite rapidly to heat input, so that it is possible to control temperature or composition at any point, including the overhead, by manipulating $B$.

Manipulation of $D$ does not elicit the same responsiveness, however. Delays consist in the time constant of the reflux accumulator and holdup of liquid in the trays throughout the column. When distillate flow is changed, the corresponding adjustment to reflux requires the level of the accumulator to respond first and its level controller to act. Thus reflux responds to distillate as a first-order lag, having a time constant equal to the residence time of reflux in the accumulator divided by the proportional gain of the level controller. To minimize this lag, the level controller should have as narrow a proportional band as practical, e.g., 10 percent. Integral action, in following behind proportional action, has little effect on the value of this lag.

The feedforward control system described in Ref. 3 and Fig. 11.7 can eliminate the lag associated with the reflux accumulator. Changes in distillate set point impose opposite changes directly on the reflux, without the need to wait for the level controller to respond. If $K = 1$, the changes are equal and the accumulator level will remain constant. But $K$ can be set $>1$, as shown, causing reflux to overshoot its steady-state value, in which case the accumulator is converted from a lag to a lead-lag function. This feature reduced the period of the composition loop for the column described in Ref. 3, from 5 h to 30 min.

Should it be necessary to control $x$ by manipulating $D$, reflux cannot be used to control accumulator level. Liquid flow into the base of the column is simply too slow in responding to reflux to make this composition loop viable. Instead, reflux flow can be fixed to set the heat load on the column, and accumulator level can manipulate the heat input to the reboiler. Because the entire column responds rapidly to heat input, this level loop can be closed satisfactorily. The feedforward loop shown in Fig. 11.7 is still desirable, although for manipulation of heat input the sign of $K$ would be reversed.

## Manipulating Heat Input

If bottom-product flow is under level control, heat input to the column may be fixed, set in ratio to the feed (as described under feedforward control), or manipulated for quality control. In the last case, there will be some interaction with the other composition loop, which deserves further consideration.

Steam is the most common heat source and also the most easily controlled. Flow control is also heat-input control, because of steam's nearly constant latent heat. Often, however, hot oil or hot water is the heating medium. Then heat flow is not linear with liquid flow, as indicated by Figs. 9.4 to 9.6, and it is also strongly influenced by liquid temperature. If heat input is under control of either base or accumulator level, these problems are not important, although an equal-percentage valve is needed to improve linearity of response.

In situations where close control of vapor flow is needed while manipulating a liquid heating medium, a column differential-pressure measurement is recommended. The trays restrict vapor flow much like an orifice, producing a proportional differential pressure. Liquid flow also affects differential pressure but to a lesser degree. Many of the newer columns are equipped with valve trays, designed to extend their operating range. The valves open as vapor flow is increased, tending to regulate differential pressure, as shown in Fig. 11.8. At very low vapor rates, the valves are fully closed, and at very high rates they are fully open. But even in the midrange where they regulate, the relationship between differential pressure and vapor flow is still reproducible and useful for control.

Furthermore, column differential pressure is a reliable index of flooding. A rising differential pressure indicates increasing vapor velocity and accompanying entrainment. A point is eventually reached where the differential pressure across a tray exceeds the head of liquid in the downcomer, thereby interfering

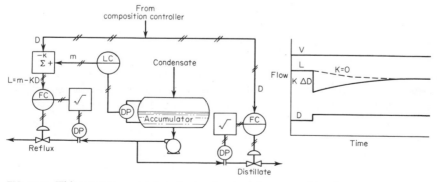

**FIG. 11.7**  This arrangement injects lead action into the composition loop.

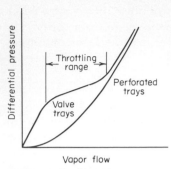

FIG. 11.8 Perforated trays produce a differential pressure proportional to the square of vapor flow, while valve trays tend to regulate it.

with liquid flow. This condition, called *downcomer flooding,* causes liquid to accumulate in the column, destroying its internal material balance and separation as well. The onset of downcomer flooding is accompanied by a sharp and steady increase in differential pressure. It can be relieved by reducing either reflux or boilup, but as much time is required to empty the column as was taken to fill it. In other words, corrective action should be applied as soon as possible following the first indication of flooding. Flooding can be avoided altogether by controlling differential pressure with heat input at a value just below the flooding point. If heat input is normally well below flooding but manipulated for temperature or composition control, differential pressure may be used as an override to protect against flooding.

Proper installation of the transmitter is the key to reliable differential-pressure control. The low-pressure connection should be close-coupled to the column or overhead vapor line, at the lowest possible elevation, to facilitate maintenance. Piping from the base of the column to the high-pressure connection should be at least 1 in (25 mm) in diameter with no horizontal sections. Vapors will tend to condense in this line, so it should drain freely back to the column. Insulation will reduce condensation, and ball valves should be used to avoid liquid traps.

### Pressure-Control Methods

The energy balance must be closed. If the sum of the vapor fed and generated exceeds that withdrawn and condensed, the pressure in the column will rise. Column pressure tends to be self-regulating because as it increases, the temperature difference across the reboiler heat-transfer surfaces diminishes and that across the condenser increases, both of which tend to restore the energy balance. Nonetheless, pressure cannot be left uncontrolled. In certain configurations of column and controls, a cycle may develop due to internal interactions which can be eliminated only by tight pressure control. Furthermore, when atmospheric cooling is used, rapid changes in weather conditions can upset column pressure to the point of initiating flooding.

Pressure may be controlled by manipulating the rate of vapor generation, removal, or condensing. One of the important factors in controlling compositions is maintaining the internal vapor or liquid rate in ratio to the feed, to fix separation. This is best accomplished by controlling the rate of heat input or internal reflux, whichever is the most accurately measured and controlled. Usually (but not always) heat input is the preferred choice for setting separation. Then the energy balance is best closed by the pressure controller acting on the less predictable rate of condensation. Most configurations then will have the pressure controller manipulating condenser cooling.

There are several common methods of adjusting the rate of condensation:

1. Changing the flow of coolant
2. Flooding the condenser
3. Bypassing the condenser
4. Injecting noncondensible gas and subsequent venting

Each has certain advantages and disadvantages. Generally, coolant manipulation is slow and nonlinear, as described in Chap. 9. In the case of fan condensers, the associated variable-speed motors, variable-pitch fans, and louver operators are also quite costly.

Pressure control through condenser flooding was also described in Chap. 9, in connection with Fig. 9.8. A hot-vapor-bypass line, shown next to the condenser in Fig. 11.4, is used with total condensers only. To understand its function, recognize that all the bypassed vapor must condense on the walls and liquid surface of the accumulator. Opening the bypass raises column pressure by reducing the pressure drop across the condenser, tending to keep the condensate from draining freely. The size of the bypass valve depends on the pressure drop available; if the condenser is mounted below the accumulator (submerged), a relatively small valve can be used. Pressure tends to respond fairly rapidly to valve motion, with natural periods of less than 2 min being common. There is a dependence on the condenser arrangement, however; a submerged condenser with flooded shell is much slower than an overhead condenser with vapor condensing in the tubes.

Figure 11.9 describes a pressure-control method which is helpful with a flooded condenser when heat input is not accurately measurable. Although the scheme shows the pressure transmitter located at the top of the column, better response is obtained when manipulating heat input if it is located near the bottom.

Scheme 4 listed above intentionally introduces noncondensable gas into the

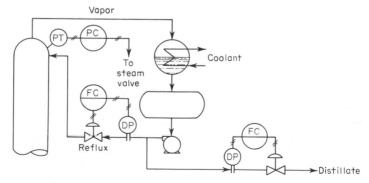

**FIG. 11.9** With rising pressure, vapor flow is reduced, thereby exposing more heat-transfer surface.

condenser, ultimately reducing its heat-transfer capability to that required by the heat input. But a path must be open for release of the gas in the event of rising pressure. Since every cubic foot of gas added to the tower will eventually sweep a certain amount of product with it out the vent, this scheme is *not*

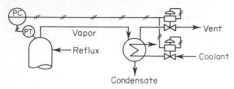

**FIG. 11.10** Operating a vent valve in split range with the coolant valve will allow for the release of noncondensables.

recommended. Noncondensables also dissolve in the reflux and tend to reduce its boiling point, affecting temperatures on the top trays of the tower. This may invalidate the use of these temperature points for quality control. There is also the danger, particularly during startup, of blanketing the condenser tubes with gas, preventing heat removal altogether. The condenser must then be vented until normal condensation begins.

Columns operated at atmospheric pressure actually are controlled in this same way. If condensation is too efficient, a vacuum will start to develop, drawing air into the condenser. Eventually, the vapor mixture in the condenser will contain enough air to limit the rate of condensation to the rate of boilup. Every increase in boilup will expel some air and product; a decrease will draw more air into the system.

To allow for the release of noncondensable gases that may be contained in the feed stream, a vent valve must be added to whatever pressure-control scheme has been chosen. This valve would operate in split range with the main control valve.

Figure 11.10 shows a typical arrangement, with valve positioners used to

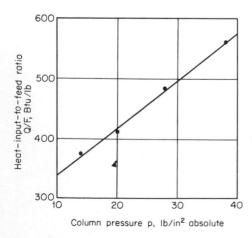

**FIG. 11.11** Reducing column pressure saves energy for most separations like this one between pentane isomers.

effect the split-range operation. Both valves must be fully open at maximum controller output, but the vent valve ought to close before the coolant valve, on decreasing output. An acceptable sequence would have the vent valve closed below 50 percent output and the coolant valve fully open at 50 percent output. The coolant valve needs equal-percentage characteristics, the vent valve, linear.

If condensing area is limited, separation can be maximized by using the pressure controller to set the rate of heat input to the reboiler. In this way, just as much heat will be introduced as the condenser is capable of removing.

In a vacuum still, pressure can be controlled by manipulating a valve in the line leading to the vacuum system. As before, introduction of a noncondensable gas to control the vacuum is not generally recommended.

Sometimes overhead product is withdrawn as a vapor under flow control. This provides an escape for noncondensables, in which case any of schemes 1 to 3 may be used with a single valve. The flow of the product must be manipulated to control composition, however, not pressure.

### Floating Pressure Control

Most pressure-control systems function by restricting heat transfer through the condenser. If the condenser is allowed to operate without restriction, column pressure and temperature will be as low as attainable consistent with conditions of heat load and coolant temperature. For most mixtures, relative volatility improves with falling temperature and pressure, thereby requiring less energy to effect the same separation. The savings to be gained amounts to nearly 1 percent per degree Fahrenheit reduction in coolant temperature, particularly attractive for columns with fan condensers [4]. Figure 11.11 shows the heat input required to separate a 50-50 mixture of normal and isopentane into products each 97 percent pure in a column of 67.7 theoretical trays. The points were found using the model of Eq. (11.16) and values of $\alpha$ and $H_v$ taken from Ref. 4.

However, pressure control needs to be maintained in the short term to provide stability and resistance to upsets in the cooling medium. Short-term control may be combined with long-term minimization using the two control loops shown in Fig. 11.12. The PI pressure controller is adjusted normally for tight control at its set point. However its set point is manipulated slowly by the integral-only valve-position controller, to drive the valve to an extreme position in the long term. An integral time of 1 h has been found satisfactory if the pressure signal is also fed forward to the separation control system.

Naturally, no energy will be saved unless the heat input or reflux is adjusted as pressure falls. If no such adjustment is made, the purity of the products will simply improve. While the task of

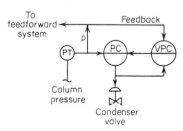

**FIG. 11.12**  The valve-position controller minimizes the pressure set point in attempting to drive the condenser valve to an extreme position.

readjustment may be left to a composition controller, there will be less upset to the products and faster response if the adjustment is made directly as a function of pressure in accordance with the relationship shown in Fig. 11.11

$$\frac{Q}{F} = m(p_0 + p) \tag{11.17}$$

where $p$ = column pressure

$m$ = a function of separation, tray efficiency, etc.

$p_0$ = constant determined by mixture being separated

If temperature is being used to indicate and control composition, it must be compensated for variable pressure

$$T_b = T + \left.\frac{\partial T}{\partial p}\right|_x (p_b - p) \tag{11.18}$$

This equation is used to calculate the temperature $T_b$ corresponding to base pressure $p_b$ from measured temperature $T$ and pressure $p$. The slope of the vapor-pressure curve for hydrocarbon mixtures can be estimated from the formula given in Ref. 5

$$\left.\frac{\partial T}{\partial p}\right|_x \approx 0.1 \frac{T_a}{p_a} \tag{11.19}$$

where subscript $a$ identifies absolute temperature and pressure.

## COMPOSITION CONTROL

The composition of products from distillation columns is quite difficult to control, even with the optimum loop configuration. Delays are substantial, particularly when analyzers must be located far from the column. The high degree of interaction between the capacities represented by tray holdup means that time constants are very long, often hours. Furthermore, sensitivities are high, and responses tend to be nonlinear. As a result, feedforward is a very worthwhile addition to any composition loop, and its contribution is easily justified.

### Feedforward Control

The feedforward system first must solve the column material-balance equations for the set point (*) of one product

$$D^* = F \frac{z_i - x_i}{y_i - x_i} \tag{11.20}$$

or

$$B^* = F \frac{y_i - z_i}{y_i - x_i} \tag{11.21}$$

Where product purities are high, these expressions can be simplified to

$$D^* = m_y F(z_l + z_{ll}) \tag{11.22}$$

and

$$B^* = m_x F(z_h + z_{hh}) = m_x F(1 - z_l - z_{ll}) \tag{11.23}$$

where $m_y$ and $m_x$ are adjusted to achieve the desired product quality. These parameters should be essentially constant if separation is held constant by control of $V/F$ (or $L/F$). This requires a second feedforward loop, as shown in Fig. 11.13. In this system, one composition controller adjusts $m_y$ or $m_x$ to maintain constant quality of its product while the other sets separation. The separation loop need not be closed if there is no specification on the second product.

The feedforward input from feed composition is probably not required for most columns. Composition tends not to change as widely or rapidly as feed rate, and analyzers are more costly and less reliable than flowmeters. Although Fig. 11.13 shows only one component being analyzed, in a multicomponent system, significant lighter-than-light-key components should be added to the light key to calculate distillate flow, as called for by Eq. (11.22). Similarly, when calculating bottom-product set point, heavier-than-heavy-key components should be added to the heavy key. Because light components are easier to detect, the composition input can be determined by difference, as in Eq. (11.23).

Requirements for dynamic compensation vary principally with the condition of the feed. If the feed is liquid, its entire influence is projected toward the bottom of the column, with delays associated with the hydraulics of the trays below the point of entry. To maintain a dynamic balance within the column, changes in bottom-product flow and heat input must be delayed until the change in feed arrives at the base. Dead time and lag compensation are

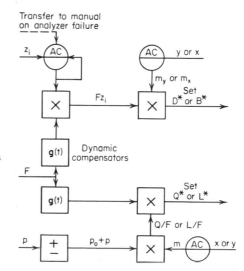

**FIG. 11.13** This system provides for control of both products with feedforward inputs from pressure, as well as feed rate and composition.

required, adjusted so that base level is not noticeably upset following a step change in feed rate. If the internal material balance is maintained, compositions will remain steady. For the case of liquid feed, the distillate or reflux compensator can be set the same as the boilup or bottom-flow compensator because of the rapid response of the overhead system to changes in boilup. In fact, a single compensator has been used for the two loops.

If the feed is completely vaporized, its variations affect the overhead balance directly. Then the distillate or reflux compensator needs lead-lag action without dead time. Variations in accumulator level can be used as a guide in setting the compensator. Then the liquid loading originates as reflux from the top of the column, requiring longer dead time and lag for the boilup or bottoms compensator than with liquid feed. A feed that is partially vaporized affects both ends of the column and therefore will require the combination of the overhead compensator adjusted as for vapor feed and the bottom compensator adjusted as for liquid feed.

Dynamic compensation for the composition input in Fig. 11.13 is applied using a PI controller whose output is fed back to its controlled-variable input. The function of this device is described in Eq. (7.21). It is necessary to provide an input signal to the feedforward multiplier should the analyzer be out of operation. In fact, if a logic signal is available from the analyzer, the controller may be transferred to manual upon a failure, holding its last output. The operator should be advised not to readjust its output to track changes in feed composition unless those changes are current, otherwise a secondary upset may be introduced after the feedback controller has already adjusted for a change in feed composition.

No compensator is needed for the pressure input because its rate of change is controllable through the time constant of the valve-position controller in Fig. 11.12. The feedforward loop from pressure actually contains some positive feedback, in that reducing heat input as pressure falls will cause pressure to fall further. But the response of the positive-feedback loop through the valve-position controller can be made much slower than that of the negative-feedback loop through the pressure controller, so that stability can be assured.

Columns operating in the maximum-recovery mode have no feedforward loop from feed to boilup or reflux. Furthermore, their recovery factor is variable, so that product flow is related nonlinearly to feed rate. The data in Table 11.1 are plotted in Fig. 11.14 to illustrate the extent of this nonlinearity. A simple parabolic model of the form

$$D^* = m_y F - kF^2 \tag{11.24}$$

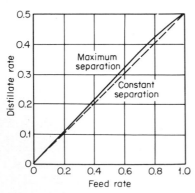

**FIG. 11.14** Distillate flow varies non-linearly with feed rate when recovery is maximized.

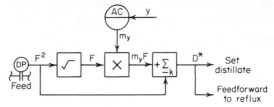

**FIG. 11.15**  This parabolic model is used to match the curve of Fig. 11.14 in the operating region.

is used to match the curve. Its slope is adjusted by $m_y$, the output of the feedback controller, and the nonlinearity set as coefficient $k$ in the subtractor. To match the curve of Fig. 11.14 at 0, 80, and 100 percent feed, $m_y = 0.62$ and $k = 0.12$. The fit at 60 percent feed is high by 0.013 unit, or 2.6 percent of $D/F$. If feed composition is measured, it should be multiplied by linear feed rate, as with the constant-separation system.

Dynamic compensation for this system differs markedly from those where boilup is manipulated. Lacking the accelerating effect of boilup, feedforward control of reflux is needed, as shown in Fig. 11.8. Lead action is definitely required and is obtainable by setting forward-loop gain $K$ greater than unity.

A maximum-recovery system manipulating bottom-product flow uses the same nonlinear function shown for distillate. Its dynamic compensation, however, is the same as for constant recovery, in that boilup is manipulated to control base level. Should $B$ be manipulated to control $y$ for maximum recovery of $D$, the sign of nonlinear coefficient $k$ is reversed.

## Feedback Control

Temperature is used more frequently than analysis to control product quality. It has the advantage of being an inexpensive, reliable, and responsive measurement. The composition of a binary mixture is virtually linear with temperature for common mixtures. The relationship is sensitive to concentration variations in off-key components, however. As an example, the percentage of propane in the bottom product from a depropanizer is typically controlled by temperature; but should the concentration of heavy components such as gasoline fractions increase, the temperature controller will allow more propane to enter the product to maintain the same bubble point.

Temperature is not a sufficiently sensitive measurement of composition for high-purity mixtures. To improve its sensitivity, the measurement is usually located several trays from the end of the column, where purity is lower. The effect of variable pressure can be reduced either through the compensation described in Eq. (11.18) or by using a differential-vapor-pressure-measuring device. In this apparatus [6], a temperature bulb containing a sample of the desired product mixture is connected to one side of a differential-pressure transmitter. The bulb is inserted into the column, with the other side of the transmitter directly connected to the same point. If the mixture in the column is exerting the same vapor pressure on the direct-connected side as the mixture in the bulb is exerting on its side of the transmitter, their differential will be zero. This indicates that the product of the desired vapor pressure is being made. This device is used widely for proof control in alcohol distilleries.

Improved sensitivity without influence from pressure can also be gained by combining two or more temperature measurements into a differential or double-differential system, as described in Ref. 7. While excellent results are reported for well-documented separations such as benzene-toluene, considerable effort may be required to calibrate such a system properly for other separations.

A temperature-control loop responds with a natural period in the range of 20 to 40 min; this short period is attributed to the location of the measuring element directly in the column. Photometric analyzers connected directly to the column exhibit similar control response. Most analyzers, however, cannot be so close-coupled, particularly when precise control must be exercised over valuable final products. The chromatograph, the most commonly used analyzer in distillation, also presents internal delays in sampling and sample transport, as well as transport delays external to the analyzer. The period of oscillation for a chromatograph control loop on a column is typically from 1 to 3 h. The long period makes feedback rather ineffective, forcing the engineer to use feedforward control or temperature feedback set in cascade from composition. To maximize speed of response, locate the analyzer as close to the column as possible, use vapor samples under low pressure, and maximize velocity in the sample line. Avoid using multistream analyzers for control.

Finally, a sampled-data PID controller is recommended for chromatographs controlling distillation columns. The analyzer can provide a contact closure when a new analysis is complete, which can be used to transfer the controller to automatic for a short interval. It then is to remain in manual until the next analysis is complete. Should there be no next analysis due to an analyzer failure or interruption for maintenance, the controller simply remains in manual. A digital control algorithm should be allowed to make only one calculation following each analysis. Its derivative action is surprisingly effective when operated in this mode.

Operating people have discovered that a column is easier to control with increasing product purity. Thus there is a tendency to overpurify products to simplify operation, although it is costly in terms of recovery and energy consumption. This variable sensitivity can be detected by differentiating the relationship between composition and $D/F$ for constant separation

$$\frac{\partial y_h}{\partial (D/F)}\bigg|_S = \frac{(x_h - y_h)^2}{x_h - z_h + \dfrac{(z_h - y_h)(1/y_l + 1/y_h)}{1/x_l + 1/x_h}} \tag{11.25}$$

The principal components effecting the sensitivity are the key impurities

$$\frac{\partial y_h}{\partial (D/F)}\bigg|_S \propto \frac{y_h}{x_l} \tag{11.26}$$

This variability can be seen in the record of distillate impurity while under control; the composition cycle has flat valleys and sharp peaks. Bottom-product composition exhibits the same variable gain.

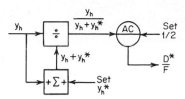

To compensate for this gain variation the input to the composition controller is computed as the impurity measurement divided by measurement plus set point [8, p. 259], as shown in Fig. 11.16. As $y_h$ decreases, the gain of the divider increases, virtually offsetting the gain reduction of the process. When measurement $y_h$ equals set point $y*_h$, the output of the divider

**FIG. 11.16** Dividing by the impurity helps compensate for the variable gain of the process. (*From F. G. Shinskey, "Distillation Control," copyright 1977 by McGraw-Hill, Inc., used by permission of McGraw-Hill Book Company.*)

is ½, at which value the set point of the composition controller must be fixed.

Nonlinearities observed in the response of temperature control in distillation are largely due to the same phenomenon. As the temperature element is moved away from the end of the column, the concentration of the key impurity rises and linearity improves. However, as the feed tray is approached, sensitivity decreases again; a discontinuity in the composition profile is common at that point. When the temperature measurement is located midway between the feed tray and the end of the column, a nonlinear controller of the type described in Fig. 5.28 is useful. If temperature is set in cascade from a composition controller, integral action can be omitted from the secondary loop, thereby improving stability.

## Interaction

Interaction between composition loops can be quite severe because of their similar long periods of oscillation. The first step in configuring a control system for two products is then to estimate relative gains. Although a distillation column is typically a $5 \times 5$ system, the relative-gain array can conveniently be broken into $2 \times 2$ subsets to find the most-effective control scheme.

The first subset to be examined is the control of $y$ and $x$ using $D$ and $V$. This assumes that accumulator level is controlled by reflux, base level by $B$, and pressure by condenser cooling. The steady-state gain of $y$ with respect of $D/F$ for constant $x$ can be determined by differentiating the material-balance equation

$$\left.\frac{\partial y_h}{\partial (D/F)}\right|_x = \frac{(x_h - y_h)^2}{x_h - z_h} \tag{11.27}$$

When this is combined with the gain for constant separation given in Eq. (11.25), the relative gain is obtained

$$\lambda_{yD} (\Lambda_{DV}) = \cfrac{1}{1 + \cfrac{(z_h - y_h)(1/y_l + 1/y_h)}{(x_h - z_h)(1/x_l + 1/x_h)}} \tag{11.28}$$

If the influence of feed composition is neglected, $\lambda_{yD}$ of the subset $\Lambda_{DV}$ approaches zero when $x_l \gg y_h$ and approaches unity when $y_h \gg x_l$. In other words, the material balance should generally be manipulated to control the composition of the less pure product and separation to control the composition of the purer product. Relative gains of the $\Lambda_{DV}$ subset always fall between 0 and 1.

Should $\lambda_{yD}$ be closer to 0 than to 1, subset $\Lambda_{DV}$ may have to be rejected because of the poor dynamic response of the alternate loop represented by favorable gain $\lambda_{xD}$. Another material-balance subset is $\Lambda_{BL}$. As it happens,

$$\lambda_{yB}(\Lambda_{BL}) \approx \lambda_{yD}(\Lambda_{DV}) \tag{11.29}$$

Consequently, if $\lambda_{yD}(\Lambda_{DV})$ is low, $\lambda_{yB}(\Lambda_{BL})$ will be too, in which case the alternate loop $\lambda_{xB}(\Lambda_{BL})$ will approach unity. Then $x$ can be controlled successfully with $B$ and $y$ with reflux $L$, leaving base level under the control of boilup and the accumulator level manipulating $D$.

Another possible material-balance subset is $\Lambda_{DL}$, one of whose elements is

$$\lambda_{yD}(\Lambda_{DL}) \approx \lambda_{yD}(\Lambda_{DV}) \tag{11.30}$$

If $\lambda_{yD}$ is low, the alternate gain $\lambda_{yL}(\Lambda_{DL})$ will approach unity and also provide reasonable dynamic response. Although $x$ would be controlled by $D$, response could be attained by placing boilup under accumulator level or pressure control, as in Fig. 11.10.

The only non-material-balance subset is $\Lambda_{LV}$, where compositions are controlled by reflux and boilup and both products are under level control. Relative gain $\lambda_{yL}$ for subset $\Lambda_{LV}$ is derived for the multicomponent column in Ref. 8 (pp. 309 to 311) and is related to the $\lambda_{yD}$ of subset $\Lambda_{DV}$ as follows:

$$\frac{\lambda_{yL}(\Lambda_{LV})}{\lambda_{yD}(\Lambda_{DV})} = 1 + \frac{\beta(y_l - x_l)^2}{z_l - x_l}\left(\frac{1}{y_l} + \frac{1}{y_h}\right) \tag{11.31}$$

where $\beta = V/F \ln S$. The dominant term in the above expression is $1/y_h$, which will vary from about 5 into the hundreds. As a result, $\lambda_{yL}(\Lambda_{LV})$ is always positive and usually much greater than unity. This is to be expected, because both manipulated variables move both controlled variables in the *same direction*. Increasing boilup, for example, will raise temperatures at both ends of the column, and increasing reflux will lower them both.

**example 11.2**    The ethylene fractionator in Example 11.1 requires control over both products. Given $V/F = 2.52$, calculate $\lambda_{yD}(\Lambda_{DV})$ and $\lambda_{yL}(\Lambda_{LV})$ and determine the most effective loop configuration.

$$\lambda_{yD}(\Lambda_{DV}) = \frac{1}{1 + \dfrac{(0.511 - 0.002)(1/0.997 + 1/0.002)}{(0.946 - 0.511)(1/0.02 + 1/0.946)}} = 0.080$$

$$\beta = \frac{2.52}{\ln 23{,}579} = 0.25$$

$$\frac{\lambda_{yL}(\Lambda_{LV})}{\lambda_{yD}(\Lambda_{DV})} = 1 + \frac{0.25(0.997 - 0.02)^2}{0.47 - 0.02}\left(\frac{1}{0.997} + \frac{1}{0.002}\right) = 266.7$$

$$\lambda_{yL}(\Lambda_{LV}) = 266.7(0.080) = 21.3$$

For this column, $\lambda_{xB}(\Lambda_{BL}) \approx 1 - \lambda_{yD}(\Lambda_{DV}) = 0.92$.

Therefore bottom-product composition should be controlled with its flow and distillate composition with reflux.

Because the subset $\Lambda_{LV}$ typically produces such high relative gains, considerable interaction can be expected when a non-material-balance scheme is used to control both products. Furthermore, attempts to improve response by the addition of decoupling bring a very real risk of instability. Figure 8.16 shows how little decoupler error is tolerable as the relative gain rises above unity. For the column in Example 11.2, the decoupler error corresponding to an infinite decoupled relative gain is only 2.4 percent. This sensitivity was demonstrated in a simulation study by Toijala and Fagervik [9], where a decoupler error of +10 percent increased the period of the composition loops by an order of magnitude. Compositions were being controlled by reflux and boilup, with a relative gain of 5.2.

## Decoupling

A partial decoupling scheme that has been found particularly effective in controlling distillation columns is shown in Fig. 11.17. Measured heat input is divided into dynamically compensated feed rate to adjust the $D*/F$ ratio

$$\frac{D*}{F} = m_y - \frac{d_{DV}F(t)}{Q} \tag{11.32}$$

Solving for the distillate set point $D*$ reveals that this decoupler mimics the parabolic model given in (11.24) for columns with constant heat input

$$D* = m_y F(t) - \frac{d_{DV}[F(t)]^2}{Q} \tag{11.33}$$

In a constrained situation, i.e., when $Q$ is limited, the column operates in the

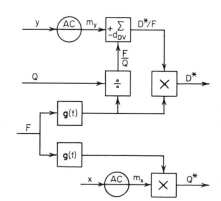

**FIG. 11.17** Partial decoupling is applied nonlinearly by dividing compensated feed rate by measured heat input.

maximum-recovery mode, and therefore the parabolic feedforward model is appropriate.

In the unconstrained mode, distillate flow is made to follow variations in heat input to allow for corresponding variations in separation. Dynamic compensation of the heat-input signal is not required, because of the rapid response of the overhead internal balance to heat flow. For this same reason this decoupling loop is more worthwhile than attempting to decouple bottom-product composition from changes in distillate flow. Bottom-product composition responds only very slowly to this source of upset, so that the effects on its variations are attenuated. If a decoupler were to be added in this direction, it would require dynamic compensation in the form of dead time plus a long lag. This second decoupler is then both less valuable and more complex than the one shown in Fig. 11.17.

From Ref. 8 (p. 321), the value of the decoupling coefficient is estimated as

$$d_{DV} = \frac{(y_l - z_l)(V/F)^2}{(y_l - x_l)^2\,\beta(1/x_l + 1/x_h)} \tag{11.34}$$

Here again, the key impurity is the controlling factor. As $x_l$ approaches zero, $1/x_l$ becomes quite high and less decoupling is required.

**example 11.3**  Calculate $\lambda_{yD}(\Lambda_{DV})$ and $d_{DV}$ for the column described in the top row of Table 11.1

$$\lambda_{yD}(\Lambda_{DV}) = \frac{1}{1 + \dfrac{(0.50 - 0.05)(1/0.95 + 1/0.05)}{(0.95 - 0.50)(1/0.95 + 1/0.05)}} = 0.50$$

Decoupling is necessary.

$$\beta = \frac{V/F}{\ln S} = \frac{3}{\ln 361} = 0.509$$

$$d_{DV} = \frac{(0.95 - 0.50)(3)^2}{(0.95 - 0.05)^2(0.509)(1/0.95 + 1/0.05)} = 0.466$$

Compare $d_{DV}/V$ with the nonlinear coefficient $k$ of 0.12, which was used to fit Eq. (11.24) to the same column

$$\frac{d_{DV}}{V} = \frac{0.466}{3} = 0.155$$

For the case where $B$ is manipulated to control $x$ and $L$ to control $y$, decoupling may be added to the bottom control loop from reflux in a similar manner

$$\frac{B*}{F} = m_x - \frac{d_{BL}F(t)}{L} \tag{11.35}$$

Again from Ref. 2 (pp. 315 and 317), the decoupling coefficient $d_{BL}$ is estimated as follows:

$$d_{BL} = \frac{(z_l - x_l)(L/F)^2}{(y_l - x_l)^2\,\beta(1/y_l + 1/y_h) + (z_l - x_l)} \qquad (11.36)$$

Dynamic compensation in the form of a lag plus dead time is required.

## BATCH DISTILLATION

Although gradually diminishing in favor, batch distillation still is an interesting process to control and deserves more than casual attention. Like most batch processes, its control system requires special consideration, ultimately bearing only faint resemblance to that of its continuous counterpart.

### The Material Balance

A batch separation will require an amount of time inversely proportional to the rate at which heat is introduced. Consequently, if processing time is to be minimized, heat input must be maintained at the maximum permissible level throughout distillation. This feature then fixes one of the variables which was subject to manipulation on a continuous tower. With vapor rate fixed, a material balance can be readily constructed for the batch still shown in Fig. 11.18.

If distillate flow is selected as the variable to be manipulated for product-quality control, reflux is then dependent. In the continuous system, product quality was affected by both $D/F$ and $V/F$. But here $F = 0$, and so it follows that product quality is a function of the ratio of the remaining variables, that is, $D/V$. In a sense, a batch still is similar to the enriching section of a continuous tower,

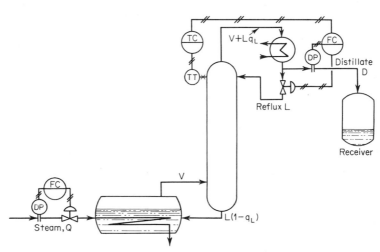

**FIG. 11.18**  With $V$ fixed, the only variable capable of independent manipulation is $D$.

part of whose vapor flow is feed. In the continuous tower if $V/F$ were maintained constant, manipulation of $D/V$ would also be manipulation of $D/F$.

Batch distillation is an unsteady-state process, because bottoms composition is continually changing as long as distillate is being withdrawn. If $D/V$ is constant, $y$ will change as $x$ changes. Constant $D/V$ is essentially constant separation and constant withdrawal of distillate. Distillate composition will then vary with time as the light component is removed from the tower. Starting with an initial charge $W_0$, containing $x_0$ mole fraction of the lighter component, at any time $t$ there remains

$$W = W_0 - Dt \tag{11.37}$$

The bottoms composition $x$ at time $t$ can be found from the material balance of the light component

$$Wx = W_0 x_0 - D \int y \, dt$$
$$x = \frac{W_0 x_0 - D \int y \, dt}{W_0 - Dt} \tag{11.38}$$

Now $y$ can be found from $x$, using Eq. (11.13).

The point to be remembered is that constant $D/V$ will produce variable distillate composition. Distillate is collected in a receiver whose final contents must meet a certain specification. Thus the average value of $y$, designated $\bar{y}$, is the controlled variable. For a constant distillate rate

$$\bar{y} = \frac{D \int y \, dt}{Dt} = \frac{\int y \, dt}{t} \tag{11.39}$$

Withdrawal of distillate is to be stopped when $\bar{y}$ falls to its desired value.

The disadvantages of constant-distillate-rate control are these: (1) If $D/V$ is relatively high, separation will be low, and withdrawal of distillate must be stopped at a relatively high value of $x$. This means that recovery of light ends will be poor. (2) If $D/V$ is reduced to enhance recovery, the distillation may consume an unreasonable amount of time and energy.

### Constant-Composition Control

A more efficient way to operate a batch still is on the basis of constant distillate composition. Because bottoms composition continually changes, separation must also change if constant distillate composition is to be maintained. Conse-

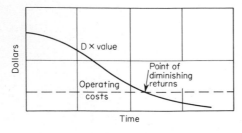

**FIG. 11.19** When the rate of remuneration equals the operating cost, distillation should be terminated.

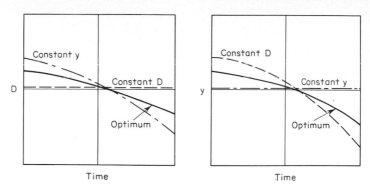

**FIG. 11.20** Optimal distillate rate is intermediate between constant-rate and constant-composition programs.

quently, $D/V$ will be high at the beginning of a batch and will gradually be reduced to zero by the distillate-composition controller when all the recoverable product has been withdrawn. A control system operating on this basis appears in Fig. 11.18. The temperature controller would require integral action to maintain constant quality with changing distillate rate.

Toward the end of the distillation, not only will the flow of product diminish but its rate of change will also diminish. As a result, all the light component cannot be removed economically, so a decision must be made when to stop withdrawal. This decision can be made on an economic basis by comparing the distillate flow times its value against operating costs. When the two are equal, the point of diminishing returns has been reached, as indicated in Fig. 11.19.

Another factor limiting the amount of recoverable distillate is the holdup of liquid in the trays of the column. Before the next higher-boiling product can be withdrawn as distillate, the mixture held on the trays must be removed as a *slop cut*. This material is collected in its own receiver and returned to the reboiler with a later batch of feed stock. The piping arrangement for withdrawal of distillate, shown in Fig. 11.18, has been designed to minimize holdup.

It can be seen that multicomponent separations can be accommodated without difficulty in a batch still. A separate receiver is necessary for each product, and manual operations are required to change receivers and to readjust the set point of the temperature (or composition) controller. But with each additional product cut there is also a slop cut. Hence as the number of products increases, the percentage of the batch recovered as product diminishes.

## Maximizing Product Recovery

Computer studies have shown that there is a program of distillate withdrawal which will recover the maximum amount of product of specified average composition in a specified time interval [10]. Figure 11.20 shows how this program falls between that of constant distillate rate and constant distillate composition. In effect, the final bottoms composition will be lowest because both $y$ and $D/V$ are low when distillation is terminated.

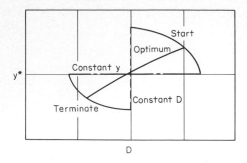

**FIG. 11.21** The optimal policy varies distillate composition with rate in a programmed manner.

Plotting distillate rate vs. composition for each of these three programs gives an indication of how the optimal program might be implemented. A typical plot is constructed in Fig. 11.21. The optimal program calls for varying the set point of the temperature (or composition) controller based on the current value of distillate flow. Although the optimal program is not linear, it can be approximated to a satisfactory degree by a simple linear equation

$$y^* = kD + y_0 \tag{11.40}$$

where $k$ is the slope and $y_0$ is the intercept. This linear expression can readily be implemented with the simple arrangement of analog devices pictured in Fig. 11.22.

## SUMMARY

Unfortunately it is impossible to cover even a sampling of the variety of distillation columns that are in service in industry. They are nearly as individualistic as people. Consequently much is left to practitioners in the way of interpreting the design rules discussed here in terms of their own problems. In this regard, a word of warning: do not attempt to make your particular separation fit the structure of the control system: take care to mold the control system to the peculiarities of the separation.

One very important class of separation is omitted from this chapter, however. It includes all the most difficult problems, i.e., extremely close-boiling mixtures and constant-boiling mixtures (azeotropes). The reason for the omission is that distillation alone is insufficient for their separation. They will be discussed in as much detail as seems reasonable after a brief treatment of extraction in the next

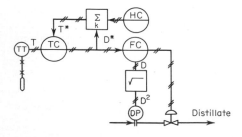

**FIG. 11.22** The set point of the temperature (or composition) controller can be adjusted automatically with a summing device and a manual set station.

chapter. For a more detailed presentation on these separations, the reader is directed to Chaps. 4 and 5 of Ref. 8.

## REFERENCES

1. Fenske, M. R.: Fractionation of Straight-Run Pennsylvania Gasoline, *Ind. Eng. Chem.*, May 1932.
2. Douglas, J. M., A. Jafarey, and T. J. McAvoy, Short-cut Techniques for Distillation Column Design and Control, pt. I. Column Design, *I & EC Process Des. and Devel.*, to be published.
3. Van Kampen, J. A.: Automatic Control by Chromatographs of the Product Quality of a Distillation Column, *Conv. Adv. Autom. Control, Nottingham*, April 1965.
4. Shinskey, F. G.: Energy Conserving Control for Distillation Units, *Chem. Eng. Prog.*, May 1976.
5. Rademaker, O., J. E. Rijnsdorp, and A. Maarleveld: "Dynamics and Control of Continuous Distillation Units," p. 71, Elsevier, Amsterdam, 1975.
6. The Foxboro Company: Differential Vapor Pressure Cell Transmitter, Model 13VA, *Tech. Inform. Sheet* 37-91a, Foxboro, Mass., April 1965.
7. Boyd, D. M.: Fractionating Column Control, *Chem. Eng. Prog.*, June 1975.
8. Shinskey, F. G.: "Distillation Control for Productivity and Energy Conservation," McGraw-Hill, New York, 1977.
9. Toijala, K., and F. Fagervik: A Digital Simulation Study of Two-Point Feedback Control of Distillation Columns, *Kem. Teollisuus*, January 1972.
10. Converse, A. O., and G. D. Gross: Optimal Distillate Rate Policy in Batch Distillation, *Ind. Eng. Chem.*, August 1963.

## PROBLEMS

**11.1** For the column with $S = 361$ at $V/F = 3$, and $z = 0.50$, calculate the $D/F$ required to raise $y$ to 0.97, and the resultant value of $x$.

**11.2** Calculate $x$ and $D/F$ with $y = 0.95$ for $V/F = 2.5$. Let $\alpha = 1.3$.

**11.3** A particular column is fed a binary mixture containing 80 to 90 percent light component. Distillate is to be controlled to a purity of 99.9 percent. Write the feedforward control equation assuming a constant $V/F$ ratio. Repeat for constant heat input.

**11.4** Feed to a tower contains 5 percent propane, 50 percent isobutane, and 40 percent normal butane, with the balance being higher-boiling components. The feed is analyzed by chromatograph for propane and isobutane. All the propane in the feed goes out in the distillate. Under normal conditions, the bottom stream contains 2 percent isobutane if the distillate composition is controlled at 5 percent normal butane. Write the feedforward control equation, evaluating all coefficients.

**11.5** How much less energy will be saved in making the separation in Prob. 11.4 if the isobutane in the bottom of the column is allowed to increase to 4 percent? The relative volatility for the separation is 1.32, and $V/F = 3.0$.

**11.6** By reducing the pressure in the column of Prob. 11.4 from 95 to 72 lb/in² absolute, relative volatility increases to 1.35. Latent heat of vaporization increases from 129 to 135 Btu/lb at the same time. Estimate the savings in energy possible with the reduction in pressure.

**11.7** Calculate $\lambda_{yD}$ for a column that is splitting feed containing 12 percent lower-boiling component into a 90 percent pure distillate and a bottom product containing only 0.6 percent lower-boiling component.

# Chapter Twelve

# Other Mass-Transfer Operations

Although distillation may be the most common mass-transfer operation, it is also the most difficult to assimilate. Indeed, the separation between components is noticeably obscure because they occupy the same phase. Other mass-transfer operations involve separation or combination of different phases:

1. Vapor-liquid: absorption, humidification
2. Liquid-liquid (immiscible): extraction
3. Liquid-solid: evaporation, crystallization
4. Vapor-solid: drying

Because of this distinction, one of the exit streams in each of the above is either pure, e.g., the vapor from an evaporator, or in an equilibrium state independent of material-balance considerations. Although material-balance control can be enforced in each of these mass-transfer operations, the separation between phases generally simplifies its formulation by eliminating one variable. This reduces the number of manipulated variables by the same amount.

The final controlled variable in every case is composition, requiring some sort of an anlytical measurement. For most of these applications, a nonspecific determination, such as density, is sufficient. But occasionally, as in a drying operation, even nonspecific analyses are not available, so other variables must be found to provide some degree of regulation.

## ABSORPTION AND HUMIDIFICATION

Mass transfer between liquids and gases depends on the vapor pressure of the components as functions of temperature. Thus appropriate selection of operating temperature and pressure allows the reverse (desorption, or stripping, and dehumidification) to be performed. The purpose of absorption and stripping operations is to remove and recover the maximum amount of a particular component from a feed stream. It is most efficiently accomplished in multiple stages, as in tray or packed columns. Humidification and dehumidification are similar in principle but are directed toward control of an environment short of equilibrium, (for example, <100 percent humidity); for them, a single stage is ordinarily sufficient.

### Equilibrium Mixtures of Vapors and Liquids

Each component in a vapor mixture exerts a partial pressure $p_i$ relative to its concentration $y_i$

$$p_i = p y_i \tag{12.1}$$

It can be seen that since the concentrations total 100 percent, the sum of the partial pressures is the total static pressure $p$ exerted by the system.

According to Raoult's law, [1a], each component in an ideal-liquid solution generates a partial pressure relative to its concentration $x_i$ in the liquid

$$p_i = p°_i x_i \tag{12.2}$$

The coefficient $p°_i$ in Eq. (12.2) is the vapor pressure of component $i$ at the prevailing temperature. Unfortunately, wide departures from the ideal situation are encountered in typical solutions; nonetheless, linearity prevails over certain ranges, allowing $p°_i$ to be replaced with an equilibrium constant $K_i$

$$p_i = K_i x_i \tag{12.3}$$

The ideal situation is most closely realized where the gaseous components are above their critical temperature.

Combining Eqs. (12.1) and (12.2) or (12.3) establishes equilibrium conditions for a single stage

$$y_i = \frac{K_i x_i}{p} \tag{12.4}$$

If it is desired that $y_i/x_i$ exceed $K_i/p$, more stages must be used.

One unusual factor in absorption is the temperature rise of the absorbing liquid due to condensation of the absorbed vapors. These vapors actually change to the liquid state and in doing so release their latent heat. If the system is adiabatic, the temperature of the absorbent rises, which shifts the equilibrium, tending to retard further absorption. If the solution is quite dilute, this heating effect may be unimportant, but interstage cooling is necessary with high concentrations. Absorption of HCl and $NH_3$ are typical of the latter situation.

In stripping and humidification, heat must be applied to counteract the cooling effect of evaporation.

## Absorption

An absorption column is like the top half of a distillation tower. Feed vapor enters at the bottom, and the depleted gas leaves the top. Figure 12.1 points out the flowing streams.

There are four streams, but vapor and liquid inventory controls manipulate two. Feed rate is the load; the only manipulated flow then available for composition control is absorbent stream $L$. The temperature of stream $L$ is also a factor, but for maximum absorption it should be as low as practicable. For the same reason, pressure should be maintained at a high value.

The uppercase letters in Fig. 12.1 represent molal flow rates, while the lowercase indicate the mole fraction of the principal absorbed component in the respective streams. An overall material balance requires that

$$F + L = V + B \qquad\qquad (12.5)$$

The balance for the absorbed component is

$$Fz + Lw = Vy + Bx \qquad\qquad (12.6)$$

If the other components in the vapor phase are not absorbed, another equation can be written to close out material balance

$$V(1 - y) = F(1 - z) \qquad\qquad (12.7)$$

The combination of Eqs. (12.5) to (12.7) permits solution for the value of the manipulated variable $L$ required to control either $y$ or $x$, the other being a dependent variable

$$\frac{L}{F} = \frac{(z - y)(1 - x)}{(x - w)(1 - y)} \qquad\qquad (12.8)$$

Notice the resemblance of Eq. (12.8) to the feedforward control equation for binary distillation.

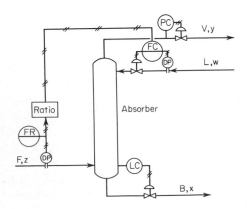

**FIG. 12.1**  The absorber features two liquid and two vapor streams.

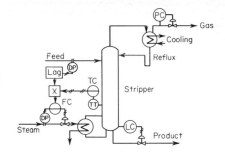

**FIG. 12.2**  In a stripping column, all the condensables are refluxed and all the noncondensables discharged.

As in distillation, there is a relationship between $y$ and $x$, of which Eq. (12.4) was a single-stage representation. Without attempting to arrive at a rigorous definition, it is important to point out that the ratio $L/F$ is the principal manipulated term, subject, however, to variations in feed composition.

Absorption is not a refining operation and is rarely the last operation conducted on a product. Consequently, close control of the concentration of either effluent stream is not paramount, and on-line analyzers are seldom used. More importance is placed on minimizing losses (such as $Vy$) or total operating costs, for which the simple optimizing feedforward system was designed at the close of Chap. 7. In that example, as in the control equation (12.8), maintenance of a designated ratio of $L/F$ applies.

### Stripping

Absorption is usually followed by a stripping operation, in which the absorbed component is removed from the solvent. Stripping may also be carried out independently, to preferentially remove lighter components as dissolved gases from a liquid product.

A stripping column looks quite like a distillation tower, equipped with both a reboiler and condenser. The reboiler raises the vapor pressure of all components, driving the most volatile preferentially up the column. A condenser is necessary to reflux whatever solvent might otherwise be carried away with the stripped vapors. A tower for removal of volatile impurities in a liquid product is shown in Fig. 12.2. Only the reflux would contain more dissolved impurities than the feed, which therefore enters near the top.

The concentration of the heaviest component (solvent) in the stripped vapor is determined by the pressure and temperature at the condenser. Ordinarily the pressure is controlled as shown in Fig. 12.2, but the temperature may be left to fall as low as the cooling medium will allow. This maximizes the recovery of solvent but increases the energy required to hold bottom quality constant. Alternately, reflux temperature could be controlled by manipulating condenser cooling, and thereby control solvent recovery.

Floating-pressure control can be applied by programming the pressure set point to follow uncontrolled reflux temperature, as shown in Fig. 12.3. This can provide constant recovery of solvent at a minimum energy requirement because of the improvement in relative volatility attainable with falling pressure. The function generator is an approximation of the vapor-pressure curve for the desired vapor composition.

Because inventory control for vapor and liquid manipulate both effluent streams, as in an absorber, heat input is the only variable remaining for bottom-composition control. In Fig. 12.2, control of temperature near the base is used to specify its initial boiling point. Feedforward is added from feed rate through a lag, because the feed enters as a liquid well above the base.

Should the column pressure be allowed to float, the base-temperature set point must be adjusted accordingly. This is most easily achieved by setting it at a fixed differential above reflux temperature or by controlling the differential temperature across the column.

When operated in conjunction with an absorber, the product becomes the vapor leaving the condenser, while the bottom stream is recycled to the absorber. A typical absorber-stripper combination for the separation of carbon dioxide and hydrogen is shown in Fig. 12.4. Monoethanolamine (MEA) is used as the solvent. Control of $CO_2$ content in the MEA leaving the stripper is important only for its influence on the equilibrium maintained with the gas leaving the top tray of the absorber; $CO_2$ is not lost. Cooling the lean MEA enhances absorption, although its control is not really warranted. In addition, the absorber usually operates at a higher pressure than the stripper.

### Humidification

Cooling towers dissipate tremendous quantities of heat into the atmosphere through the process of humidification. Water circulated countercurrently to a stream of air is reduced in temperature since atmospheric air is ordinarily far from saturated with water vapor. The latent heat of the evaporated water is converted into a change in sensible heat of the remainder.

Humidification and dehumidification also apply to environmental control where a certain moisture content is desired in the air. As pointed out earlier, an operation of this sort is generally conducted in a single stage, so control is actually not difficult. Yet the significance of the terms and principles is sufficiently confusing to deserve a general review and definition:

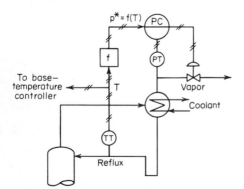

**FIG. 12.3**  Pressure is made to follow variations in reflux temperature to control vapor composition while saving energy.

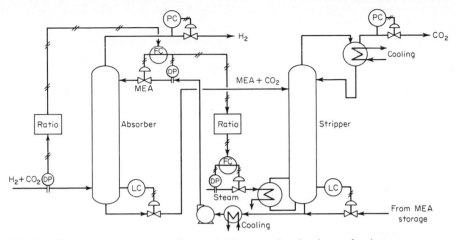

**FIG. 12.4** The solvent is continuously recycled between the absorber and stripper.

1. The vapor pressure of water in atmospheres varies with its temperature

$$\log p^\circ_w = \begin{cases} 6.69 - \dfrac{4407}{T} & {}^\circ\mathrm{R} \\[2mm] 6.69 - \dfrac{2446}{T} & \mathrm{K} \end{cases}$$
(12.9)

2. Partial pressure $p_w$ was defined by Eq. (12.1). With regard to humidification, the liquid is essentially pure, so $x$ in Eq. (12.2) is 1.0. At equilibrium (100 percent saturation), the partial pressure of water vapor is equal to its vapor pressure at the prevailing temperature, that is, $p_w = p^\circ_w$.

3. Absolute humidity is the ratio of the mass of water vapor to the mass of air or gas in the mixture

$$\text{Lb water/lb dry air} = \frac{18p_w}{29(1 - p_w)}$$
(12.10)

4. The mass of water per unit volume of humid air is sometimes used. Its units are typically [1b]

$$\text{gr/ft}^3 = 1.73 \times 10^5 \frac{p_w}{T} \quad \text{or} \quad \text{g/m}^3 = 4.25 \times 10^5 \frac{p_w}{T}$$
(12.11)

where gr stands for grains, $p_w$ is in atmospheres, and $T$ is in degrees Rankine in the left-hand equation and in Kelvin on the right.

5. Relative humidity is the percent saturation at prevailing temperature and pressure and is exactly defined as $100 p_w/p^\circ_w$.

6. Dew point is the temperature at which a mixture becomes saturated when cooled out of contact with liquid at constant pressure. It is often used to determine the moisture content of gases, by converting the temperature to

vapor pressure by Eq. (12.9). Below 32°F (0°C) the dew point is actually a frost point.

7. Wet-bulb temperature is the equilibrium temperature reached by a small amount of liquid evaporating adiabatically into a large volume of gas. Equilibrium exists when the rate of heat transfer from the gas to the cooler liquid equals that consumed by evaporation. It is affected by heat- and mass-transfer coefficients as well as humidity and therefore is dependent on maintaining turbulent gas flow around the bulb. Humidity can be determined from wet-bulb $T_w$ and dry-bulb $T$ temperatures by following the adiabatic-saturation curves on a psychrometric chart, or by

$$T - T_w = 2.58 H_v \left( \frac{p°_w}{1 - p°_w} - \frac{p_w}{1 - p_w} \right)$$  (12.12)

where $H_v$ is the latent heat of evaporation and $p°_w$ is the vapor pressure at the wet-bulb temperature.

Humidity measurements may be made by several different means, wet-bulb temperature being but one. Some instruments are equipped with a hair element which is sensitive to changes in relative humidity. Though dew point can be measured directly, a more reliable instrument [2] uses a hygroscopic salt whose conductivity varies with moisture content. The salt is self-heated simply by application of an ac voltage, and its temperature is an indication of the absolute humidity. The measured temperature is not the dew point but is related to it such that scales are available for direct reading in dew point or units of absolute humidity.

Choice of the type of measurement to be used for control depends on the process. Under isothermal conditions, the moisture content of solid materials varies with relative humidity, but in adiabatic processes, a determining factor is wet-bulb temperature. An exact analysis of moisture content can best be found by an absolute-humidity measurement, however.

Usually it is necessary to control both humidity and temperature in an environmental chamber. This presents an opportunity for interaction. Spraying water into the air raises its humidity while reducing temperature. Heating

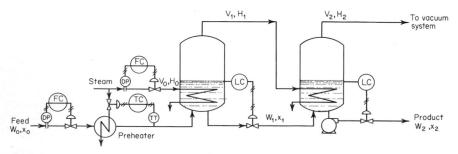

**FIG. 12.5**  A double-effect evaporator with forward feed.

the air raises temperature and reduces relative humidity. However, wet-bulb temperature is unaffected by spraying, and so it can be used to control heat input. Then control of dry-bulb temperature by spraying will produce only partial coupling. Alternately, dew point can be used to control spraying because it is unaffected by heat input.

Dehumidification requires cooling of the humid air below its dew point, with or without compression, depending on the dryness required. Manipulation of cooling under constant pressure is effective.

## EVAPORATION AND CRYSTALLIZATION

These operations may be conducted separately or in combination in an effort to separate a solid from its solvent. The product from an evaporator is a concentrated solution, whereas a crystallizer discharges a slurry of crystals in a saturated solution. These two operations may not be technically classified as mass transfer, in that no equilibrium exists between the composition of the two phases (the vapor leaving an evaporator and the crystals in the crystallizer are both essentially pure); yet the control of both these operations depends heavily on the material balance.

### Multiple-Effect Evaporation

To conserve steam, evaporation is usually carried out in two or more stages, each stage being heated by the vapors driven from the previous stage. To maintain a temperature difference across each heat-transfer surface, a pressure difference must be maintained between stages. The most economic operation is realized with low-pressure steam heating, requiring each stage to be maintained under a different vacuum. A double-effect evaporator is shown in Fig. 12.5; although the arrangement could be extended indefinitely, there is a practical limit.

The arrangement shown in Fig. 12.5 is *forward feed,* in that the feed stream enters the first effect only. *Reverse feed,* i.e., entering the last effect first, is another possibility. In addition, each effect can receive fresh feed, an arrangement called *parallel feeding.* The first described is the most common.

The controlled variable is product concentration. It can be determined by density measurement, electrolytic conductivity, refractive index, or by measuring the elevation in boiling point or the depression in freezing point of the solvent.

In the past, control of product composition typically entailed manipulation of the discharge valve. The level controllers for each effect were left to manipulate each inflow, ultimately affecting feed rate. This arrangement results in a series of interactions between flows and compositions from the last effect to the first and back again. Furthermore, production rate can be adjusted only by altering the heat input, which constitutes a prime source of disturbance. These deficiencies prompted the investigation of material-balance control.

## Material-Balance Control

A certain amount of solvent is evaporated in each effect relative to its heat input; all the solids in the feed are discharged with the product. Let $W_1$ represent the mass flow of a feed whose solids content is $x_1$ (weight fraction) such that $X$ is the mass flow of solids in the feed

$$X = W_1 x_1 \tag{12.13}$$

The total flow of solution leaving the effect $W_2$ contains $x_2$ weight fraction of solids

$$X = W_2 x_2 \tag{12.14}$$

The rate of evaporation is designated $V_2$

$$V_2 = W_1 - W_2 \tag{12.15}$$

By combining Eqs. (12.13) to (12.15) it is possible to calculate the rate of evaporation required to convert a feed of known composition into a specified product composition

$$V_2 = W_1 \left( 1 - \frac{x_1}{x_2} \right) \tag{12.16}$$

The heat input to the effect, in the form of vapor or steam, will flow at a rate $V_1$ with a latent heat $H_1$ in order to cause the evaporation of $V_2$, whose latent heat is $H_2$, if the feed is preheated to the boiling point

$$V_1 H_1 = V_2 H_2 \tag{12.17}$$

Combining the last two expressions gives the relationship between the input variables necessary to maintain a desired output quality

$$V_1 H_1 = W_1 \left( 1 - \frac{x_1}{x_2} \right) H_2 \tag{12.18}$$

To apply this to a double-effect evaporator, let Eq. (12.18) represent conditions existing in the second effect. The material balance for the two effects can be derived in the same way as Eq. (12.16), relating total evaporation to inflow rate $W_0$ and weight fraction solids $x_0$

$$V_1 + V_2 = W_0 \left( 1 - \frac{x_0}{x_2} \right)$$

Relating first-effect vapor inflow $V_0$, of latent heat $H_0$, to $V_1$ and $V_2$, as was done in Eq. (12.17), permits elimination of the last two variables

$$V_0 H_0 = \frac{W_0 (1 - x_0/x_2)}{1/H_1 + 1/H_2}$$

Extension to an $n$-effect evaporator follows directly

$$V_0 H_0 = \frac{W_0(1 - x_0/x_n)}{\sum\limits_{i=1}^{i=n} \frac{1}{H_i}} \tag{12.19}$$

Equation (12.19) can be implemented for control of product quality by manipulating either heat input or feed rate in relation to the other. The choice depends on the relative availability of each. If short-term reductions in steam availability are common, feed rate should be manipulated accordingly. But if feed is coming from another processing unit without intermediate surge capacity, the alternate arrangement is favored.

Actually none of the terms in Eq. (12.19) is directly measurable. Steam flow is usually measured by an orifice meter, which requires correction for pressure variations, as described below. The customary device for feed-rate determination is a magnetic flowmeter; since it is a volumetric device, it must be corrected for feed density. Feed density is also used to infer feed composition.

Heat input is related to steam-flow differential $h$ as a function of latent heat and specific volume $v$

$$V_0 H_0 = k \sqrt{\frac{hH_0^2}{v}}$$

The ratio $H_0^2/v$ is linear with pressure $p$ over a reasonable operating range, so that

$$V_0 H_0 = \sqrt{h(a + bp)} \tag{12.20}$$

The mass-flow term in Eq. (12.19) can be converted into volumetric flow and desnity $\rho$

$$W_0\left(1 - \frac{x_0}{x_n}\right) = F\rho\left(1 - \frac{x_0}{x_n}\right)$$

Since $x_0$ is also a function of $\rho$, the two terms can be combined into a single function of density, which when plotted against density for an aqueous system forms a linear relationship

$$\rho\left(1 - \frac{x_0}{x_n}\right) = 1 - m(\rho - 1) \tag{12.21}$$

The intercept (1, 1) of Eq. (12.21) indicates that when feed density is 1.0, one unit of vapor must be driven away from each unit of feed. Slope $m$ of the line is a function of the density of the solid and of its final concentration $x_n$.

When Eq. (12.20) and (12.21) are substituted into (12.19), the feed rate needed to satisfy a given steam flow can be calculated

$$F^* = \frac{K\sqrt{h(a + bp)}}{1 - m(\rho - 1)} \tag{12.22}$$

Coefficient $K$ will be recognized as the denominator in $K$ (12.19), and it is predictable within a close tolerance for any evaporator. Because slope $m$ varies with product concentration, it is essentially the set point for the feedforward system and therefore should be adjusted by the feedback controller. The combination of feedforward and feedback for the evaporator is shown in Fig. 12.6. Steam flow is set from the measurement of feed-tank level, to maximize the utilization of its surge capacity.

In low-capacity evaporators of the falling-film type, feed-rate changes may affect product composition before steam-flow changes do. Then the dynamic compensation used in the system of Fig. 12.6 will be a dominant lag. Should the system be rearranged to set steam flow as a function of feed rate, the lead must dominate, as in a heat exchanger. Still another possibility is the reverse-feed evaporator, where steam enters at the point where product leaves. Then product composition may respond more rapidly to steam flow, and dynamic compensation must be opposite of that used for the forward-feed evaporator.

### Control of Crystallizers

A solution is saturated when an equilibrium exists between dissolved and undissolved molecules of a solute in a solvent. The concentration of undissolved solute present as crystals does not affect the equilibrium state. The concentration of solute in solution is fixed by the equilibrium state, which varies with temperature.

As a result, crystals can be deposited from solution by either of two mechanisms, evaporation of solvent or reduction in solution temperature. Evaporation can be caused by the application of heat or vacuum or both, but if vacuum alone is used, the temperature of the solution is reduced along with the evaporation.

A usual requirement is control over the concentration of crystals in the discharge slurry. In many cases, however, crystal size is important as well. Crystal concentration is customarily measured as density, if the crystals are

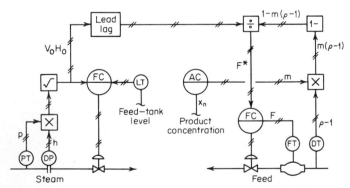

**FIG. 12.6**   Feed-flow set point is calculated from steam flow and feed density, with feedback correction of the density function.

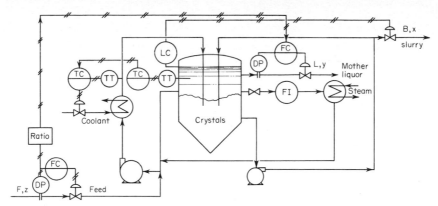

**FIG. 12.7**  Control of crystal content involves manipulation of mother-liquor and slurry flows.

uniformly dispersed across the sensitive span of the detector. Crystal-size determination unfortunately does not lend itself to on-line analysis.

Figure 12.7 shows a typical cooling crystallizer. Temperature of the solution is maintained by circulating the slurry through a chiller, which removes sensible heat in the feed stream and heat of fusion of the crystals. The crystal slurry must be kept in motion to avoid plugging. A centrifuge or filter subsequently separates the crystals and returns the mother liquor to the process.

Since fine particles settle slowly, they accumulate at the top of the mass of crystals. By withdrawing a sidestream from this region, a token amount of fine crystals can be redissolved, increasing the average size of the crystals remaining. Increasing the density of the slurry tends to increase crystal size by raising the level of the mass relative to the side-stream tap.

Examination of the material balance across a crystallizer gives an indication of how it ought to be controlled. Mass flow of feed $F$ is separated into saturated mother liquor $L$ and crystal slurry $B$

$$F = L + B \qquad (12.23)$$

Weight fractions of the solute in each stream are represented by $z$, $y$, and $x$, respectively,

$$Fz = Ly + Bx \qquad (12.24)$$

Load variables are $F$ and $z$; $y$ is the weight fraction of solute in solution at saturation as fixed by temperature, so it is not a variable; $L$ and $B$ remain to be manipulated so as to control $x$ and the crystallizer level.

Following the usual procedure for material-balance control, $L$ is selected to be manipulated for density ($x$) control because its flow is readily measurable, whereas that of the slurry is not. Eqs. (12.23) and (12.24) are therefore solved for $L$, eliminating $B$

$$L = F\frac{x-z}{x-y} = F\left(1 - \frac{z-y}{x-y}\right) \tag{12.25}$$

Notice the similarity to the control equation for distillation, but in this case $y$ is fixed, considerably simplifying matters. The content of crystals in the slurry is $(x - y)/(1 - y)$, whereas total solute content is $x$.

The most significant outcome of the above derivation is that mother-liquor flow should be set in ratio to feed rate. Feedback from crystal density is not mandatory because the process is self-regulating and extreme accuracy is not usually warranted.

Evaporative crystallizers can be treated just like evaporators since pure solvent is driven off by a proportional flow of heat. The only difference is that the part of the product which remains in solution is a function of solution temperature. The latter is not an independent variable, however, because the solution is boiling.

## EXTRACTION AND AZEOTROPIC DISTILLATION

Extraction is defined as the transfer of a dissolved material between two immiscible solvents. The material being transferred may be solid, liquid, or gas in its ordinary state. The purpose of the extraction is to permit its separation from the first solvent or some contaminant in it. Many liquids whose boiling points are nearly identical, preventing their distillation, can be separated by extraction, often in conjunction with distillation.

A serious problem area in distillation technology involves substances which form constant-boiling mixtures, called *azeotropes*. Azeotropes occur so often in chemical systems that their separation deserves special mention apart from the conventional distillation practices discussed in the previous chapter. Because the principles of extraction play a major role in azeotropic distillation, the subject will be discussed in detail.

### Extraction and Decantation

In a single extraction stage, an equilibrium will be approached between the concentration of product in the two exit streams. Consider a product dissolved in solvent A being extracted into solvent B. If $x_1$ is used to represent the concentration of product remaining unextracted and $y_1$ that of the product leaving in $B$, then

$$x_1 = Ky_1 \tag{12.26}$$

where $K$ is the equilibrium constant.

Similarly, a material balance can be drawn across the stage

$$A(x_0 - x_1) = B(y_1 - y_0) \tag{12.27}$$

A and B are mass flow rates of the two solvents, and $x_0$ and $y_0$ are the concentrations of product in each influent stream.

It is usually convenient to manipulate the flow of extractant B in response to changes in feed rate A and composition $x_0$, in order to control the concentration of extracted product. If this is indeed the case, Eqs. (12.26) and (12.27) can be solved for B in terms of $y_1$

$$B = A \frac{x_0 - Ky_1}{y_1 - y_0} \qquad (12.28)$$

Note once more the familiarity of the material-balance equation, particularly the ratio existing between A, the feed rate, and B, the manipulated variable. Whether single or multistage, control of extraction always involves manipulation of this ratio.

Multistage extraction can be carried out in a series of mixers, each followed by a settling chamber, where the solvents are separated and allowed to flow countercurrently. However, extraction is more commonly conducted in packed towers, with extractant flowing countercurrently to the feed. The less dense solvent must enter at the bottom, flowing upward at a rate determined by its difference in density from the heavier solvent.

Not only must liquid level be controlled but interface as well, in order to maintain inventory of both solvents. Figure 12.8 shows how the control loops would be arranged for a typical extraction column. The location of the interface between two solvents is easily measured as differential pressure between two taps or buoyant force on a displacer. In either case, the measurement reflects the average density across the vertical span of the instrument.

Decanters are used to separate two immiscible solvents following extraction, mixing, or condensation from the vapor phase. As in extractors, both level and interface height must be regulated, but fortunately in many cases this can be accomplished simply by proper equipment design. Figure 12.9 points out the dimensions which are important.

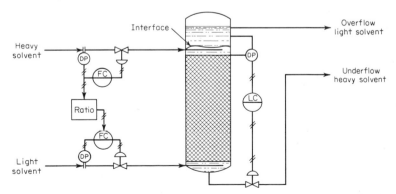

**FIG. 12.8**  Both liquid level and interface must be regulated in an extractor.

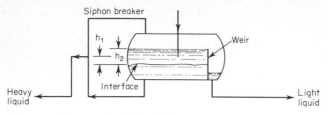

**FIG. 12.9** A properly designed decanter can function without controls.

The static differential pressure between the top of the light liquid and the highest point in the underflow loop is zero. The vertical distance from the interface to each of these points is a function of the densities of the heavy and light liquids, designated $\rho_1$ and $\rho_2$

$$\rho_1 h_1 = \rho_2 h_2 \qquad\qquad (12.29)$$

For finite flow rates of heavy liquid $h_1$ will decrease; i.e., the interface will rise. This piping arrangement is not directly applicable to control of the extraction tower shown in Fig. 12.8 because of the variable pressure drop encountered in the packing and discharge line.

### Azeotropic Distillation

An azeotrope is a mixture of two or more materials that cannot be separated by distillation; the vapor and liquid in equilibrium are of the same composition, and there is no difference between the boiling point and the dew point of an azeotrope. The individual components may have entirely different boiling points, but their azeotropic mixture will exhibit a higher or lower boiling point than either, the latter being more common.

Azeotropes act like pure substances. Ethanol and water form an azeotrope containing about 89 mole percent ethanol. Any mixture of ethanol and water containing more than 89 percent ethanol can be fractionated into ethanol and the azeotrope; a mixture containing less than 89 percent ethanol can be fractionated into water and the azeotrope. The composition of an azeotrope and its boiling point at a given pressure are characteristics peculiar to the system.

Heterogeneous azeotropes separate into two immiscible layers of different composition when condensed. This is a considerable advantage, for it permits separation into the pure components, by decantation followed by a second distillation. Figure 12.10 shows that both columns use the same condenser, since their vapors are both of the azeotropic composition. Steam-flow rates are the only variables which can be manipulated for composition control.

Occasionally a binary azeotrope which is homogeneous can be broken by adding a third component which forms a ternary heterogeneous azeotrope. The third component is called an *entrainer;* it is not intentionally removed from

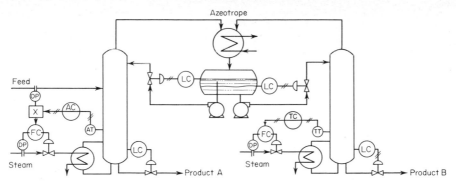

**FIG. 12.10** A heterogeneous azeotrope can be separated by two columns with a single condenser.

the system but circulates in the reflux loop. The ternary azeotrope must boil at a lower temperature than the binary in order for the operation to be successful. Figure 12.11 shows the similarity of the process to separation of a binary heterogeneous azeotrope. Again, removal of the light component is directly proportional to the vapor flow, so heat input must be manipulated for bottoms-composition control.

Although entrainer is intended to circulate continuously through the reflux loop, a certain amount will be lost with the light component. Because the entrainer is a third component, its inventory cannot be detected by liquid level. However, a deficiency of entrainer resulting in the loss of the ternary azeotrope will cause temperatures in the top of the tower to rise. Therefore top temperature can be used to manipulate its addition.

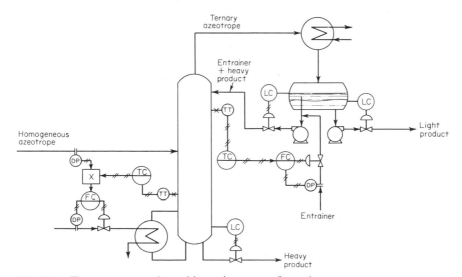

**FIG. 12.11** Top temperature is sensitive to inventory of entrainer.

## DRYING OPERATIONS

The drying of solids has historically defied control principally because a continuous measurement of product moisture has been lacking. Any kind of measurement on solids—even flow rate—is fraught with problems, but on-line analytical determinations are virtually impossible. Consequently environmental measurements must be relied upon, but their successful employment hinges entirely on how capably they represent the true state of the process.

### The Rate of Drying

The rate of drying of a wet solid is a function of driving force, surface area, mass-transfer coefficient, and the moisture content of the material. If the particles are uniformly wet, moisture will evaporate from the surface as from any wet surface. This condition is called *constant-rate drying* because the rate is not influenced by the actual amount of moisture present. As dry areas appear however, the rate of evaporation begins to diminish; the *falling-rate zone,* shown in Fig. 12.12, has been entered. The transition point between the two zones is known as the *critical moisture content* $x_c$. The rate eventually falls to zero when the solid moisture content reaches $x_e$, in equilibrium with the drying air.

To facilitate calculations, it will be assumed that the rate of drying in the falling-rate zone is proportional to the ratio of solids moisture to critical moisture. Then the rate of evaporation $dW$ from a particle of solid having a surface $dA$ can be represented as

$$dW = dA \; \gamma(T - T_w) \frac{x}{x_c} \tag{12.30}$$

where $\gamma$ is a mass-transfer coefficient. $T$ and $T_w$ are dry-bulb and wet-bulb temperatures, whose difference act as the driving force for heat and mass transfer. The wet-bulb temperature is essentially the temperature of the solid, while the dry-bulb temperature is that of the air.

Evaporation of moisture into a flow $G$ of air results in a decrease in dry-bulb temperature proportional to the ratio of the heat capacity $C$ of the air and the latent heat $H_v$ of the water

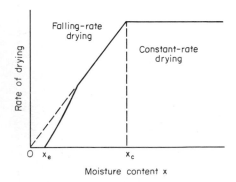

**FIG. 12.12** In the falling-rate zone, the rate of drying is roughly linear with product moisutre.

$$dW = -GC \frac{dT}{H_v} \tag{12.31}$$

Substituting (12.31) into (12.30) and integrating from inlet ($i$) to outlet ($o$) gives a solution for temperature as a function of total surface area

$$\int_0^A dA = \frac{GC/H_v}{\gamma x/x_c} \int_{Ti}^{T_o} \frac{-dT}{T - T_w}$$

$$A = \frac{GC/H_v}{\gamma x/x_c} \ln \frac{T_i - T_w}{T_o - T_w}$$

Then solids moisture can be related to inlet, outlet, and wet-bulb temperatures

$$x = \frac{x_c GC}{H_v \gamma A} \ln \frac{T_i - T_w}{T_o - T_w} \tag{12.32}$$

In effect, (12.32) states that moisture $x$ can be controlled if the other parameters are constant or if numerators and denominators are maintained in a constant ratio.

In practice, the terms ahead of the logarithm tend to be constant, with the possible exception of $G$ and $\gamma$; however, $\gamma$ tends to vary nearly linearly with $G$, in which case their ratio is nearly constant.

Conventional dryer control systems attempt to hold outlet air temperature $T_o$ constant by manipulating heat input or inlet temperature $T_i$ in response to load changes. This action also affects $T_w$. If $T_o$ is held constant by varying $T_i$ and $T_w$, the ratio of temperature differences in Eq. (12.32) must change with load. Increasing loads, requiring higher values of $T_i$ (and hence $T_w$), will result in increasing product moisture. Moisture content can be regulated only by increasing $T_o$ at the same time, to keep the ratio of the temperature differences constant.

### Inferential Controls

The model given in Eq. (12.32) cannot be implemented directly for control, principally because of the difficulty in measuring $T_w$. While wet-bulb temperature is easy to measure at ambient conditions, air in a dryer is very hot and contaminated with solids, resulting in failure of the bulb wick. Fortunately, wet-bulb temperature is principally a function of inlet dry-bulb temperature, with a lesser influence from humidity, particularly at higher temperatures. Therefore, it is possible to establish a program relating $T_o$ to $T_i$ for any particular ratio of temperature differences. This has been done for three different ratios in Fig. 12.13, using air with a dew point of 50°F.

Implementation consists in approximating these curves with a set of straight lines having both an adjustable slope $R$ and intercept $b$

$$T^*_o = b + RT_i \tag{12.33}$$

where $T^*_o$ is the desired set point for outlet temperature. A control system embodying this concept is shown in Fig. 12.14, applied to a fluid-bed dryer.

It is entirely possible to estimate appropriate values of $b$ and $R$ from design conditions. However, adjustments may also be made in the field to calibrate the system to actual operating conditions. Raising either $b$ or $R$ will cause product moisture to fall.

The generation of $T^*_o$ forms a positive-feedback loop. Raising heat input will raise $T_i$, causing a proportional increase in $T^*_o$, calling for more heat. Concurrently, the outlet-air temperature measurement begins to respond, applying negative feedback. For the system to be stable, both the dynamic and steady-state gains of the positive feedback loop must be lower than those of the negative loop. Proper adjustment of $R$ will assure steady-state stability. A lag, shown in Fig. 12.14, is needed for dynamic stability. Its value must exceed the integral time of the outlet temperature controller since it parallels the internal feedback loop generating integral action (see Fig. 4.3). If the lag is too short, a slow cycle will develop, identical to that caused by too short a value of integral time.

### Limitations and Extensions

The relationship shown in Fig. 12.13 is somewhat sensitive to ambient humidity, and correction has been applied where extreme variations have been encountered [3, pp. 226 to 228]. When the inlet air temperature exceeds 400°F, this sensitivity diminishes to the point of being negligible.

Direct-fired dryers are heated by the combustion of fuel, which adds some moisture to the air. However, the moisture added is in direct proportion to the inlet temperature achieved, and the resultant influence over wet-bulb temperature is predictable [3, pp. 231, 232].

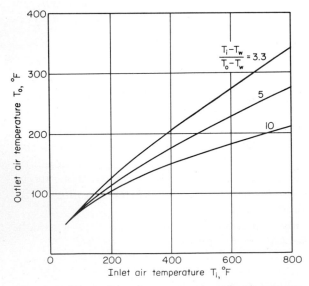

**FIG. 12.13**  To control moisture constant, outlet-air temperature must be programmed as a function of inlet-air temperature.

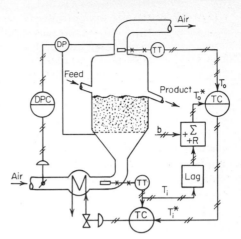

**FIG. 12.14** A fluid-bed dryer controlled by the inferential system.

In a longitudinal dryer, the moisture content of the solid varies from inlet to outlet, in contrast to the fluid-bed dryer, whose contents are uniformly mixed. As a result, the mathematical model needs to be integrated with respect to both temperature and composition to obtain a representative solution. However, Eq. (12.32) has been found to fit reasonably well if the moisture profile remains essentially constant. Should feed moisture tend to increase, the rate of drying improves and $T_o$ does not require as much elevation as needed for a feed-rate increase. If a measurement of feed rate is available, it can be used to augment the inferential control system

$$T^*_o = b + R_T T_i + R_F F \tag{12.34}$$

For a constant feed rate, $T^*_o$ needs to respond to $T_i$ for only a change in feed moisture, such that $R_T < R$. With a change in feed rate at constant moisture, both $F$ and $T_i$ will change, resulting in a change in $T_o$ equivalent to $R\ dT_i$.

This concept is described in detail in Ref. 3, where it is applied further to batch and steam-tube dryers.

## REFERENCES

1. Treybal, R. E.: "Mass Transfer Operations," McGraw-Hill, New York, 1955, (a) p. 204; (b) p. 158.
2. The Foxboro Company, Foxboro Dew Point Recording System, Using the Dewcel Element, *Tech. Inform. Sheet* 19–30a.
3. Shinskey, F. G.: "Energy Conservation through Control," Academic, New York, 1978.

## PROBLEMS

**12.1**  What are the upper and lower limits of $y$ and $x$ in the streams leaving the absorption tower? What will make $y$ approach its lower limit? What will happen to $x$ under these conditions?

**12.2**  Saturated air at 60°F is being heated to 72°F. What is the relative humidity after heating?

**12.3**  The three effects of a triple-effect evaporator are operated at 12, 7, and 2 lb/in² absolute, respectively. How many pounds of water will 1 lb of 20 lb/in² absolute saturated steam evaporate, excluding losses? How many pounds of solution can be concentrated from 35 to 70 weight-percent solids from 1 lb of steam?

**12.4**  Feed to a crystallizer contains 44 weight-percent solute. It is chilled to a controlled temperature where saturation represents only 17 weight-percent solute. What is the ratio of mother liquor to feed flow (weight basis) needed to control the discharged slurry at 70 weight-percent crystals?

**12.5**  A dryer is designed to operate at an inlet dry-bulb temperature of 1000°F, wet-bulb temperature of 153°F, and an outlet dry-bulb temperature of 260°F. Calculate the outlet dry-bulb temperature needed to hold the same moisture at inlet conditions of 600°F dry-bulb and 134°F wet-bulb. Estimate $R$ and $b$.

# Appendix A

# Notation

| | | | |
|---|---|---|---|
| $A$ | Amplitude, area | $M$ | Mass |
| A | Component | $N$ | Speed |
| $B$ | Bottom-product flow | $P$ | Proportional band |
| B | Component | $Q$ | Heat flow |
| $C$ | Flow coefficient, specific heat | $R$ | Rangeability, ratio, gas constant |
| $D$ | Derivative, determinant, distillate flow | $S$ | Separation factor |
| | | $T$ | Temperature |
| $E$ | Integrated error, reaction-rate exponent | $U$ | Heat-transfer coefficient |
| | | $V$ | Volume, vapor flow |
| $F$ | Flow, feed rate | $W$ | Weight, mass flow |
| $G$ | Dynamic gain, gas flow | $X$ | Variable |
| $H$ | Enthalpy, reciprocal gain | X | Component |
| $I$ | Integral time | $Y$ | Variable |
| $K$ | Steady-state gain | Y | Component |
| $L$ | Length, reflux flow | $Z$ | Variable |

| | | | |
|---|---|---|---|
| $a$ | Valve opening, nonlinearity, constant | $f$ | Flow, function, feedback |
| | | $g$ | Gravitational acceleration |
| $b$ | Constant, bias | $\mathbf{g}$ | Dynamic-gain vector |
| $c$ | Controlled variable | $h$ | Head, differential pressure |
| $d$ | Differential, decoupler gain | $k$ | Constant, reaction rate |
| $e$ | Deviation, 2.718 | $l$ | Loss |

| | | | |
|---|---|---|---|
| $m$ | Manipulated variable, slope | $v$ | Volume, value, specific volume |
| $n$ | Number | $w$ | Molecular weight, composition |
| $p$ | Pressure | $x$ | Constant, variable, composition |
| $q$ | Load, enthalpy | $y$ | Constant, variable, conversion, composition |
| $r$ | Set point | | |
| $t$ | Time | $z$ | Constant, variable, dead zone, composition |
| $u$ | Velocity | | |

| | | | |
|---|---|---|---|
| $\alpha$ | Degree of mixing, angle, relative volatility | $\Lambda$ | Relative-gain array |
| | | $\lambda$ | Relative gain |
| $\beta$ | Angle, column characterization factor | $\Pi$ | Product |
| | | $\pi$ | 3.1416 |
| $\gamma$ | Mass-transfer coefficient | $\rho$ | Density |
| $\Delta$ | Difference | $\Sigma$ | Sum |
| $\delta$ | Error | $\tau$ | Time constant |
| $\partial$ | Partial differential | $\phi$ | Phase angle |

| | | | |
|---|---|---|---|
| ——//—— | Control signal (pneumatic) | —⊣⊢— | Orifice |
| ——✕—— | Control signal (capillary) | | Nozzle or venturi |
| — — — — | Control signal (electric) | | Magnetic flowmeter |
| | Control valve | | Turbine flowmeter |
| | Control damper | ✕ | Multiplier |
| | Control valve with positioner | ÷ | Divider |
| | Solenoid or on-off valve | √ | Square-root extractor |
| (LC) | Controller, locally mounted (liquid level) | Σ | Summer (adding and subtracting) |
| (TC) | Controller, board-mounted (temperature) | < | Low selector |
| | | > | High selector |
| (FFC) | Controller, ratio, board-mounted (flow) | f | Function generator |
| | | f(t) | Dynamic compensator |
| (PS) | Switch (pressure) | ± | Bias station |
| (DP) | Transmitter (differential pressure) | K | Ratio station |
| | | ∫ | Integrator |
| | | 1− | Reversing device |

# Appendix B

# Solutions to Problems

## Chapter 1

**1.1** $\tau_d = \dfrac{4 \text{ ft}}{12 \text{ ft/min}} = 0.33 \text{ min}$

$\tau_o = 4\tau_d = 1.33 \text{ min}$     $I = \dfrac{4}{\pi}\tau_d = \underline{0.42 \text{ min}}$

**1.2** $\phi_d = -180 - (-60) = -120°$     $\tau_o = \dfrac{360}{120}\tau_d = 3(0.33) = \underline{1.0 \text{ min}}$

$\dfrac{\tau_o}{2\pi I} = -\tan(-60°) = 1.732$     $I = \dfrac{\tau_o}{1.732(2\pi)} = \underline{0.092 \text{ min}}$

$K_c G_c = \dfrac{100}{P}\sqrt{1 + (1.732)^2} = 0.5$     $P = 200\sqrt{4} = \underline{400\%}$

**1.3**

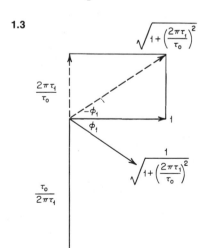

**1.4**

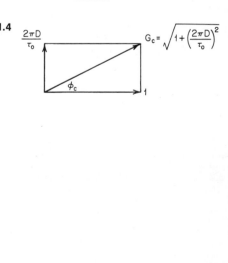

**1.5** $\phi_d = -360\dfrac{\tau_d}{3\tau_d} = -120°$    $\phi_1 = -180 - (-120) = -60°$

$\dfrac{2\pi\tau_1}{\tau_o} = -\tan(-60°) = 1.732$    $G_1 = \dfrac{1}{\sqrt{1 + (1.732)^2}} = \underline{0.5}$

$\dfrac{2\pi\tau_1}{3\tau_d} = 1.732$    $\dfrac{\tau_d}{\tau_1} = \dfrac{2\pi}{3(1.732)} = \underline{1.21}$

**1.6** PD control:

$\phi_d = -90 - 45 = -135°$    $\tau_o = \dfrac{360}{135}\tau_d = 2.67\tau_d$

$G_1 = \dfrac{2.67\tau_d}{2\pi\tau_1} = \dfrac{2.67(1)}{2\pi(30)} = 0.0142$    $G_c = \sqrt{1 + (\tan 45)^2} = 1.41$

$P = 200G_cG_1 = \underline{4.0\%}$    $D = (\tan 45)\dfrac{2.67\tau_d}{2\pi} = \underline{0.42\text{ min}}$

PI control:

$\phi_d = -90 + 45 = -45°$    $\tau_o = \dfrac{360}{45}\tau_d = 8.0\tau_d$

$G_1 = \dfrac{8\tau_d}{2\pi\tau_1} = \dfrac{8(1)}{2\pi(30)} = 0.0424$    $G_c = \sqrt{1 + (\tan 45)^2} = 1.41$

$P = 200G_cG_1 = \underline{12.0\%}$    $I = (\tan 45)\dfrac{8.0\tau_d}{2\pi} = \underline{1.27\text{ min}}$

## Chapter 2

**2.1** Noninteracting:

$\phi_i = \dfrac{180°}{15} = 12°$    $\tan\phi_i = \dfrac{2\pi\tau}{\tau_o} = 0.212$

$\tau_o = \dfrac{2\pi\tau}{0.212} = \underline{29.56\tau}$

$G_i = \dfrac{1}{\sqrt{1 + (\tan\phi_i)^2}} = 0.978$    $\Pi G_i = (0.978)^{15} = \underline{0.718}$

Interacting: from Fig. 2.6,

$\dfrac{\tau_d}{\tau_1} \approx 0.22$    $\tau_d + \tau_1 = \dfrac{15^2 + 15}{2}\tau = 120\tau$

$\tau_d + \tau_d/0.22 = 120\,\tau$    $\tau_d = \dfrac{120\tau}{1 + 1/0.22} = 21.6\tau$

$\tau_o = 4\tau_d = \underline{86.6\tau}$    $G_1 = \dfrac{\tau_o}{2\pi\tau_1} = \dfrac{4\tau_d}{2\pi(\tau_d/0.22)} = \underline{0.140}$

**2.2** 50 noninteracting lags + integrating capacity:

$\phi_i = \dfrac{90}{50} = 1.8°$    $\tan\phi_i = 0.0314$

$\tau_o = \dfrac{2\pi(6\text{ s})}{0.0314} = 1200\text{ s} = \underline{20\text{ min}}$

$G_1 = \dfrac{\tau_o}{2\pi\tau_1} = \dfrac{20}{2\pi[200\text{ gal}/(100\text{ gal/min})]} = 1.59$

$G_n = \left[\dfrac{1}{\sqrt{1 + (0.0314)^2}}\right]^{50} = 0.976$    $P = 200(1.59)(0.976) = \underline{310\%}$

Very ineffective control!

**2.3**  $h = F^2$   $\dfrac{dh}{dF} = 2F$

Loop gain will double when flow doubles. For equal-percentage valve, $G_v = F \ln R$

$\dfrac{dh}{dm} = 2F^2 \ln R$

Gain doubles at $\sqrt{2}(60\%) = 85\%$ flow

**2.4**  At $m = 0.5$, $a = 50^{-0.5} = 0.141$

Set $f = 0.5 = \dfrac{1}{\sqrt{1 + (1/0.141^2 - 1)\,\Delta p_{min}/\Delta p_{max}}}$

$\dfrac{\Delta p_{min}}{\Delta p_{max}} = \dfrac{1/0.5^2 - 1}{1/0.141^2 - 1} = 0.061$

Adjust manual valve so that $\Delta p$ across control valve at full flow is only 6 percent of $\Delta p$ at no flow. Valve size is increased by $\sqrt{1/0.061} = 4.04$. Rangeability reduced to $50/4.04 = 12.4$.

**2.5**  At $a = 0.50$: $m = 1 + \dfrac{\ln 0.5}{\ln 50} = 0.823$

$z + (1 - z)(0.5) = \dfrac{0.5}{0.823}$     $z = \dfrac{1}{0.823} - 1 = \underline{0.215}$

At $m = 0.25$: $f(m) = \dfrac{0.25}{0.215 + (0.785)0.25} = 0.608$

$a = 50^{0.608-1} = 0.216$     $\underline{error = -3.4\%}$

At $m = 0.75$: $f(m) = \dfrac{0.75}{0.215 + (0.785)0.75} = 0.933$

$a = 50^{0.933-1} = 0.770$     $\underline{error = +2.0\%}$

**2.6**  $\dfrac{\tau_n}{\tau_d} = \dfrac{1.5}{23/60} = 3.91$

Dead time and steady-state gain vary directly with flow; use equal-percentage valve to compensate.

## Chapter 3

**3.1**  First estimate: $\tau_o = (3 \text{ s}) \left(\dfrac{180}{150}\right) = 3.6$ s     Second estimate: $\tau_o = 4$ s

| | |
|---|---|
| $\phi_1 = -23.6°$ | $\phi_1 = -21.4$ |
| $\phi_2 = -15.6$ | $\phi_2 = -14.1$ |
| $\phi_3 = -29.2$ | $\phi_3 = -26.7$ |
| $\phi_4 = -8.0$ | $\phi_4 = -7.2$ |
| $\phi_5 = -8.0$ | $\phi_5 = -7.2$ |
| $\phi_6 = -79.2$ | $\phi_6 = -78.0$ |
| $\Sigma\phi = -163.6°$ | $\Sigma\phi = -154.7°$ |

Third estimate: $\tau_o = \underline{4.25}$ s

| | |
|---|---|
| $\phi_1 = -20.3°$ | $G_1 = 0.938$ |
| $\phi_2 = -13.3$ | $G_2 = 0.973$ |
| $\phi_3 = -25.3$ | $G_3 = 0.904$ |
| $\phi_4 = -6.8$ | $G_4 = 1.0$ |
| $\phi_5 = -6.8$ | $G_5 = 1.0$ |
| $\phi_6 = -77.3$ | $G_6 = 0.22$ |
| $\Sigma\phi = -149.8°$ | $\Pi G = 0.182$ |

$$P = 200\ (0.182)\ \sqrt{1 + (\tan 30)^2}\ (0.346)(8.89) = 129\%$$

$$I = \frac{\tau_o}{2\pi \tan 30} = \underline{1.17\ \text{s}}$$

**3.2** Estimate $\tau_n = 2.7$ s        Estimate $\tau_n = \underline{2.6}$ s

| | | |
|---|---|---|
| $\phi_1 = -30.2°$ | $\phi_1 = -31.1°$ | $G_1 = 0.856$ |
| $\phi_2 = -20.4$ | $\phi_2 = -21.1$ | $G_2 = 0.932$ |
| $\phi_3 = -36.7$ | $\phi_3 = -37.7$ | $G_3 = 0.791$ |
| $\phi_4 = -10.7$ | $\phi_4 = -11.1$ | $G_4 = 1.0$ |
| $\phi_5 = -10.7$ | $\phi_5 = -11.1$ | $G_5 = 1.0$ |
| $\phi_6 = -66.7$ | $\phi_6 = -67.5$ | $G_6 = 0.382$ |
| $\Sigma\phi = -175.4°$ | $\Sigma\phi = -179.7°$ | $\Pi G = 0.241$ |

**3.3** $C_R^2 = \dfrac{F^2}{p - p_0} = \dfrac{(10\ \text{gal/min})^2}{8\ \text{lb/in}^2}$

$K_p = \dfrac{dp}{dF} = \dfrac{2F}{C_R^2} = \dfrac{2(10)8}{(10)^2} = 1.6\ \text{lb/(in}^2\text{)(gal/min)}$

$K_t = \dfrac{100\%}{25\ \text{lb/in}^2} = 4\%/(\text{lb/in}^2)$

$P = 200(0.110)(0.346)(1.6)(4) = \underline{48.8\%}$

**3.4** $\tau_o = 2\pi\sqrt{\dfrac{30/12}{64.4}} = 2\pi\sqrt{0.0388} = 1.24$ s

It will raise $\tau_n$ for the flow loop.

**3.5** At 10 gal/min: $\tau_d = \dfrac{(0.4\ \text{gal/ft})(20\ \text{ft})}{10\ \text{gal/min}} + \dfrac{15\ \text{s}}{60\ \text{s/min}} = 1.05$ min

At 80 gal/min: $\tau_d = \dfrac{0.4(20)}{80} + \dfrac{15}{60} = 0.35$ min

$\tau_n$ varies from 2(0.35) to 2(1.05) or $\underline{0.70}$ to $\underline{2.10}$ min

At 100 gal/min: $\tau_d = \dfrac{0.4(20)}{100} + \dfrac{15}{60} = 0.33$ min

$\tau_n = 2(0.33) = \underline{0.66\ \text{min}}$

**3.6** $K_p = \dfrac{dx}{dF} = \dfrac{1}{F} = \dfrac{1}{100\ \text{gal/min}} = 1\%/(\text{gal/min})$

$K_t = \dfrac{100\%}{1\%}$        $K_v = \dfrac{1.2\ \text{gal/min}}{100\%}$        $G_c = \sqrt{1 + (\tan 60)^2} = 2$

$P = 200(2)(1)(100)\ \dfrac{1.2}{100} = \underline{480\%}$

## Chapter 4

**4.1**

| $\phi_c$ | $PI/100$ |
|---|---|
| $-30°$ | $1.53\tau_d$ |
| $-45$ | $1.19$ |
| $-60$ | $\underline{1.12}$ |
| $-75$ | $1.15$ |

**4.2**  $e = \dfrac{P \, \Delta m}{100} = 2(5) = \underline{10\%}$

$e = \dfrac{PI \, \Delta m}{100 \Delta t} = 2(10) \dfrac{5}{30} = \underline{3.3\%}$

$G_c = \dfrac{100}{P} \sqrt{1 + \left(\dfrac{\tau_o}{2\pi I}\right)^2} = 0.5 \sqrt{1 + \left(\dfrac{120}{2\pi 10}\right)^2} = 0.5\sqrt{1 + 3.67} = 1.8$

$e = \dfrac{\Delta m}{G_c} = \dfrac{5\%}{1.08} = \underline{4.63\%}$

**4.3**  $I_{eff} = 0.9 + 0.45 = 1.35\tau_d$      $D_{eff} = \dfrac{0.9(0.45)}{1.35} = 0.30\tau_d$

Assume $\tau_o = 4\tau_d$:

$\dfrac{\tau_o}{2\pi I} = \dfrac{4\tau_d}{2\pi(1.35\,\tau_d)} = \underline{0.47}$

$\dfrac{2\pi D}{\tau_o} = \dfrac{2\pi(0.30\tau_d)}{4\tau_d} = \underline{0.47}$

$\tan \phi = 0$

$P_{eff} = 200 K_p \dfrac{0.64\tau_d}{\tau_1} = \dfrac{128 K_p \tau_d}{\tau_1}$

$P = P_{eff}\left(1 + \dfrac{D}{I}\right) = \dfrac{128 K_p \tau_d}{\tau_1}\left(1 + \dfrac{0.45}{0.9}\right) = \dfrac{191 K_p \tau_d}{\tau_1}$

$\dfrac{PI}{100} = \dfrac{1.72 K_p \tau_d^2}{\tau_1}$

**4.4**  $\tau_o = 5.81\left(\tau_d + \dfrac{\Delta t}{2}\right) = 5.81\left(2 + \dfrac{5}{2}\right) = 26.1 \text{ min}$

$G_p = \dfrac{26.1}{2\pi 30} = 0.139$

$P = \dfrac{209 K_p \tau_d}{\tau_1} = 209 K_p \dfrac{4.5}{30} = \underline{31.4 K_p \%}$

$I = 1.74\left(\tau_d + \dfrac{\Delta t}{2}\right) = 1.74(4.5) = \underline{7.83 \text{ min}}$

**4.5**  Let $\Delta t_c = 0.1$ min, $\Delta t = \tau_d + \Delta t_c = 2.1$ min

$I = K_p \, \Delta t_c = 2.5(0.1 \text{ min}) = \underline{0.25 \text{ min}}$

$\Delta m = e \dfrac{\Delta t_c}{I}$      $\dfrac{e}{\Delta m} = \dfrac{I}{\Delta t_c} = K_p$

$E = e \, \Delta t = K_p \, \Delta m$      $\dfrac{E}{\Delta m} = K_p \, \Delta t = 2.5(2.1 \text{ min}) = \underline{5.25 \text{ min}}$

## Chapter 5

**5.1**  $A = 200 \dfrac{K_p G_p}{\pi}$      $K_p G_p = \dfrac{20}{100} = 0.2$

$A = \dfrac{200(0.2)}{\pi} = \underline{\pm 12.7\%}$

**5.2**  $\tau_o = 4\tau_d = 40$ s     $G_p = \dfrac{\tau_o}{2\pi\tau_1} = \dfrac{40/60}{2\pi5} = 0.0212$

$K_p = \dfrac{60°F}{100\%}$     $A = 200\dfrac{K_pG_p}{\pi} = \dfrac{200(60°F)(0.0212)}{100\pi} = \pm0.81°F$

**5.3**  $\dfrac{a\tau_1}{K_p\tau_d} = \dfrac{(2°F)(5\text{ min})}{(60°F/100\%)(10/60)} = 100\% = 1.0$

From Fig. 5.14,     $\tau_o = 11.7\tau_d = \underline{1.95\text{ min}}$

$A = 1.17\dfrac{K_p\tau_d}{\tau_1} = \dfrac{1.17(60°F/100\%)(10/60)}{5} = \underline{\pm2.34°F}$

**5.4**  Loop gain must be $<1$ at $A/a = 1$. Then

$\tan\phi_3 = \dfrac{1 - z/a}{\cos\beta}$     $G_3 = \dfrac{2}{\pi a}\sqrt{(\cos\beta)^2 + \left(1 - \dfrac{z}{a}\right)^2}$

Let $z/a = 0.5$, $\tan\phi_3 = 0.5/(\cos 30) = 0.577$; $\phi_3 = -30°$

$\phi_1 = -60$     $\tau_o = \dfrac{2\pi(10)}{\tan 60} = 3.62$ s

$G_p = \dfrac{3.62}{2\pi10}\sqrt{1 + (\tan 60)^2} = 0.029$

$G_3a = \dfrac{2}{\pi}\sqrt{(\cos 30)^2 + (0.5)^2} = \dfrac{2}{\pi} = 0.64$

$G_c = \dfrac{0.64}{a}$     $G_3G_p = \dfrac{0.0186}{a}$

$a = \text{dead band} + z = 0.02 + 0.5a$     $a = 0.04$

$G_cG_p = \dfrac{0.0186}{0.04} = 0.464 \text{ (stable)}$

Let $z/a = 0.3$, $\tan\phi_3 = 0.7/(\cos 17.46) = 0.734$; $\phi_3 = 36.3°$

$\phi_1 = -53.7°$     $\tau_o = \dfrac{2\pi(1\text{ s})}{\tan 53.7} = 4.61$ s

$G_p = \dfrac{4.61}{2\pi10}\dfrac{1}{\sqrt{1 + (\tan 53.7)^2}} = 0.0434$

$G_3a = \dfrac{2}{\pi}\sqrt{(\cos 17.46)^2 + (0.7)^2} = 0.753$

$G_3 = \dfrac{0.753}{a}$     $G_3G_p = \dfrac{0.0327}{a}$

$a = 0.02 + z = 0.02 + 0.3a$     $a = \dfrac{0.02}{0.7} = 0.028$

$G_3G_p = \dfrac{0.0327}{0.028} = \underline{1.17 \text{ (unstable)}}$     $z = a - \text{dead band} = 0.028 - 0.02 = 0.008$

Conclude $z$ must be set at $\sim\underline{0.01}$; $\tau_o \approx \underline{4\text{ s}}$

**5.5**  From Eq. (5.11):

$t_d = -\tau_2 \ln\dfrac{q}{100} = -3\ln 0.3 = \underline{3.6\text{ min}}$

$e_l = [100(3 + 2) - 30(3 + 2 + 3.6)]\dfrac{1°F/\text{min}}{50\%} = (500 - 258)\dfrac{1°F}{50\%} = \underline{4.85°F}$

Preload = normal load = 30%

**5.6**  $G_p = \dfrac{50}{100} = 0.5$

$f|e| = 0.2 + \dfrac{0.8|e|}{100} = 0.2 + \dfrac{0.8(20)}{100} - 0.36$

$P = 100 G_p f|e| = 50(0.36) = \underline{18\%}$

## Chapter 6

**6.1**  Original $\tau_n = 35$ min

Effective dead time $= \dfrac{35}{4} = 8.8$ min

Subtract jacket $\tau_d$ and $\tau_3$  $\underline{-6.2}$

Remaining in primary loop 2.6 min

$\tau_{n2} = 4\tau_d = 8$ min

First estimate: $\tau_{o1}/\tau_{n2} = 2$; then $\tau_{o1} = 16$ min. From Fig. 6.3, $\phi_{o2} = -75°$. $\phi_d$ in

primary loop then should be $-15°$, but $360\dfrac{2.6}{16} = -58°$

Second estimate: $\tau_{o1}/\tau_{n2} = 3$; then $\tau_{o1} = 24$ min. From Fig. 6.3, $\phi_{o2} = -40°$. $\phi_{d1}$

should be $-50°$; $360\dfrac{2.6}{24} = -39°$

Third estimate: $\tau_{o1}/\tau_{n2} = 2.7$; $\tau_{o1} = 21.6$ min. $\phi_{o2} = -47°$; $\phi_{d1}$ should be $-43°$

$360\dfrac{2.6}{21.6} = \underline{43.3°}$

$G_{o2} = 1.2 \qquad G_1 = \dfrac{21.6}{2\pi 48} = 0.072$

If the ranges of the two temperature transmitters are identical,

$K_t K_v = 1.0 \qquad K_p = \dfrac{dT}{dT_{c2}} = 1.0$

$P_{eff} = 200(1.2)(0.072) = 17.3\% \qquad P = \underline{34.6\%}$

$D = I = \dfrac{\tau_{o1}}{2\pi} = \dfrac{21.6}{2\pi} = \underline{3.44\ min}$

**6.2**  Moving the secondary measurement to the jacket inlet places jacket dynamics back in the primary loop. Thus only the steady-state gain is different from that given in Example 3.6.

$G_{o2} = 1.0 \qquad K_t K_v K_p = 1.0 \qquad G_1 = 0.116$

$P_{eff} = 200(1.0)(0.116) = 23.2\% \qquad P = \underline{46.4\%}$

$D = I = \dfrac{\tau_{o1}}{2\pi} = \dfrac{35}{2\pi} = \underline{5.57\ min}$

**6.3**

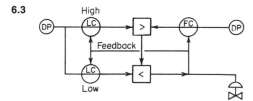

**6.4**  At 50% flow, $\tau_o = 3.6\tau_d$, $D = I = 0.68\tau_d$, $G_c = 1.015$

$\phi_c = \tan^{-1}\left(\dfrac{\pi 0.68\tau_d}{\tau_o} - \dfrac{\tau_o}{4\pi 0.68\tau_d}\right) = \tan^{-1}\left(\dfrac{2.14\tau_d}{\tau_o} - \dfrac{\tau_o}{8.55\tau_d}\right)$

Let dead time under undamped conditions $= \tau_u$

Estimate: $\tau_o = 6\tau_d$     $\phi_c = \tan^{-1}\left(\dfrac{2.14}{6} - \dfrac{6}{8.55}\right) = -19°$

$\tau_u = \dfrac{90 - 19}{360}\, 6\tau_d = 1.18\tau_d$     $G_c = \sqrt{1 + (\tan 19)^2} = 1.058$

$\dfrac{(K_p G_p)_u}{K_p G_p} = \dfrac{\tau_o}{3.6\tau_d} = \dfrac{6}{3.6} = 1.67$     $\dfrac{(G_c)_u}{G_c} = \dfrac{1.058}{1.015} = 1.04$

$\dfrac{(\Pi KG)_u}{\Pi KG} = 1.67(1.04) = 1.74$

Second estimate: $\tau_o = 7\tau_d$     $\phi_c = \tan^{-1}\left(\dfrac{2.14}{7} - \dfrac{7}{8.55}\right) = -27°$

$\tau_u = \dfrac{90 - 27}{360}\,(6\tau_d) = 1.22\tau_d$     $G_c = \sqrt{1 + (\tan 27)^2} = 1.12$

$\dfrac{(K_p G_p)_u}{K_p G_p} = \dfrac{7}{3.6} = 1.944$     $\dfrac{(G_c)_u}{G_c} = \dfrac{1.12}{1.015} = 1.105$

$\dfrac{(\Pi KG)_u}{\Pi KG} = 1.94(1.105) = 2.15$ (loop gain doubled)

$f_u = \dfrac{50\%}{\tau_u/\tau_d} = \dfrac{50\%}{1.22} = \underline{41\%}$

**6.5**  With an equal-percentage valve, gain is reduced proportional to flow.

Estimate: $\tau_o = 9\tau_d$     $\phi_c = \tan^{-1}\left(\dfrac{2.14}{9} - \dfrac{9}{8.55}\right) = -39°$

$\tau_u = \dfrac{90 - 39}{360}\, 9\tau_d = 1.27\tau_d$     $G_c = 1.29$

$\dfrac{(\Pi KG)_u}{\Pi KG} = \dfrac{9}{3.6}\dfrac{1.29}{1.015}\dfrac{1}{1.27} = 2.50$

Second estimate: $\tau_o = 8\tau_d$     $\phi_c = \tan^{-1}\left(\dfrac{2.14}{8} - \dfrac{8}{8.55}\right) = -33.7°$

$\tau_u = 1.25\tau_d$     $G_c = 1.20$

$\dfrac{(\Pi KG)_u}{\Pi KG} = \dfrac{8}{3.6}\dfrac{1.20}{1.015}\dfrac{1}{1.25} = 2.11$ (loop gain doubled)

$f_u = \dfrac{50\%}{1.25} = \underline{40\%}$

## Chapter 7

**7.1**  $\Delta t = K \dfrac{W_s}{W_p}$     $K = 100$     $dT_2 = d\Delta T = \dfrac{K}{W_p}\, dW_s$

$dT_2 = \begin{cases} \dfrac{100(0.02)}{0.5} = 4°F & \text{at 50\% flow} \\[2mm] \dfrac{100(0.02)}{0.25} = 8°F & \text{at 25\% flow} \end{cases}$

**7.2**  $y = \dfrac{kzF}{L}$     $L = \dfrac{Fkz}{y}$

**7.3**

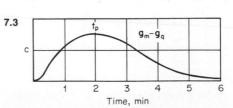

$\tau_1 = 1.8\ \text{min};\ \tau_2 = 2.2\ \text{min}$

**7.4** $dc = dq \dfrac{r}{q}(e^{-t/\tau_q} - e^{-t/\tau_m})$

$\displaystyle\int \dfrac{dc}{dq} = \dfrac{0.5}{0.5} \int (e^{-t/\tau_q} - e^{-t/\tau_m}) = \tau_m - \tau_q = 3 - 1 = \underline{2 \text{ min}}$

$\tau_1 - \tau_2 = 2.5 - 1 = \underline{1.5 \text{ min}}$  $\qquad$ Area $= 2 - 1.5 = \underline{0.5 \text{ min}}$

**7.5** See Table 4.3 for comparison of the performance of PI and PID controllers. Integrated area is reduced by 2:1 by using derivative, 4:1 with noninteracting PID, and 10:1 using feedforward control having ±10 percent accuracy.

## Chapter 8

**8.1** $m_1 + m_2 = F$  $\qquad$ $m_1 x_1 + m_2 x_2 = Fx$

$\dfrac{\partial F}{\partial m_1}\Big|_m = 1$  $\qquad$ $m_1 x_1 + (F - m_1)x_2 = Fx$  $\qquad$ $F(x - x_2) = m_1(x_1 - x_2)$

$\dfrac{\partial F}{\partial m_1}\Big|_x = \dfrac{x_1 - x_2}{x - x_2}$  $\qquad$ $\lambda_{F1} = \dfrac{x - x_2}{x_1 - x_2}$

**8.2** $\dfrac{\partial T}{\partial B}\Big|_m = 0$, therefore $\lambda_{TB} = 0$. Then $\lambda_{L_1 B} = \lambda_{L_2 B} = 0.5$, and $\lambda_{TQ_1} = \lambda_{TQ_2} = 0.5$. Rows and columns must sum to 1. Therefore

|       | B    | $Q_1$   | $Q_2$   |
|-------|------|---------|---------|
| $T$   | 0    | 0.5     | 0.5     |
| $L_1$ | 0.5  | a       | 0.5-a   |
| $L_2$ | 0.5  | 0.5-a   | a       |

Two loops can be closed having relative gains of 0.5. The third loop must have a relative gain of either $a$ or $0.5 - a$. The initial responses of the two levels to a change in heat input are approximately equal and opposite, in which case $a$ is apparently a large positive number. Neither $a$ nor $0.5 - a$ is then favorable.

**8.3** Determinant $= -0.5734$

$$\mathbf{H} = \begin{bmatrix} 1.064 & & 0 \\ & -1.639 & \\ & & 0.159 \end{bmatrix} \qquad \mathbf{\Lambda} = \begin{bmatrix} 0.617 & -0.458 & \underline{0.841} \\ 0 & 1.0 & 0 \\ \underline{0.383} & 0.458 & 0.159 \end{bmatrix}$$

The best control-loop pairs are underlined. Decoupling should be provided because the best choice is not particularly favorable. If the second row and column are removed by decoupling, the remaining gains are more favorable

$$\mathbf{\Lambda} = \begin{bmatrix} \underline{0.617} & 0.383 \\ 0.383 & \underline{0.617} \end{bmatrix}$$

Note that the optimum pairing has been changed by decoupling.

**8.4**

$$\mathbf{K} = \begin{matrix} & T_c & F \\ T & 1 & 12 \\ p & [0.4 & 4.8] \end{matrix}$$

$\lambda_{TT} = \dfrac{1}{1 - 12(0.4)/1(4.8)} = \infty$

Pressure and temperature cannot be controlled independently.

**8.5**  $F_1 = m_A c_2 \qquad c_2 = m_2 + F_1$

$$c_1 = \frac{F_1}{c_2} = \frac{m_A c_2}{c_2} = m_A \qquad \frac{\partial c_1}{\partial m_A}\Big|_{c_2} = \frac{\partial c_1}{\partial m_A}\Big|_{m_2} = 1.0$$

$$\lambda_{1A} = 1.0$$

**8.6**
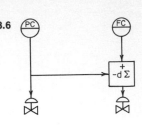

## Chapter 9

**9.1**
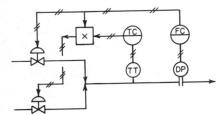

**9.2**  Rearrange Eq. (9.14) to yield

$$W_H C_H = \frac{W_C C_C}{2(T_{H1} - T_{C1})/(T_{C2} - T_{C1}) - 2W_C C_C/UA - 1}$$

$$UA = \frac{Q}{\Delta T_m} \qquad Q = W_C C_C(T_{C2} - T_{C1}) = 100(0.8)(200)$$

$$Q = 16{,}000 \qquad T_{H1} - T_{H2} = \frac{Q}{W_H C_H} = \frac{16{,}000}{400(0.8)} = 50$$

$$T_{H2} = T_{H1} - 50 = 500 - 50 = 450$$

$$\Delta T_m = \frac{(T_{H1} - T_{C2}) + (T_{H2} - T_{C1})}{2} = \frac{(500 - 400) + (450 - 200)}{2}$$

$$\Delta T_m = 175 \qquad UA = \frac{16{,}000}{175} = 91.5$$

For 120 lb/min

$$W_H C_H = \frac{120(0.8)}{2(300/200) - 2[120(0.8)/91.5] - 1} = \frac{96}{3 - 3.1} \text{ (no solution)}$$

For 80 lb/min

$$W_H C_H = \frac{80(0.8)}{3 - 2[80(0.8)/9.15] - 1} = \frac{64}{3 - 2.4} = 107$$

$$W_H = \frac{107}{0.8} = \underline{133 \text{ lb/min}}$$

For $T_{C1} = 240°F$

$$W_H C_H = \frac{100(0.8)}{2(260/160) - 2[100(0.8)/91.5] - 1} = \frac{80}{3.25 - 2.75} = 160$$

$$W_H = \frac{160}{0.8} = \underline{200 \text{ lb/min}}$$

For $T_{C1} = 160°F$

$$W_H C_H = \frac{80}{2(340/240) - 2.75} = \frac{80}{2.84 - 2.75} = 890$$

$$W_H = \frac{890}{0.8} = \underline{1110 \text{ lb/min}}$$

**9.3** From Eq. (9.14)

$$T_{C2} = T_{C1} + \frac{T_{H1} - T_{C1}}{W_C C_C/UA + \frac{1}{2}(1 + W_C C_C/W_H C_H)}$$

$$\frac{dT_{C2}}{dW_H} = \frac{C_H(T_{H1} - T_{C1})W_C C_C/2}{[W_H C_H(W_C C_C/UA + \frac{1}{2}) + W_C C_C/2]^2}$$

For $W_C = 100$ lb/min

$$\frac{dT_{C2}}{dW_H} = \frac{(0.8)(300)(80/2)}{[320(80/91.5 + \frac{1}{2}) + 80/2]^2} = \frac{9600}{(470)^2} = \underline{0.0435°F/(lb/min)}$$

For $W_C = 80$ lb/min

$$\frac{dT_{C2}}{dW_H} = \frac{(0.8)(300)(64/2)}{[107(64/91.5 + \frac{1}{2}) + 64/2]^2} = \frac{7670}{(160.5)^2} = \underline{0.297°F/(lb/min)}$$

$$\frac{dT_{C2}}{dm} = \begin{cases} 400k(0.0435) = \underline{17.5k} & \text{for 100 lb/min} \\ 133k(0.297) = \underline{39.6k} & \text{for 80 lb/min} \end{cases}$$

**9.5** Assume inversion can be approximated by an equal dead time. From Table 4.3

$$P = \frac{209K_p\tau_d}{\tau_1} = 209(\frac{12}{50}) = \underline{50\%}$$

$$I = 1.74\tau_d = \frac{1.74(12)}{60} = \underline{0.35 \text{ min}}$$

**9.6** $\frac{p}{\rho} = k_1 N^2 - k_2 F^2$

$$100 = k_1(3600)^2 - 0 \qquad k_1 = \frac{100}{3600^2} = \underline{7.70 \times 10^{-6}}$$

$$0 = k_1(3600)^2 - k_2(100)^2 \qquad k_2 = k_1\left(\frac{3600}{100}\right)^2 = \frac{100}{100^2} = \underline{0.01}$$

$$N^2 = \frac{p/\rho + k_2 F^2}{k_1} = \frac{50 + 0.01(50)^2}{7.70 \times 10^{-6}} = \frac{75}{7.70 \times 10^{-6}} = 9.75 \times 10^6 \qquad N = 3120 \text{ r/min}$$

$$\text{hhp} = \frac{Fp}{1714} = \frac{50(50)0.4335 \text{ (lb/in}^2)/\text{ft}}{1714} = \underline{0.632 \text{ hp}}$$

At 3600 r/min

$$\frac{p}{\rho} = 75 \text{ ft; hhp} = \frac{50(75)(0.4335)}{1714} = \underline{0.948 \text{ hp}}$$

**Chapter 10**

**10.1** $\frac{\partial y}{\partial T}\big|_x = \frac{E}{RT^2}y \qquad T - T_c = \frac{y}{\partial y/\partial T|_x} = \frac{RT^2}{E}$

$$T_c = T - \frac{RT^2}{E} = 230 - \frac{(460 + 230)^2}{20,000} = 230 - 23.8 = \underline{206.2°F}$$

**10.2** $UA - \frac{Q}{\Delta T} - \frac{H_r F x_0 y}{T - T_c} = \frac{5000(4000)(0.2)(0.5)}{230 - 190} = 50,000 \text{ Btu/°F}$

$$\tau_T = \frac{4000(0.4)}{50,000 + 4000(0.4) - [(5000)(4000)(0.2)(20,000)(0.5)]/(460 + 230)^2}$$

$$= \frac{1600}{50,000 + 1600 - 84,000} = \frac{1600}{-32,400} = -0.0495 \text{ h} \approx \underline{-3 \text{ min}}$$

Estimate: $\tau_0 = 4.5\tau_d = 4.5$

$$\phi_T = -180° + \tan^{-1}\left(2\pi\frac{3}{4.5}\right) = -180° + 76° = -104°$$

$$\phi_d = -76° \qquad \tau_0 = \frac{360}{76} = 4.75$$

Second estimate: $\tau_o = 4.8$

$$\phi_T = -180° + \tan^{-1}\left(2\pi\frac{3}{4.8}\right) = -180° + 75.7°$$

$$\phi_d = -75.7° \qquad \tau_o = \frac{360}{75.7} = \underline{4.77}$$

$$K_T = \frac{UA}{-32,400} = \frac{50,000}{-32,400} = -1.54$$

$$G_T = \frac{1}{\sqrt{1 + (2\pi\,3/4.77)^2}} \qquad K_T G_T = \frac{-1.54}{\sqrt{16.7}} = -0.378$$

$$P = 200K_T G_T = \underline{75\%} \qquad \frac{P}{100K_T} = \frac{75}{100(-1.54)} = -0.49 \text{ (marginally stable)}$$

**10.3** $\quad \tau_T = \dfrac{1600}{50,000 + 800 - 42,000} = \dfrac{1600}{8800} = 0.182 \text{ h}$

$\tau_T = 0.182 \times 60 = \underline{+10.9 \text{ min}}$

Estimate: $\tau_o = 3.8$ min

$$\phi_T = -\tan^{-1}\left(2\pi\frac{10.9}{3.8}\right) = 87° \qquad \tau_o = \frac{360}{180 - 87} = \underline{3.9 \text{ min}}$$

$$K_T G_T = \frac{3.9/60}{6.28(1600)/(50,000)} = 0.324 \qquad P = 200K_T G_T = \underline{65\%}$$

For $x_0 = 0.1$, $\tau_T = \dfrac{1600}{50,000 + 1600 - 42,000} = \dfrac{1600}{9600} = 0.167 \text{ h}$

$\tau_T = \underline{+10 \text{ min}} \qquad \tau_o = \underline{3.9 \text{ min}}$

$K_T G_T = 0.324 \qquad P = \underline{65\%}$

**10.4** A feedback optimizing controller can hold minimum conductivity, but the process must be made self-regulating first. This can be accomplished by feeding solvent from one storage tank while flowing into another, thus breaking the positive-feedback loop. Tanks are switched when their capacity is reached.

**10.5** $\quad 10^{-\text{pH}}\left(1 + \dfrac{1}{K_A}\right) = 10^{\text{pH}-14} \qquad 1 + \dfrac{1}{K_A} = 10^{2\text{pH}-14}$

$$2\text{pH} = 14 + \log\left(1 + \frac{1}{K_A}\right) \qquad \text{pH} = 7 + 2.37 = \underline{9.37}$$

$$x_A = x_B + [\text{H}^+]\left(1 + \frac{1}{K_A}\right) - \frac{10^{-14}}{[\text{H}^+]} \qquad \frac{dx_B}{d[\text{H}^+]} = -\left(1 + \frac{1}{K_A} + \frac{10^{-14}}{[\text{H}^+]^2}\right)$$

$$\text{pH} = -\log[\text{H}^+] \qquad \frac{d\text{pH}}{d[\text{H}^+]} = -\frac{\log e}{[\text{H}^+]} = -\frac{0.434}{[\text{H}^+]}$$

$$\frac{d\text{pH}}{dx_B} = \frac{0.434/[\text{H}^+]}{1 + 1/K_A + 10^{-14}/[\text{H}^+]} = \frac{0.434}{10^{-9.37}(1 + 0.54 \times 10^5) + 10^{-4.63}}$$

$$= \frac{0.434}{10^{-9.37}(10^{4.74}) + 10^{-4.63}} = \frac{0.434}{2 \times 10^{-4.63}} = \frac{0.434}{0.468 \times 10^{-4}} = \underline{9300 \text{ pH/N}}$$

**10.6** Equation (10.22) and the answers to Probs. 10.2 and 10.3 indicate that $G_T$ is independent of $T$ and $x$.

## Chapter 11

**11.1**

$$x = \frac{0.97}{0.97 + 361(0.03)} = 0.082 \qquad \frac{D}{F} = \frac{0.5 - 0.082}{0.97 - 0.082} = 0.471$$

**11.2**

$$nE = \frac{\ln 361}{\ln \dfrac{1.3}{\sqrt{1 + \dfrac{1}{[(3/0.5) - 1]\, 0.5}}}} = 62.6$$

Estimate: $\dfrac{D}{F} = 0.45$ $\qquad S = \left[\dfrac{1.3}{\sqrt{1 + 1/[(2.5/0.45) - 1]0.5}}\right]^{62.6} = 153$

$$x = \frac{0.95}{0.95 + 153(0.05)} = 0.110 \qquad \frac{D}{F} = \frac{0.50 - 0.11}{0.95 - 0.11} = 0.464$$

Second estimate: $\dfrac{D}{F} = 0.46$ $\qquad S = \left[\dfrac{1.3}{\sqrt{1 + 1/[(2.5/0.46) - 1]0.5}}\right]^{62.6} = 118$

$x = 0.38 \qquad \dfrac{D}{F} = 0.446$

Third estimate: $\dfrac{D}{F} = 0.455$ $\qquad S = 134.6$

$x = \underline{0.123} \qquad \dfrac{D}{F} = \underline{0.455}$

**11.3** $B = \begin{cases} mF(1 - z) & \text{constant } V/F \\ mF(1 - z) + kF^2 & \text{constant } V \end{cases}$

**11.4** $\dfrac{D}{F} = \dfrac{z_i - x_i}{y_i - x_i} \qquad \dfrac{D}{F} = \dfrac{z_p}{y_p}$

$y_i = 1 - y_n - y_p \qquad \dfrac{D}{F} = \dfrac{z_i - x_i}{1 - y_n - z_p/(D/F) - x_i}$

$\dfrac{D}{F}(1 - y_n - x_i) - z_p = z_i - x_i$

$\dfrac{D}{F} = \dfrac{z_i + z_p - x_i}{1 - y_n - x_i} = \dfrac{z_i + z_p - 0.02}{1 - 0.05 - 0.02} = \underline{1.075(z_i + z_p - 0.02)}$

**11.5** $\dfrac{D}{F} = 1.075(0.50 + 0.05 - 0.02) = 0.570 \qquad y_p = \dfrac{0.05}{0.570} = 0.088$

$y_i = 1 - 0.05 - 0.088 = 0.862 \qquad S = \dfrac{0.862/0.02}{0.05/0.98} = 845$

New $\dfrac{D}{F} = \dfrac{0.5 - 0.04}{0.862 - 0.04} = 0.560 \qquad$ New $S = \dfrac{0.862/0.04}{0.05/0.96} = 414$

$\dfrac{V}{D} = 1 + \dfrac{1}{z(\alpha/S^{1/nE})^2 - 1} \qquad nE = \ln \dfrac{\dfrac{\ln 845}{1.32}}{\sqrt{1 + 1/[(3/0.57) - 1]0.55}} = 67.4$

New $\dfrac{V}{D} = 1 + \dfrac{1}{0.55[(1.32/414^{1/67.4})^2 - 1]} = 4.98$

New $\dfrac{V}{F} = 0.56(4.98) = 2.78$

Savings $= \dfrac{3 - 2.78}{3} = 0.071 = \underline{7.1\%}$

$\dfrac{V}{D} = 1 + \dfrac{1}{0.55[(1.35/8.45^{1/67.4})^2 - 1]} = 4.695$

$\dfrac{V}{F} = 0.57(4.695) = 2.676 \qquad \dfrac{\text{New } Q}{\text{Old } Q} = \dfrac{2.676(135)}{3.0(129)} = 0.934$

Savings $= \underline{6.6\%}$

**11.7** $\lambda_{yD} = \dfrac{1}{1 + \dfrac{0.006(0.994)(0.9 - 0.12)}{0.9(0.1)(0.12 - 0.006)}} = \underline{0.688}$

## Chapter 12

**12.1**  At zero feed: $y = Kw$     $x = w$
At zero absorbent: $y = z$     $x = z/K$

**12.2**

At 60°F: $\log p_w° = 6.69 - \dfrac{4407}{520}$     $p_w° = 0.0164$ atm

At 72°F: $\log p_w° = 6.69 - \dfrac{4407}{532}$     $p_w° = 0.0255$

RH $= 100 \dfrac{0.0164}{0.0255} = \underline{64.4\%}$

**12.3**  $V_0 H_0 = V_1 H_1 = V_2 H_2 = V_3 H_3$

$V_1 + V_2 + V_3 = V_0 H_0 \left( \dfrac{1}{H_1} + \dfrac{1}{H_2} + \dfrac{1}{H_3} \right)$

$\dfrac{\Sigma V_i}{V_0} = H_0 \Sigma \dfrac{1}{H_i}$

From steam tables
$H_0 = \quad 960.1$
$H_1 = \quad 976.7$
$H_2 = \quad 992.1$
$H_3 = 1022.1$

$\dfrac{\Sigma V_i}{V_0} = 2.89$

$W_0 = \dfrac{V_0 H_0 \Sigma(1/H_i)}{1 - x_0/x_n} = \dfrac{2.89}{1 - 0.35/0.70} = \underline{5.78 \text{ lb}}$

**12.4**  Crystal content $= \dfrac{x - y}{1 - y} = 0.7$     $x = 0.17 + 0.7(0.83) = 0.751$

$\dfrac{L}{F} = \dfrac{x - z}{x - y} = \dfrac{0.751 - 0.44}{0.751 - 0.17} = \underline{0.535}$

**12.5**  $\dfrac{T_i - T_w}{T_o - T_w} = \dfrac{1000 - 153}{260 - 153} = 7.92$

$T_o = T_w + \dfrac{T_i - T_w}{7.92} = 134 + \dfrac{600 - 134}{7.92} = \underline{193°F}$

$T_o = b + RT_i$     $R = \dfrac{\Delta T_o}{\Delta T_i} = \dfrac{260 - 193}{1000 - 600} = \underline{0.168}$

$b = T_o - RT_i = 260 - 0.168(1000) = \underline{92°F}$

# Index

Absorption, 307–311
  optimizing control of, 192, 193
Accumulator, reflux, 286, 287
Acids, 264–270
Adaptive control, 156–164
  dynamic, 157–160
  in feedforward systems, 188–190
  programmed, 157, 158, 271, 272
  self-adaptive, 158–160, 267
  steady-state, 160–164
Agitation (*see* Mixing)
Analyzers, 78, 79
  sampling, 103, 104
Antisurge control (*see* Surge protection)
Antiwindup (*see* Batch control)
Auctioneering, 151, 152
Automanual transfer, 93, 106, 108, 153, 154
Averaging level control, 69
Azeotropic distillation, 320, 321

Bases, 263–270
Batch control, 92, 93
  (*See also* Reactors, batch)

Bias in proportional control, 10–12
  in multiple-output systems, 148
Blending systems, 175–178
  decoupling in, 219, 220, 227
  digital, 175–178
  interaction in, 205
Boilers, 235–239
Boilup, 282, 285, 287, 288
Bristol, E. H., 188, 197
Buffering, 266, 267, 269
Burst cycling, 160

$C_v$, 23, 47
Capacity, 18–24
  dead time and, 24–28
  double, 31–33
  multiple, 31–42
  single, 18–24
Cascade control, 139–151
  of flow, 118, 145, 146, 148, 149
  of ratio, 174, 175, 219, 268
  of temperature, 146, 239, 257, 261
  of valve position, 118, 144, 145

Catalyst, 249, 262
Closed loop:
    feedback, 4–8
    feedforward, 167
Combustion, 233–235
Complementary feedback, 100–103
    in cascade systems, 143
Composition control, 75–79
    in blending systems, 173–178, 205, 218
    in distillation, 292–304
    in reactors, 258–260, 275
    (See also pH control)
Compressors, 242–245
    selective control of, 152–154
Condensers, 232
    in distillation, 155, 156, 288–291
Conductivity control, 163
Considine, D. M., 114
Constraints, 215, 216
    (See also Selective control)
Contour plot, 193
Control algorithms, digital, 109, 110, 154,
    155, 183
Control interval, 106, 107
Control valves, 44–50
Controller action, 5
Controller adjustment, 96–99
    for coupled systems, 204–209
    for dead time, 17
        plus capacity, 27, 28
    in dual-mode systems, 131
    interaction in, 94–96
Controllers, 4, 5
    complementary feedback, 100–103
    integral, 13–17
    linear, 85–111
    nonlinear, 120–136
    on-off, 120–132, 149, 150
    PI, 16–18, 91, 92
    PID, 94–99
    proportional, 10–13
    proportional-plus-derivative, 27, 28, 90
    sampling, 105–107
    second integral, 177
    self-optimizing, 161–164
    self-tuning, 158–161
    three-state, 123–125
Conversion in reactors, 250–253
    computing of, 261, 262
Conveyors, 8
Coupling (see Interaction)
Crystallization, 316–318
Cycle (see Limit cycle; Oscillation)

Damping, 7, 8
    amplitude dependent, 112, 113
    critical, 101, 105, 106, 125
    quarter-amplitude, 8, 11
DDC (direct digital control), 107–109
Dead band, 116–118
    in on-off control, 121–124
    in valves, 116, 178
Dead-beat response, 106, 107, 110
    (See also Damping, critical)
Dead time, 8–18
    and capacity, 24–28
    complementary feedback with, 101–103
    effective: in mixing, 77
        in multicapacity processes, 37
    variable, 40, 230, 261, 262
Dead zone, 115, 116
    in dual-mode systems, 128
    in pH control, 135
    in PI controllers, 135, 136
    in three-state controllers, 123–125
Debits in optimization, 192, 193
Decanters, 319, 320
Decomposing multivariable systems, 204,
    297–299
Decoupling, 212–221
    nonlinear, 299–301
    using three-way valve, 227
Degrees of freedom, 139
Dehumidification (see Humidification)
Delay (see Dead time)
Density control, 316
Derivative, 27, 28, 90, 91
    in digital control, 110, 111
Describing functions, 113, 114
Determinant, 203
Deviation from set point, 5
Dew point, 311–313, 323
Difference equations, 109, 110
Differential equations:
    first-order, 21
    second-order, 67
Differential gap (see Dead band)
Differential-pressure control in
    distillation, 287, 288
Differential vapor pressure (see Vapor
    pressure)
Differentiators, 162, 163
Digital control, 107–111
Direct digital control, 107–109
Distillation, 277–305
    azeotropic, 318, 320, 321
    batch, 301–304

Distributed lag, 36, 38
Dither, 118
Dividers:
  in adaptive control, 162, 163
  in decoupling, 299
  in feedforward systems, 238
  for gain compensation, 46, 47, 296, 297
  in process model, 170, 179, 180
  in ratio control, 173
Douglas, J. M., 282
Droop of a pressure regulator, 64, 65
  (*See also* Offset)
Dryers, 322–325
Dual-mode control, 124–132
  for batch reactors, 274
Dynamic compensation, 178–186
  for boilers, 238
  in decoupling, 220, 221
  for distillation, 293–295
  for dryers, 324
  for evaporators, 316
  for heat exchangers, 178, 186
  in heat-flow calculation, 262
  in optimizers, 162
Dynamic gain, 4
  variable, 21, 40–42


Economics, 190, 191
End-point control, 263
  in batch reactors, 275
  in continuous reactors, 258–260
Energy conversion, 225
Enthalpy, 63
  in boilers, 236
  in distillation, 285, 286
  in pressure control, 65
Entrainer, 320, 321
Equal-percentage valves, 44–47, 49, 50,
  55, 56, 73
Equilibrium:
  chemical, 248, 249
  liquid-liquid, 318
  vapor-liquid, 279, 307
  vapor-solid, 322
Error criteria, 86–89
Error-squared control, 133
Evaporators, 312–316
Extraction, 318–320


Fagervik, K., 209, 299
Feedback, 4–8
  (*See also* Positive feedback)

Feedforward control, 166–194
  of absorbers, 192, 193, 308–310
  of boilers, 168, 169, 236, 238
  of crystallizers, 317, 318
  of distillation, 292–295, 321
  of evaporators, 314–316
  of extractors, 319
  of heat exchangers, 170–172, 190
  of liquid level, 168, 169, 286, 287
  optimizing, 161, 192, 193
  of pH, 268, 270–272
Fenske equation, 279
Flame temperature, 233
Floating pressure control, 291, 309, 310
Flooding, 287, 288
Flow calculations, 235, 236
Flow control, 58–63
  cascade, 145, 146, 148, 149
  interactions in, 198–201, 205
  nonlinear, 134
Flow measurement, 43
Flow-ratio control (*see* Ratio control)
Fourier series, 114
Fractionation (*see* Distillation)
Fuel-air ratio control, 160, 163, 233, 234
Furnace (*see* Heaters, fired)


Gain, 4
  amplitude dependent, 112, 113
  controller, 4, 5
  dynamic, 4
  loop, 7, 8
  process, 4, 50–52
  relative, 197–204
  steady-state, 42–52
    variable, 21, 24
  transmitter, 42, 43
  valve, 44–50
Gain compensation, 296, 297
Gain matrix (*see* Gain, relative)
Gas-pressure control, 63, 64
  interaction with flow control, 198–200
Gibson, J. E., 123
Gilbert, L. F., 234


Head-box control, 210–212, 220, 221
Heat exchangers, 227–232
  feedback control of, 40, 41, 50, 51, 190,
    230–232
  feedforward control of, 170–172
Heat-flow calculation, 261, 262

Heat transfer, 225–232
    in reactors, 72, 254, 255, 260–263
Heaters, fired, 234, 235
Horsepower (*see* Power)
Humidification, 310–313
Hydraulic horsepower, 241
Hydraulic resonance, 66–68
Hysteresis (*see* Dead band)


IAE (integrated absolute error), 86
Inertia, 18, 58–60
Inferential controls, 323–325
Integral control, 13–17
    in adaptive systems, 159, 162–164
    in blending systems, 175–178
    of integrating processes, 20
    sampled, 108, 109, 163, 164
    of valve position, 150, 151, 291
Integral feedback, 91, 92
    in cascade control, 142–144
    in PID controllers, 95–97
    in selective control, 153, 154, 156
Integrated absolute error, 86
Integrated error, 87
    with feedforward control, 184–187
    with PID control, 95, 97, 166
Integrated square error, 86
Integrating process, 19, 20, 260
Interaction:
    between capacities, 33–38
    between controller settings, 94–99
    between loops, 196–212
Interface control, 319, 320
Inverse response:
    in boilers, 236, 237
    with interacting loops, 209, 210
Ionization, 263
Ionization constants, 264, 265
ISE (integrated square error), 86


Jafarey, A., 282


Lag:
    distance-velocity (*see* Dead time)
    distributed, 36, 38
    first-order, 20–23
    multiple, 31–42
    negative, 255, 256
    second-order, 67
    transport (*see* Dead time)
    variable, 21, 40–42

Lag compensation, 189
    in positive-feedback loops, 324
Lead, 22, 27
    negative, 236
Lead-lag compensation, 181–186
Level control (*see* Liquid-level control)
Limit cycle, 113
    with cascade flow control, 145
    through dead zone, 135
    due to dead band, 117, 118
    due to valve rangeability, 268
    due to variable dead time, 230, 261, 262
    in exothermic reactors, 257
    with on-off control, 121, 122
    in transitional boiling, 120
    in vacuum systems, 120
Limiters, 114, 115
Liquid-interface control, 319, 320
Liquid-level control, 18–24, 31, 32, 66–69,
        144, 146
    averaging, 68
    in boilers, 69, 168–170, 236, 237
    in distillation, 213, 214
    feedforward, 168–170
    nonlinear, 134, 135
Liquid-pressure control, 65, 66
Load, 5
Load response, 86–89
    with feedforward control, 186
    with integral control, 16, 18
    with nonlinear controllers, 134
    with PID control, 97–99
    with proportional control, 12, 13, 18
    sinusoidal, 88, 89, 99
Lockup (*see* Dead band)
Locus of minimum debit, 193
Loop gain, 7, 8
    variable, 112, 113


McAvoy, T. J., 282
Manual control, 93, 108
Manual reset, 12
Material-balance control:
    in absorption, 308
    in crystallizers, 317, 318
    in distillation, 286, 287
    in evaporators, 314–316
    in extractors, 319
    in feedforward systems, 167, 168
Matrix arithmetic, 202–204
Median selector, 152, 153
Minimum-time control, 127–130

Mixing:
   dynamics of, 75–78
   hot and cold fluids, 226, 227
   in reaction vessels, 74, 250
Modes of control (*see* Controllers)
Modulation (*see* Proportional-time
   control)
Moisture control, 322–325
Mother liquor, 317, 318
Motor:
   diaphragm, 116, 118
   electric: constant speed, 123–125
     variable speed, 151, 240–242
Multicapacity processes, 31–42
Multiple-output systems, 147–151
Multipliers:
   in adaptive systems, 158, 159, 190
   in feedforward systems, 171, 190
   for gain compensation, 189, 190
   in ratio control, 174
Multivariable processes (*see* Interaction)

Natural period, 7
Negative feedback, 4
Negative resistance, 119, 120
Neutralization (*see* pH control)
Noise:
   in analytical measurements, 78
   in flow measurements, 62
   in liquid-level measurements, 68,
    69
   rejection of, 125, 126, 134
   in selective control, 155
Nonlinear controllers, 120–136
   dual-mode, 124–132
   on-off, 120–124, 149, 150
   PID, 132–136
   proportional-time, 122, 123
   three-state, 123, 124
Nonlinear dynamic elements, 116–120
Nonlinearity, 30, 31, 112, 113
   in cascade flow loops, 145
   in controllers, 120–136
   in distillation, 296, 297
   dynamic, 116–120
   in heat transfer, 72, 73, 226–231
   in transmitters, 43
   in valves, 44–50
Non-self-regulation, 19, 20, 255, 260
Notation, 327–328

Objective function, 156, 157

Offset, 12, 13
   in cascade control, 143
   in digital control, 110
   in feedforward systems, 171, 172, 188
   integral, 176, 177
   proportional, 12, 13
   in selective control, 155
On-off control (*see* Controllers, on-off)
Optimal switching, 127–130
Optimizing control, 160–164
   feedback, 161–164
   feedforward, 160, 161, 191–194
Orifice meter, 43
Oscillation, 5–8
   constant-amplitude, 113
   damped, 7, 8
   regenerative, 113
   resonant, 7, 66–68

Pairing variables, 197, 205–212
   in distillation, 283, 284
Payout (*see* Economics)
Pendulum, 6, 7, 32
Performance criteria, 86–89
Period of oscillation, 6, 7
pH control, 51, 54–56, 263–272
   adaptive, 267, 271, 272
   batch, 275
   feedforward, 270–272
   limit cycles in, 135, 268, 269
   nonlinear, 135, 267
Phase shift, 5–7
PI control, 16–18, 91
PID control, 94–96
   nonlinear, 132–136
Plug-flow reactor, 250
Pneumatic controllers, 92, 107
Pneumatic transmission, 38, 39, 62
Positioners, valve (*see* Valve positioners)
Positive feedback:
   in controllers, 91
   in decoupling systems, 215, 219, 220
   in dryer control, 324
   in exothermic reactors, 255, 256, 262
   in interacting loops, 208, 209
Power:
   electric, 225
   hydraulic, 241
   thermal, 235, 236
Preheaters, 262, 263
Preload, 92, 93, 130, 131
Pressure control, 63–66
   in boilers, 65, 237, 238

Pressure control (*Cont.*):
  in distillation, 65, 288–292
  interaction with flow control, 198–200
  in reactors, 263
Pressure regulator, 64, 65
Primary loop, 140–142
  (*See also* Cascade control)
Process gain, 50–55
Proportional band, 10
Proportional control, 10–13
  in cascade systems, 147
Proportional-plus-derivative control, 27, 28
  for batch processes, 275
  as split function, 239
Proportional-time control, 122, 123
Pumps, 239–242
  centrifugal, 240–242
  metering, 18, 31, 240, 268
  positive-displacement, 239, 240

Rangeability:
  of flowmeters, 268
  of metering pumps, 240
  in pH control, 267–270
  in ratio control, 173
  of valves, 45, 49, 50, 268
Raoult's law, 307
Ratio control, 172–178
  of fuel and air, 160, 163
Ratio stations, 173, 174
Reaction rate, 249–252
Reactors, batch, 92, 93, 272–275
  continuous, 257–263
  exothermic, 253–257
  temperature control in, 70–75, 146, 152, 253–257
  temperature profiles in, 152
Reboilers, 231, 287, 288
  parallel, 209, 210
Recovery factor in distillation, 283
Redundant instrumentation, 152, 153
Reflux, 283–287
Regenerative oscillations, 113
Regulation, 5
Relative gain, 197–204
Relative volatility, 279, 282, 291
Reset (*see* Integral control)
Residence time, 250
Resistance to flow, 66
Resonance, 7
  hydraulic, 66–68
Root-mean-square (rms) error, 86

Sampling controller, 105–107
  optimizing, 163, 164
Sampling element, 103–105
Saturation, 114, 115
  of controllers, 90–93
Secondary loop, 140–142
  (*See also* Cascade control)
Selective control, 151–156
  for fuel and air, 234
Self-regulation, 19, 20, 23, 24
Separation factor in distillation, 279–282
Sequencing on-off operators, 149, 150
Sequencing valves, 268–270
Set-point control, 107–109
Set-point response, 141–143
  with batch control, 92, 93
  with complementary feedback, 101
  with derivative on output, 90, 91
  with dual-mode control, 124–126
  with feedforward control, 188, 189
  in interacting systems, 207–211
  with nonlinear control, 133, 134
  with sampling control, 104–106
Shrink and swell in drum boilers, 237
Sine wave, 5, 9, 10
Single-capacity process, 18–24
Speed control, 151, 240–242
Square-root extractor, 43
Stability, 5
  of decoupled systems, 216–218
  of negative resistance, 119
  of reactors, 253–257
  variable, 112, 113
Steam-plant controls, 235–239
Stripping, 309–311
Surge protection, 243–245
Surge vessels, 68, 69, 135

Temperature control, 69–75
  in boilers, 238, 239
  in distillation, 295–297, 301–304, 321
  in drying, 323–325
  in heat exchangers, 40, 41, 170–172, 186–190, 229–232
  in reactors, 70–75, 146, 152, 253–257, 260–263, 273–275
  in stripping, 309–311
Test procedures, 52–56
Thermodynamics, 225
Three-element level control, 169
Three-mode controller (*see* Controllers, PID)
Three-state controller, 123–125

Time constant, 19, 21
  effective, 33, 37
    in mixing, 77
  negative, 255, 256
  variable, 21, 40–42, 59
Time-shared control (*see* Digital control)
Toijala, K., 209, 299
Transmission lines, 33, 38, 39
Transmitter gain, 42, 43
Two-capacity processes, 31–33
Two-mode control (*see* Controllers, PI)

U-tube, 66–68

Valve characteristics, 44–50
  in flow control, 49, 151
  in liquid-level control, 41, 49, 146
  in pH control, 55, 56, 268–271
  in temperature control, 41, 46, 73, 261
Valve gain, 44–46
Valve-position control, 150, 151
  in distillation, 291
  in reactors, 275
Valve positioners, 118, 144, 145
  in blending systems, 178
  in pH control, 268–271

Valve rangeability, 45, 49, 50, 268
Valve response, 116, 118
Valve sequencing:
  using positioners, 268–270, 290
  using selectors, 155, 156
Valves:
  control, 44–50
  electric-motor driven, 123
  pneumatic, 60–62, 116, 118
  solenoid, 123
  three-way, 227
Vapor pressure, 65, 295, 296, 307, 312, 313
Variable structuring, 155, 156
Vector arithmetic, 16, 17, 22, 94
Velocity limiting, 118, 119
Volume booster, 39, 62
Volume control, 175–178

Weight control, 8
Wet-bulb temperature, 312, 322-324
Windup, 91–93, 142, 143

Ziegler-Nichols method, 37